AF371624

Watercolour Botanical Art

RHS

First published in Great Britain in 2026
by Mitchell Beazley, an imprint of
Octopus Publishing Group Ltd,
Carmelite House, 50 Victoria Embankment,
London EC4Y 0DZ
www.octopusbooks.co.uk

An Hachette UK Company
www.hachette.co.uk

The authorized representative in the EEA is Hachette Ireland,
8 Castlecourt Centre, Dublin 15, D15 XTP3, Ireland
(email: info@hbgi.ie)

© 2026 Quarto Publishing plc
Published in association with the Royal Horticultural Society

ISBN 978-1-78472-984-4
eISBN 978-1-78472-985-1

A CIP record for this book is available from the
British Library

Printed and bound in Huizhou, Guangdong, China.
TT/Nov/25

10 9 8 7 6 5 4 3 2 1

This book was conceived, designed and produced
by The Bright Press, an imprint of the Quarto Group,
1 Triptych Place, London, SE1 9SH, United Kingdom.
T (0)20 7700 6700
www.Quarto.com

Publisher: James Evans
Editorial Director: Isheeta Mustafi
Managing Editor: Lucy Tipton
Art Director: James Lawrence
Commissioning Editor: Sorrel Wood
Senior Editor: Joanna Bentley
Project Editor: May Corfield
Picture Research: Sarah Howard, Kathy Steeden
Design: Lindsey Johns
Cover Design: Marcia Pedraza Sierra
Production Controller: George Li

Mitchell Beazley Publisher: Alison Starling
Mitchell Beazley Editor: Ellen Sleath
RHS Publisher: Helen Griffin
RHS Head of Editorial: Tom Howard
RHS Books Editor: Simon Maughan
RHS Consultant: Mike Grant

The Royal Horticultural Society is the UK's leading gardening
charity dedicated to advancing horticulture and promoting
good gardening. Its charitable work includes providing expert
advice and information in print, online and at its five major
gardens and annual shows, training gardeners of every age,
creating hands-on opportunities for children to grow plants,
and sharing research into plants, wildlife, wellbeing and
environmental issues affecting gardeners.

For more information visit www.rhs.org.uk
or call 020 3176 5800

Front cover pictures, clockwise from top left: *Juglans regia*
© Sarah Howard; Orchid 'Copper Queen' © Mary Crabtree;
Tulipa © Reinhild Reistrick SBA Fellow, reinhildraistrick.co.uk;
Kniphofia schimperi © Sarah Howard; *Iris foetidissima* fruit
© Linda Pitkin; *Chaenomeles japonica* © Roger Reynolds;
Magnolia liliflora © Reinhild Reistrick; *Iris* 'Benton Evora'
© Mayumi Hashi; *Gentiana sino-oranto* © Kathleen Munro;
Cerinthe minor © A. P. Brown, by kind permission of Chelsea
Physic Garden Florilegium Society; wheel image © Getty
Images / Anastasia Shemetova.

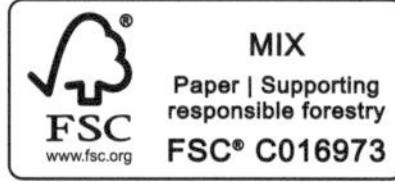

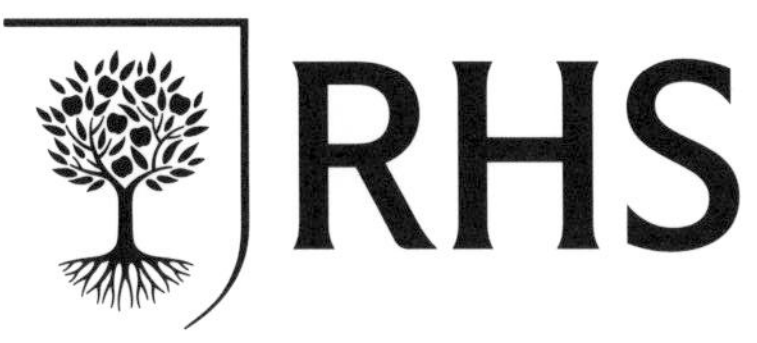

Watercolour Botanical Art

The Complete Practical Guide for Artists

SARAH HOWARD

with

RACHEL PEDDER-SMITH

MITCHELL BEAZLEY

Contents

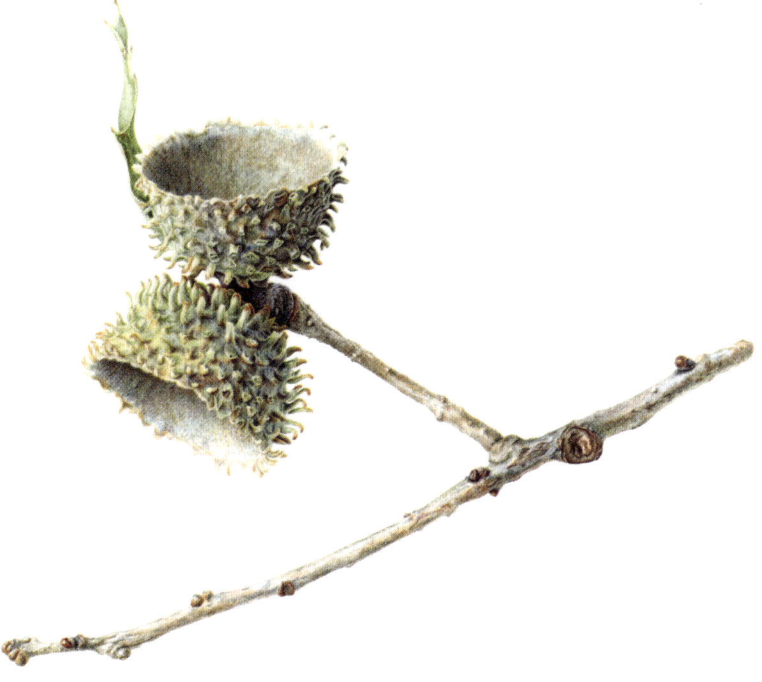

Foreword

Scientific illustration, botanical art and flower painting are terms that have often been used interchangeably. However, an artist with a talent for painting flowers is not automatically a botanical artist. The scientific understanding underpinning successful botanical art and illustration is often only gained when an artist turns to tackle a new subject. By developing this specialist knowledge, the botanical artist can ensure their close observation of the finest plant detail translates into an accurate and beautiful plant portrait.

Over the past 20 years or so, we have seen a burgeoning interest across the world, with more people than ever engaging with and wanting to create their own botanical art, fuelled in part by some excellent exhibitions, the encouragement of talented teachers and, of course, access to images via the internet and social media. A wealth of historic works, readily available online for inspiration, can be viewed alongside contemporary drawings that offer new ways of regarding traditional subjects. This, coupled with a renewed urgency to the importance of protecting our natural environment, means our attention is more focused than ever on all our wild and cultivated spaces. For many of us though, the closest we may come to the rare or endangered plants native to islands far across the world is via their representation in botanical painting.

We can all learn about and share in the beauty of our global flora through the work of talented artists.

But what many aspiring botanical artists discover upon answering the call to paint a well-loved flower, is that it may not be quite as simple as it seems. Those pristine pictures viewed at 4cm high on a screen, often belie many hours of work. The botany might seem daunting for some, but here Sarah Howard ensures it is easy to understand.

Sarah offers reassurance via comprehensive insights for both new practitioners and those wanting further guidance to support their creative process. She shows that by building a rapport with the plant, you start to gain an essential knowledge of it. Using a range of exemplary pictures from some of the most highly regarded contemporary artists, Sarah explores the various measures you can adopt with different painting techniques. No two artists will tackle the same subject in exactly the same way. Take inspiration from the many artists featured here and absorb the tips she shares. Start by putting pencil to paper and enjoy the process. Whether you are new to botanical art or seeking to hone your skill, there is always pleasure to be had in getting lost in the detail!

Charlotte Brooks, Art Curator, RHS Lindley Library
Botanical Art Judging Secretary and Curator of the RHS
Botanical Art & Photography Show

RIGHT Common Hornbeam (*Carpinus betulus*) by Pauleen Trim; one painting in a series of six entitled 'A Year in the Life of Six UK Native Deciduous Trees and their Galls', which won a Gold Medal in the RHS Botanical Art & Photography Show in 2025.

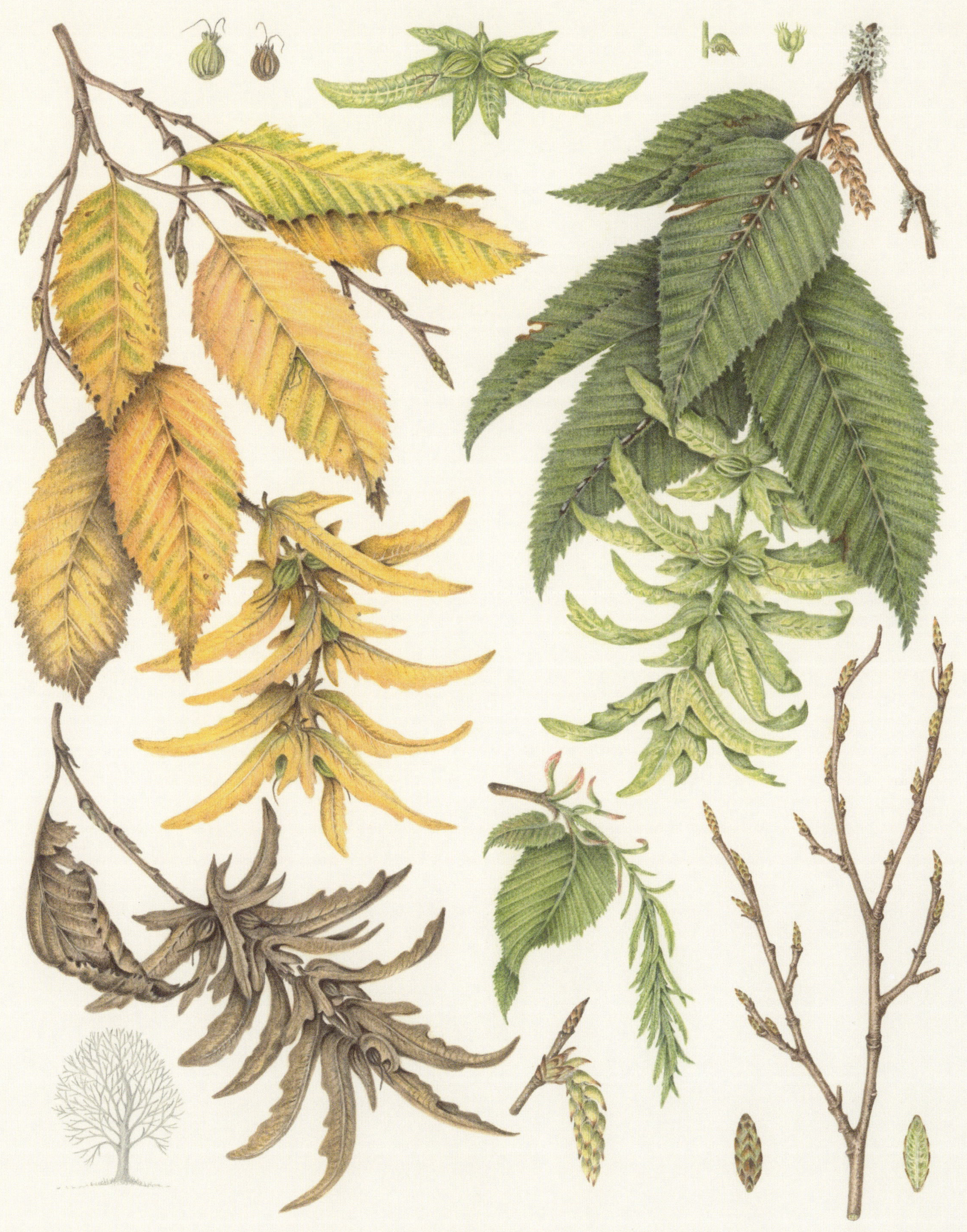

Introduction

If you want to paint a beautiful and accurate picture of a plant, and yet find the language of botany baffling, this book is for you. It is an invitation to go deeper, see further and paint with a little more understanding.

When the 19th century philosopher-artist John Ruskin exhorted his art students to observe and to feel, he meant observing with the eye of a scientist and feeling with the heart of a poet. Of course, most people are drawn to a painting's beauty rather than its detail, but to John Ruskin, great art is related to the truth of what is depicted in nature.

To draw a plant with integrity means closely observing its morphological features, especially ones you might miss or think unimportant. For this reason, botanists have built up a huge vocabulary of terms that describe the tiniest of structures that might identify a plant or express its botany.

This book cannot provide every botanical term – the plant kingdom is too diverse, and every plant is unique – but it can heighten your awareness: there is usually a reason why a bump or a hair needs attention. Every painting involves a process of deconstruction, followed by reconstruction of the different elements into a pleasing composition. Moreover, considered observation will provide a rapport with the plant, which will shine through in the result.

When it comes to painting, experience is vital, for no one begins as an expert. Everyone must practise, experiment, correct mistakes and even start again. There is often no single way to achieve one's ends. A brief explanation of techniques is provided, and Rachel Pedder-Smith, herself an experienced teacher, provides some projects to set you going.

For inspiration, beautiful portraits of plants by artists from all over the world, many of which have won awards, are dotted throughout the book. They have been painted for different purposes, some scientific, some to showcase a particular genus, others to describe the ecology and life story of a plant. All the artists have combined the craft of painting with the science of botany to provide us with artworks of which Ruskin would approve.

The brief history of botanical art suggests that its value is linked with time and culture. We are privileged to be living through a new flourishing of the genre that is technically excellent and global in reach. Ever since Walter Fitch broke with the director of Kew Gardens over pay in 1877, leaving both Kew and *Curtis's Botanical Magazine* without an illustrator, many artists – mainly women – have risen to the challenge of accurately painting plants.

The Royal Horticultural Society (RHS) has been at the forefront of the modern renaissance, mainly through its annual art show. Other institutions are augmenting their modern collections for posterity, such as Australia's state botanical art collections, or the Hunt Institute at Carnegie Mellon University in Pittsburgh, USA, which holds perhaps the largest collection of modern and historical botanical art. Private conservationists, such as the owners of the Transylvania School of Botanic Art & Illustration in Romania, Grootbos in South Africa or Oak Spring Garden Foundation in Virginia, USA, offer courses and residencies based on the rich flora of their parts of the world.

Florilegia, too, are flourishing. The Eden Project, Chelsea Physic Garden (both in the UK) and Singapore Botanic Gardens are only a few of the centres that host artists who get together to paint living plants in their collections. And HM King Charles III has commissioned florilegia of his British garden, Highgrove, and the meadow plants around his house in Romania.

Subjects for exhibition paintings are no longer simply medicinal or trophy plants. They are increasingly likely to be wildflowers, heritage plants or those that reflect themes such as our changing habitat and climate.

I hope this book opens a new interest in painting the treasures of the natural world for you.

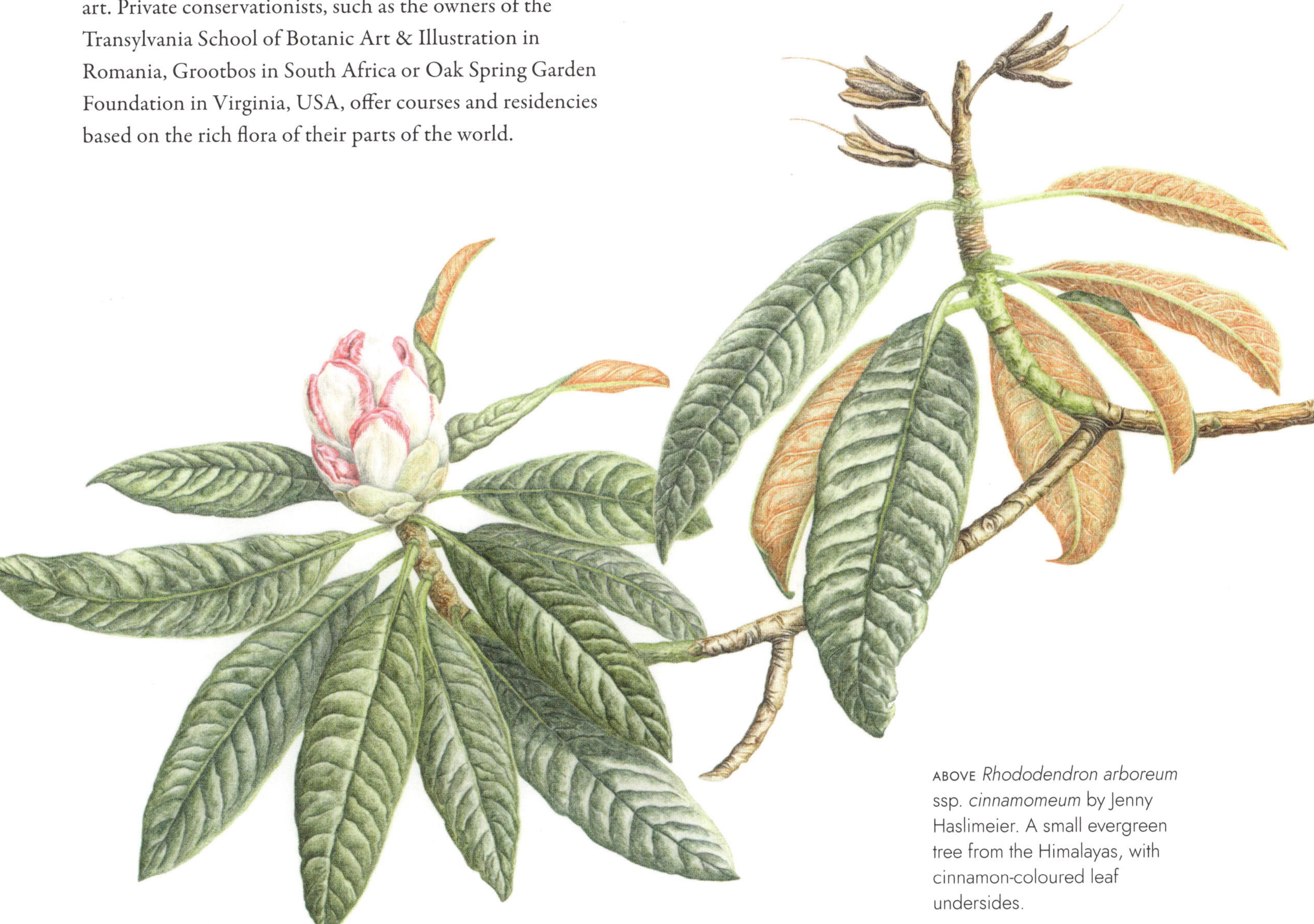

ABOVE *Rhododendron arboreum* ssp. *cinnamomeum* by Jenny Haslimeier. A small evergreen tree from the Himalayas, with cinnamon-coloured leaf undersides.

Using this book

This book is intended to be an essential guide for botanical artists at all levels. For artists new to botany, comprehensive information on plant forms, along with practical tips and step-by-step tutorials, provide the perfect starting point. But experienced botanical artists will also find the detailed level of scientific terms and explanations, and the range of beautiful examples included, make this a unique guide to a wide variety of plants, structures and techniques.

After an introduction looking at the history of botanical painting, the practicalities are covered in Chapter 1, looking at everything from paints, brushes and sketchbooks to composition and painting techniques.

Chapter 2 introduces the essential elements of botany, helping artists who 'paint what they see' to understand the significance of what they are looking at.

Chapters 3 to 6 combine plant descriptions with painting tips for all the key parts of a plant, working from roots and stems through leaves and flowers to fruits and seeds. We finish by looking at vegetative (or asexual) reproduction and non-flowering plants, which include ferns and conifers (Chapters 7 and 8). Within these chapters, there are pointers on what to look for, some advice on parts that are tricky to paint, and examples of terms that describe features in botanical terminology

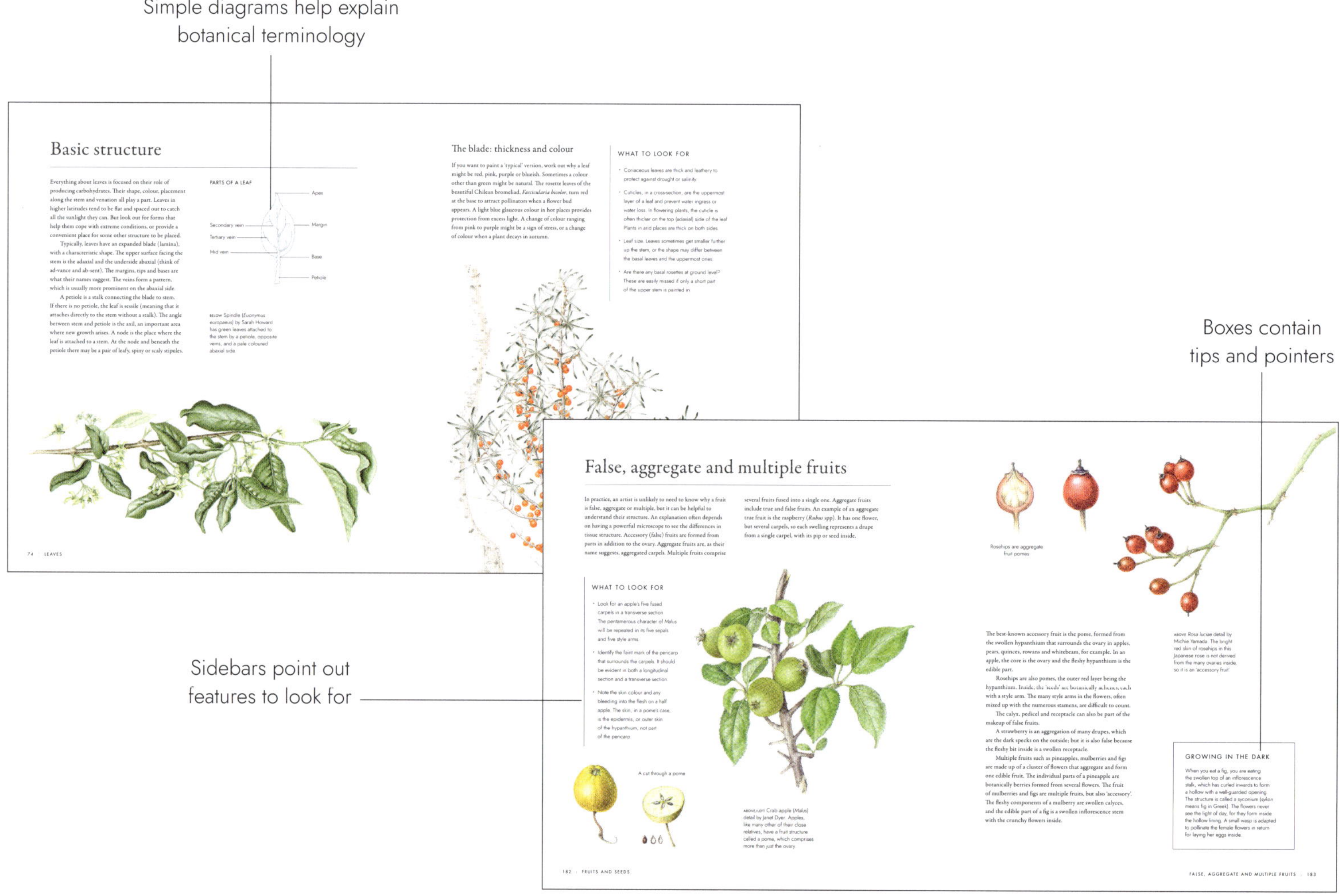

(also listed at the end in the Glossary on page 228). Simple line drawings are included to show different forms or structures in a clear style.

If you are painting a tangle of roots (Chapter 3), say, or a pine cone (Chapter 8), the relevant chapter is your guide to features that make them distinctive, and terms that indicate variations. You may want to find out some of the ways leaves are attached to a stem, in which case read Chapter 3. You can avoid traps such as painting the lie of petals incorrectly, or failing to notice that stamens can sometimes be of different lengths, by reading Chapter 5.

Within the chapters, Rachel Pedder-Smith gives you practical projects to try, preferably with a live plant in front of you. These step-by-step tutorials include a suggested colour palette and tips on how to meet the challenges of the plant in question. All the colours that Rachel uses are in the Winsor & Newton artists' range.

The lovely illustrations in the book may be motivation enough, but for added inspiration look at the full paintings in the artist profiles. These are by established, often award-winning, artists, and each painting is analyzed in terms of the chapter in which it appears.

At the end of the book we have provided a glossary of botanical terms for easy reference. A comprehensive bibliography provides suggestions for how to take your knowledge further. For further practical guidance, it is an excellent idea to do a short course with one of the many artist-tutors who have contributed to this book. They operate in person or online and their websites are listed in the further resources on page 232.

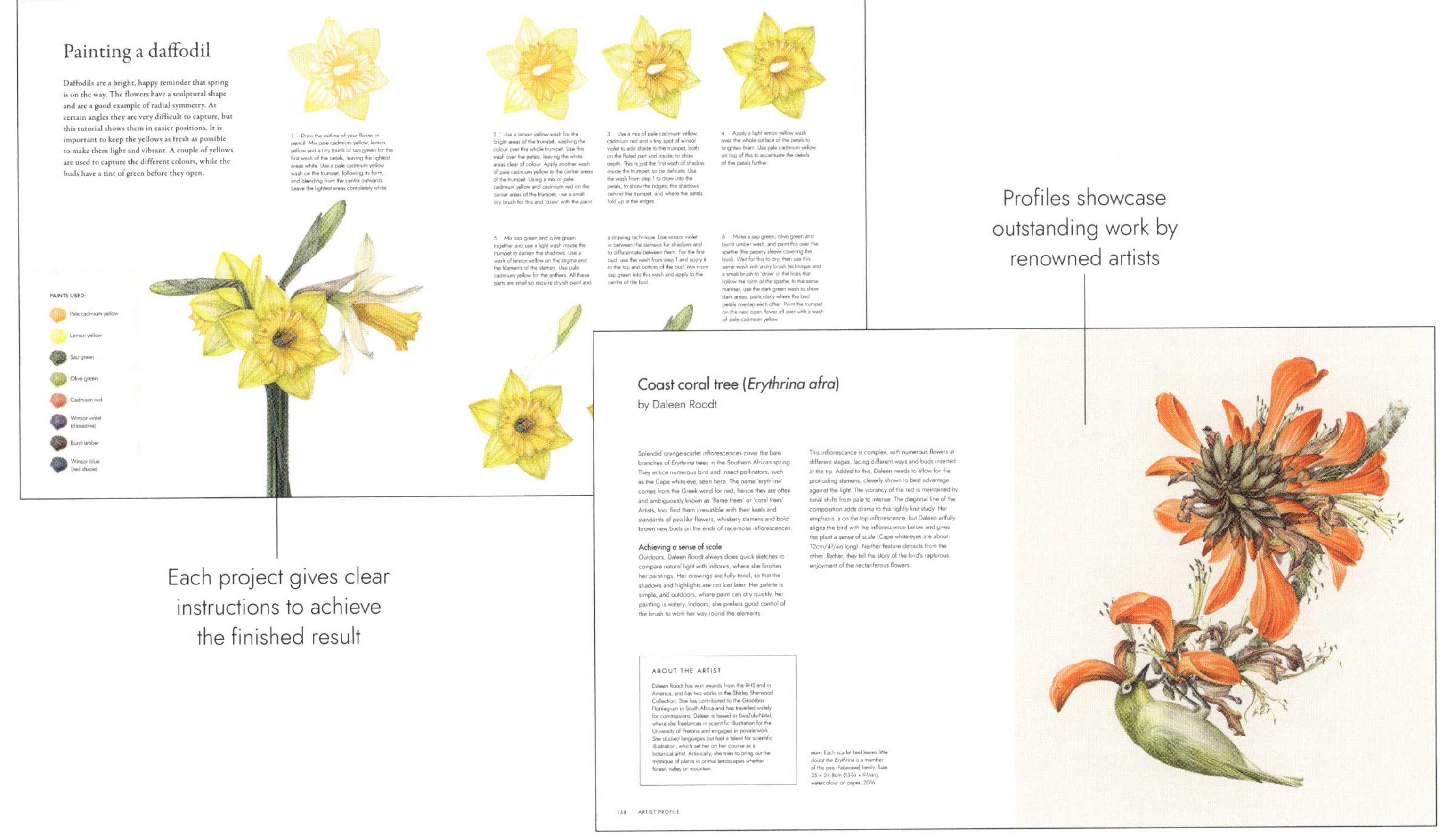

Each project gives clear instructions to achieve the finished result

Profiles showcase outstanding work by renowned artists

Two thousand years of botanical art

Today's aspiring botanical artists find inspiration as much from past masters who owed their skills to patronage, scientific curiosity, exploration, and printing techniques, as from their award-winning descendants in the 21st century. The current renaissance springs from renewed collecting, the global reach of communications, copious study opportunities, international juried exhibitions and redefining what the art form is for.

For this look at the history of botanical painting, we have used a loose definition that includes drawing, illustration and aesthetic expression, whether the medium used is paper or a screen. The important thing is to see the genre as demonstrating an increasing skill in depicting the natural form and highlighting the aesthetics of plants.

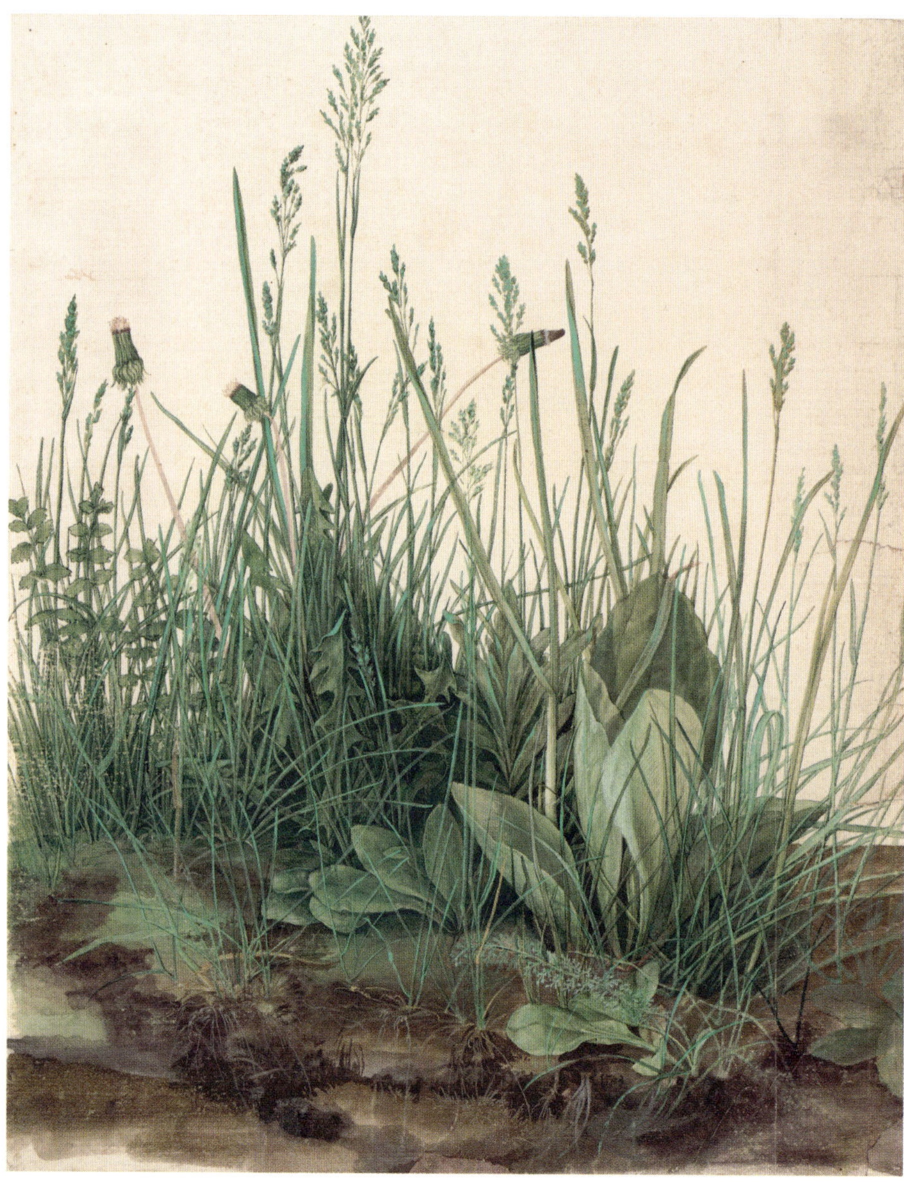

LEFT *The Great Piece of Turf* (1503) by Albrecht Dürer. One of many detailed studies of plants by Dürer, this represents a shift towards painting the natural world, from life, with a sense of realism.

An East Asian sense of beauty

Chinese, then Japanese, artists approached their art with contemplation and a sense of beauty. No work has survived from the 7th century, but we know that patrons delighted in their delicacy, grace and realism, especially for culturally significant plants such as chrysanthemums and bamboo. In Japan, a distinctive style was apparent in woodcuts and screen paintings. The greatest masters of flower painting in Japan, such as Hokusai, flourished in the 17th and 18th centuries. By the end of the 19th century, scientific enquiry and scientific illustration were well established, with Kawahara Keiga learning western techniques from Philipp Franz von Siebold, a Dutch East India surgeon.

Japanese influence still runs deep today. The refinement and seriousness of the country's artists can be seen at any international exhibition and in major collections.

The natural look in Europe

In classical Greece, Dioscorides was the main influencer of plant illustration in Europe. His 1st-century herbal (*De Materia Medica*) with illustrations of the medicinally useful parts of plants – often roots – was copied again and again over the subsequent 1,500 years to the point where the plants became unrecognizable. Then the Silk Route brought a new spirit of scientific observation, and mathematical ideas of balance, perspective and harmony redefined aesthetics. Humanist philosophy provided the impetus to explore and study the natural world. Artists learnt to produce more realistic, three-dimensional plants and painted them for their own sake.

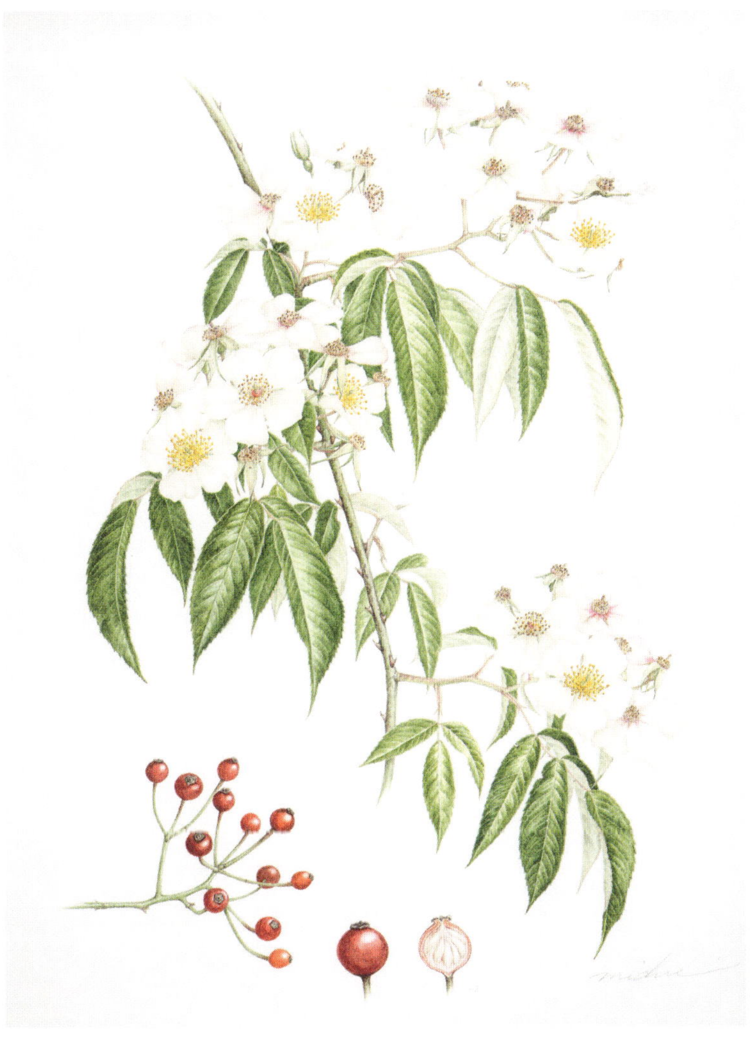

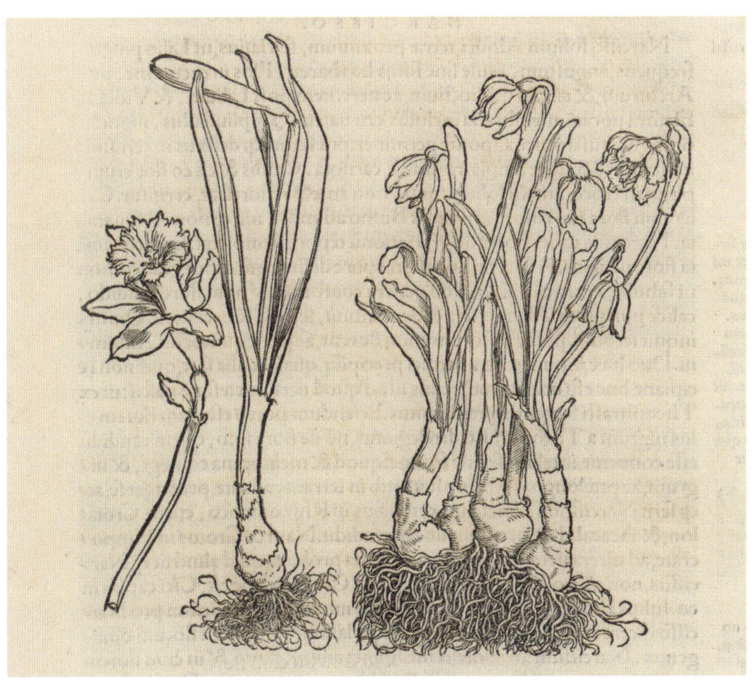

TOP RIGHT *Rosa sambucina* by Michie Yamada exemplifies the light and careful touch of modern Japanese artists.

RIGHT In the 16th century, woodcuts such as *Narcissus* by Hans Weiditz the Younger set new standards in portraying plants from life.

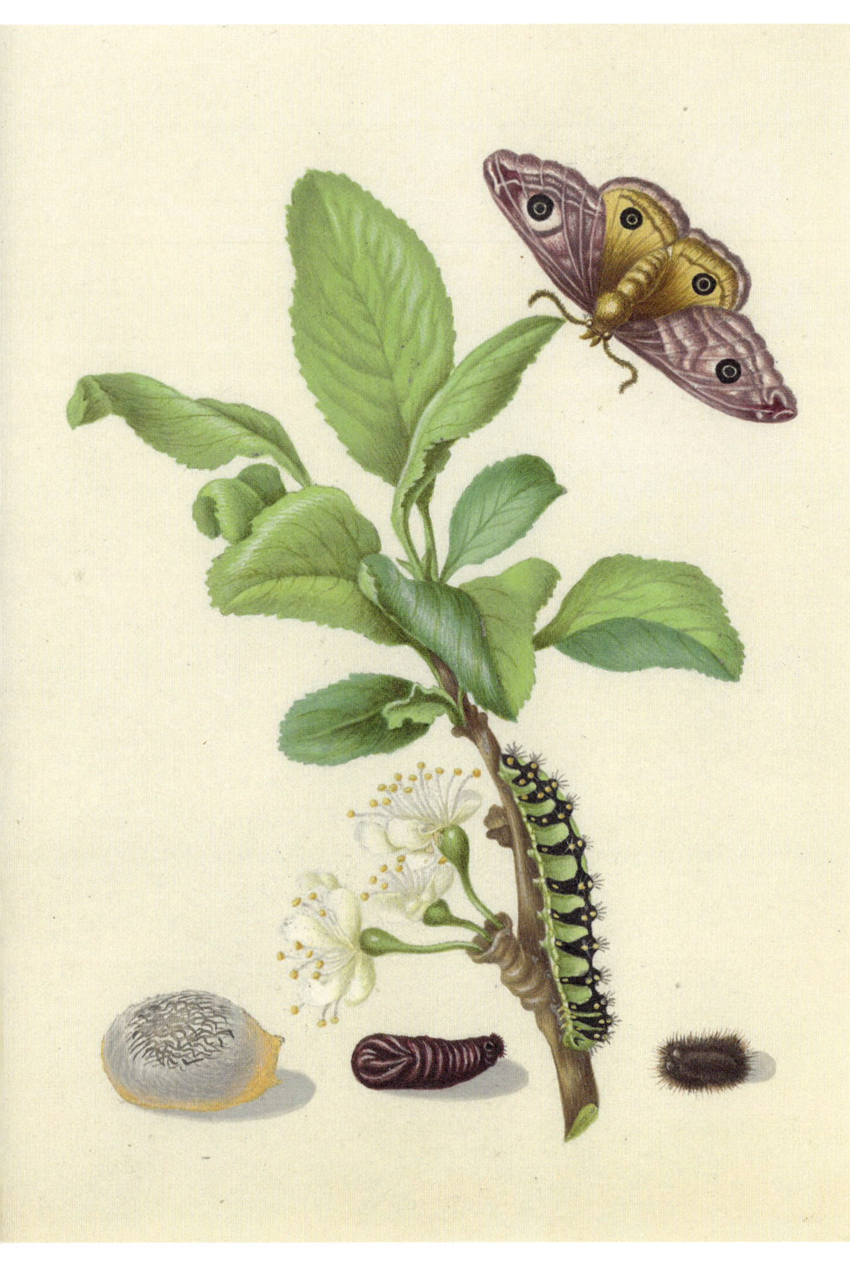

Printing arrived in Europe around 1450, and new standards were set for text and illustration. In Strasburg, Otto Brunfels, a 'Father of Botany', produced *Herbarum Vivae Eicones*, illustrated by Hans Weiditz. The content was based on Dioscorides's selection, but Weiditz created compact, accurate drawings from live material. They were cut on wood, with lines and hachures for shading, a technique that was used well into the 19th century.

Botanical illustration received another boost in the early 17th century, when western travellers brought home exotic plants, which were held in growing numbers of private gardens. Some plants, such as tulips, caused a sensation in the 1620s and 1630s, and this was reflected in contemporary paintings.

The Latin titles of illustrated books of plants of the time give a clue as to how plants were regarded in Western Europe, either as medicinal (*herbal*), for botanical study (*herbarium*), for enjoyment (*florilegium*), or in garden collections (*hortus*).

Etchings and engravings were by now superseding woodcuts, making the printing of treatises, catalogues and florilegia easier.

The 'golden age'

The mid-18th century to mid-19th century, seen as a 'golden age' of botanical art, is a period that still influences botanical artists today. Not only had skills improved, but the science was also transformed by Swedish biologist Carl Linnaeus. He classified plants using their sexual parts (an early form of taxonomy), and established a two-word naming system and rationalized descriptions.

Artists, too, began to look inside plants to draw the parts that mattered for taxonomic purposes, so pictures of the beautiful habit, leaves and flowers, rather than the roots, were now often accompanied by drawings of other parts.

The 'golden age' of botanical art is defined by the work of
four men: Georg Ehret and brothers Franz and Ferdinand
Bauer in England, and Pierre-Joseph Redouté in France.

Humbly born in Heidelberg, Ehret became gardener
to the Elector of Heidelberg until he was headhunted by the
Margrave of Baden-Durlach for his garden. The Margrave
(marquess), who prized his tulips, noticed Ehret's drawings
and favoured him above the other gardeners. Their jealousy
drove Ehret to Regensberg, where he was employed by the
botanist, Johann Wilhelm Weinmann, who was
commissioning illustrations for his *Phytanthoza
Iconographia*, one of the first botanical books to use colour
mezzotint, a method of creating halftones without lines.
Unfortunately, Ehret felt financially exploited and moved
again. He travelled from garden to garden throughout
Europe, all the while perfecting his techniques and being
paid by a wealthy patron in Nuremberg, Dr Jacob Trew.
Their friendship was a lifelong and fruitful collaboration.

In 1736 Ehret met Carl Linnaeus in Holland, where
Linnaeus taught him how to study stamens at the garden of
George Clifford in Haarlem. Twenty of Ehret's illustrations
were used in *Hortus Cliffortianus* (1737), a descriptive
catalogue by Linnaeus of Clifford's garden. Ehret then
relocated to London, where he moved in the highest social
circles, taught aristocrats how to paint, and illustrated the
new plants arriving in the country.

Britain was by now a centre of botanic energy. Ehret's
influential patrons included Sir Hans Sloane, a President of
the Royal Society, and Philip Miller, who was the best-
known nurseryman of his time and author of *The Gardeners
Dictionary* (1731–1768).

Ehret's output — especially in the 1740s — was
prodigious. It included spectacular plants such as brightly
coloured auricula primulas, irises and magnolia, and exotic
plants new to Britain and housed in the Chelsea Physic
Garden, where Phillip Miller was curator.

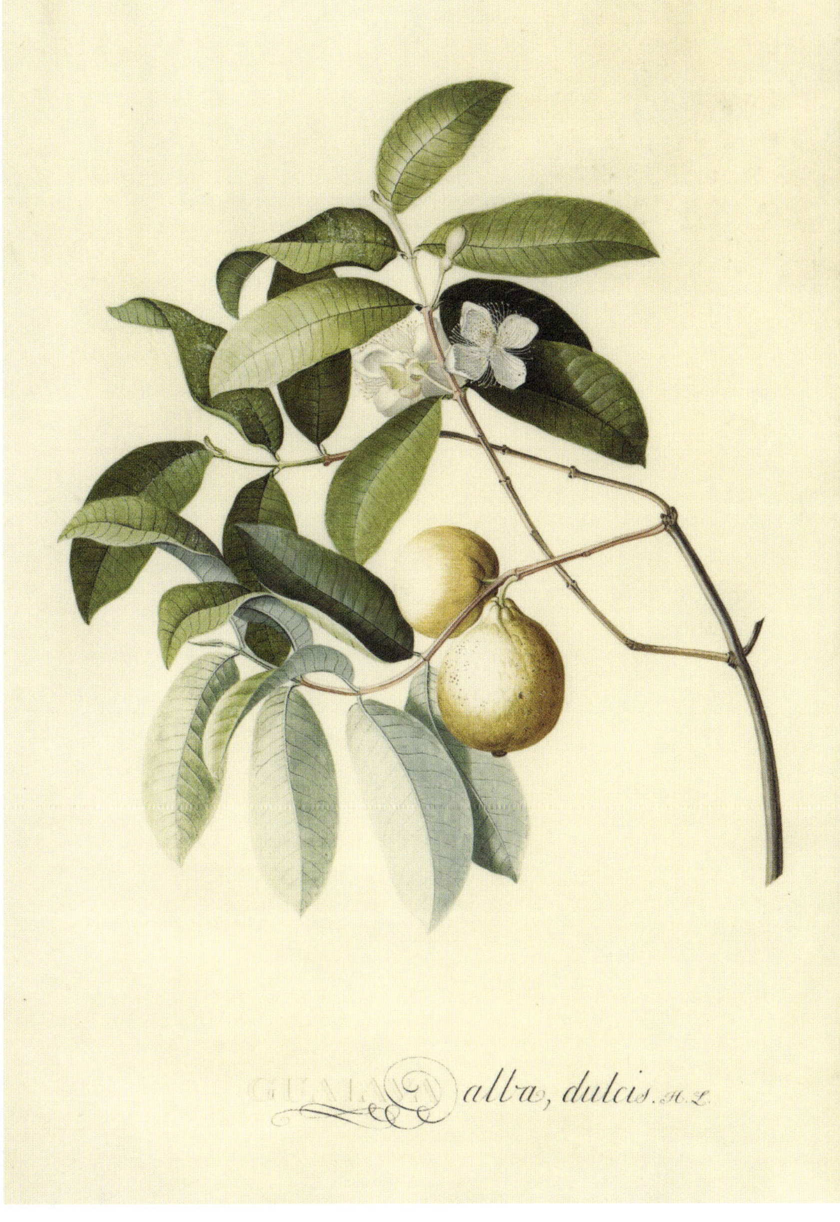

ABOVE Georg Ehret's portrayal of
opposite veins and the front and back
of leaves and flowers made him an
exceptional scientific illustrator as well
as artist. His work *Guava*, painted in
the 1750s, used gouache and
graphite on vellum.

In the 18th century, books were illustrated using engravings based on original paintings (rarely made by the same person). The new lithographic process meant that black-and-white prints could be hand coloured to create subtle shading, and colour began to be used in the printing process. Ehret himself engraved and hand coloured his own published work, *Plantae et Papiliones Rariores* (1748–9). One hundred of his paintings were collated and etched for Trew's *Plantae Selectae* (1750–73), some of them tropical fruits such as bananas and pineapples.

Interestingly for modern botanical artists, who favour the transparent properties of watercolour, Ehret preferred using opaque body-colour and painting on vellum rather than paper for his important works. Sometimes, smaller plants were painted slightly larger than life size.

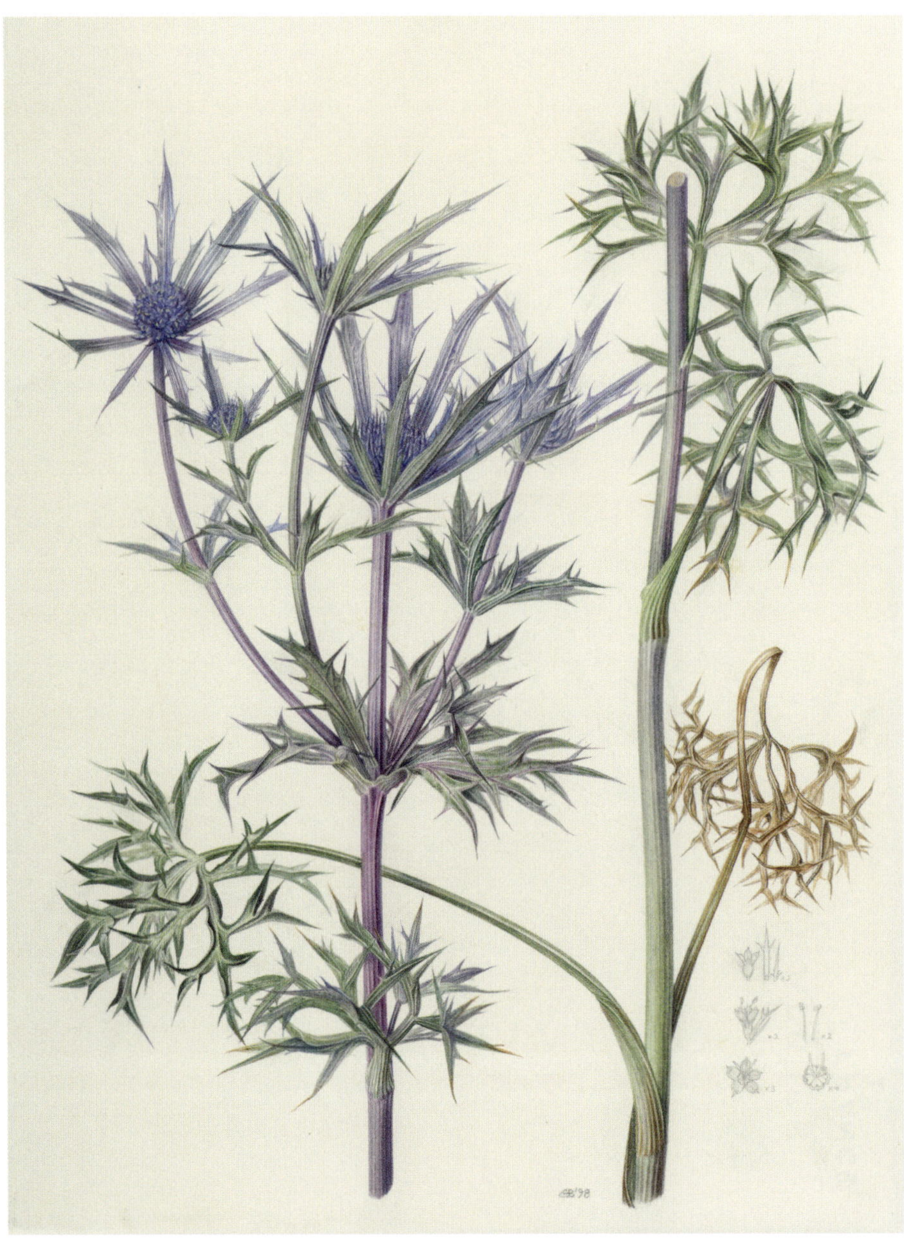

LEFT *Eryngium bourgatii* by Gillian Barlow. This Mediterranean sea holly in the Chelsea Physic Garden is part of a living collection being painted by a team of volunteer botanical artists for the Chelsea Physic Garden Florilegium, founded in 1995.

FLORILEGIA

Wealthy patrons have long commissioned florilegia to show off their plants and gardens. The word comes from 'Flos' (Latin for 'flower') and 'legere' (Latin for 'to gather'). The term florilegium now describes a collection of original flower paintings, usually on a theme. This may be a geographical area, such as *Flora of Louisiana*, illustrated by Margaret Stones; a genus, such as Celia Rosser's *The Banksias*; or a collection, such the Chelsea Physic Garden or the Royal Botanic Garden Edinburgh (RBGE).

Developments in botanical art

Shortly after Ehret died, Ferdinand and Franz Bauer, two Austrian brothers from an artistic family, were taught botanical painting by the supervisor of a royal garden in Vienna. Their studies included the Linnaean classification system and how to use a microscope. The brothers then visited London, where renowned botanists were delighted to use their talents.

Franz settled in London and became Joseph Banks' illustrator at Kew Gardens, while Ferdinand travelled widely, eventually settling back in Vienna. Between 1786 and 1787 Ferdinand travelled with an Oxford botanist, John Sibthorpe, to the eastern Mediterranean, where he produced a thousand sketches and drawings of plants. The resultant *Flora Graeca* was one of the finest florilegia ever made. It was published in instalments during the first half of the 19th century, each volume containing one hundred plates engraved by Bauer himself on copper. A full set can be seen online at the Bodleian Library, Oxford.

Ferdinand achieved so much so quickly by using a 'painting by numbers' technique, devising his own chart of colours and inserting their numbers onto his sketches for working on later.

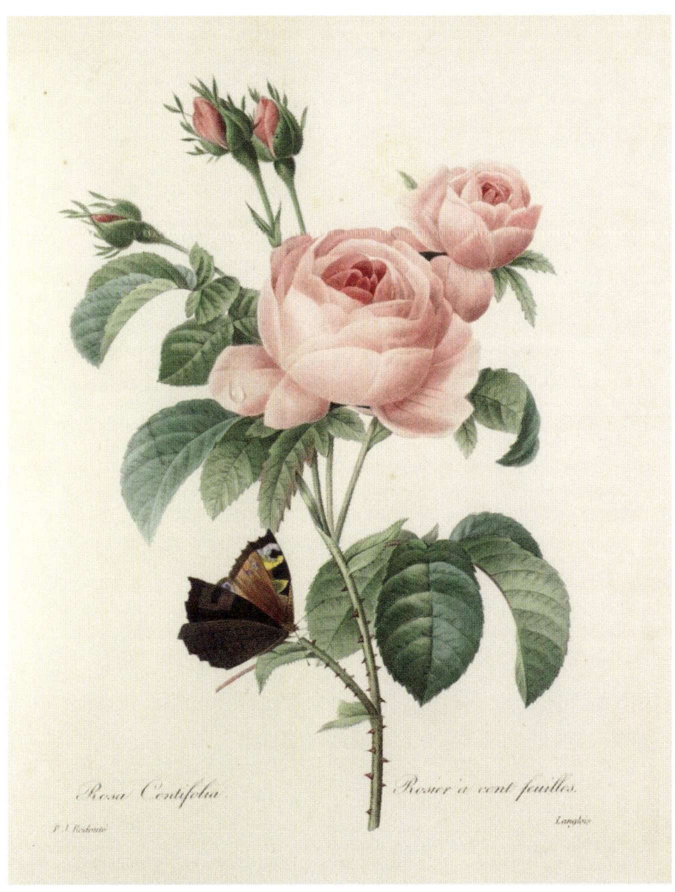

ABOVE *Rosa* x *centifolia*, engraved by renowned French artist Pierre-Joseph Redouté. His famed *Les Roses* and *Les Liliaceés* draw admiration to this day.

RIGHT *Alyogyne hakeifolia* by Ferdinand Bauer includes all the information Europeans needed to identify this southern Australian plant in the Mallow family.

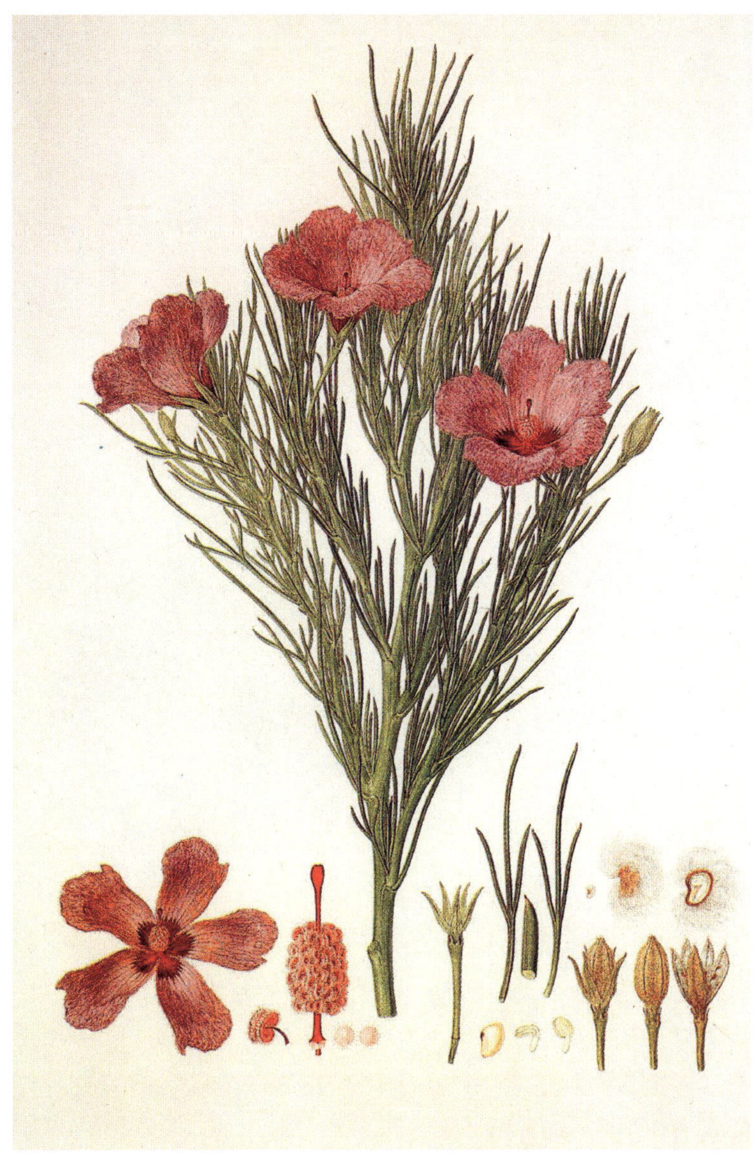

The global reach of botanical gardens

Before photography, illustrations were an essential tool. Descriptions of many plants, new to western scientists, were made from illustrations, rather than a type specimen (one that is kept as the representative of its type). The illustrations, if not made in situ, were made from flat, dried material brought back to herbaria. Even now, a published scientific description needs to be accompanied by an illustration. In India, the first scientific flora, completed in 1693, was Henrick Van Rheede's collaboration with local herbalists and draftsmen for an illustrated description of economic and medicinally useful local plants. They were named in four different scripts and proved invaluable to Linnaeus a century later.

In the 19th century, British East India Company surgeon-botanists also collaborated with Indian artists, who were often trained in the Mughal miniaturist tradition. Recent studies by Dr Henry Noltie have made the names of Vishnupersaud, Govindoo and Rungiah better known, and their distinctive style more apparent.

Sooner or later, every major botanic garden in the west employed illustrators to draw taxonomically pertinent plant features for scientific descriptions.

A new flowering

Photography became easier in the early decades of the 20th century and botanical painting experienced a lull. However, in the early 1990s institutions started to collect original botanical paintings again, fortunately before too many skills were lost. The Hunt Institute in Pittsburgh, USA, and the RHS Lindley Library in London, now have healthy modern collections. Wealthy benefactors, such as Shirley Sherwood in the UK, and Alisa and Isaac Sutton in the US, began a trend for private collections around the same time. A botanist herself, Sherwood paid for a new gallery at Kew Gardens and has exhibited her collection using an astonishing variety of curatorial themes, with catalogues to match.

Juried art shows sponsored by the Royal Horticultural Society and the American Society of Botanical Artists (ASBA) based at New York Botanical Garden, have raised standards in painting and aesthetic skills.

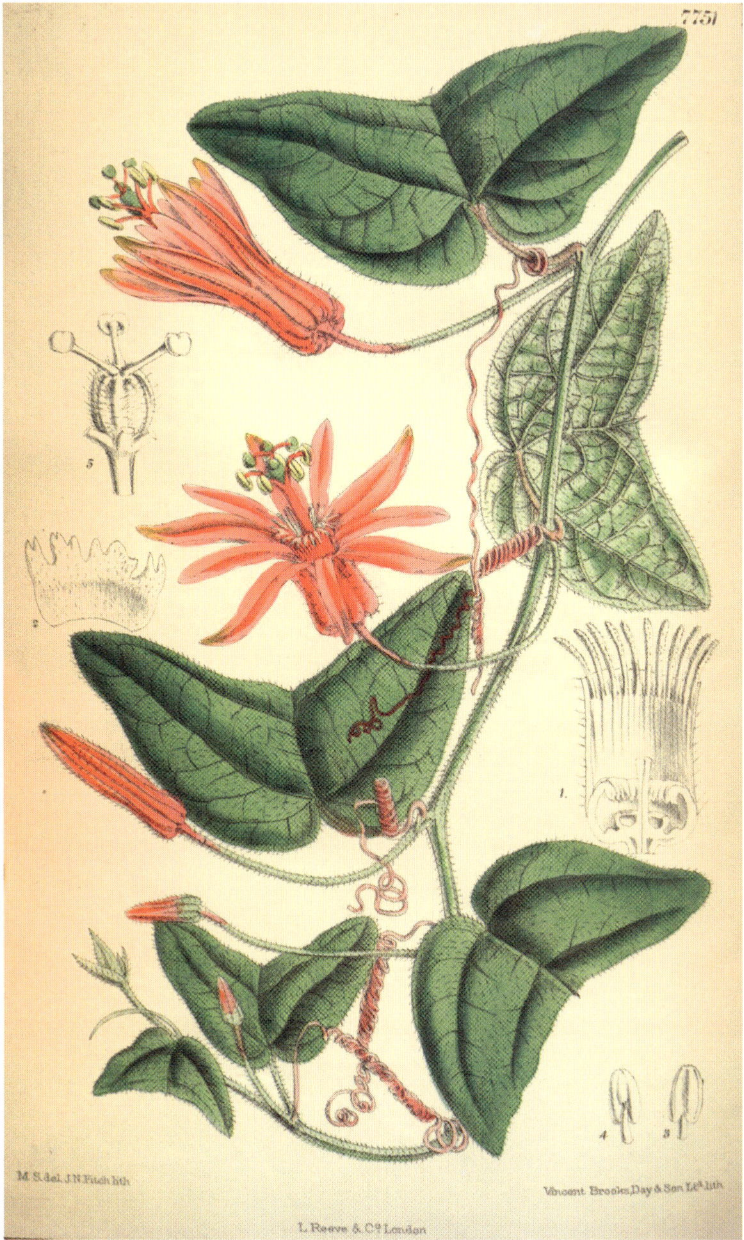

LEFT *Passiflora capsularis* (1900) by Matilda Smith, sole artist for *Curtis's Botanical Magazine* from 1898–1923. Founded in 1787, the magazine is still published today. Among its many other distinguished illustrators were Walter Fitch, Lilian Snelling and Stella Ross-Craig.

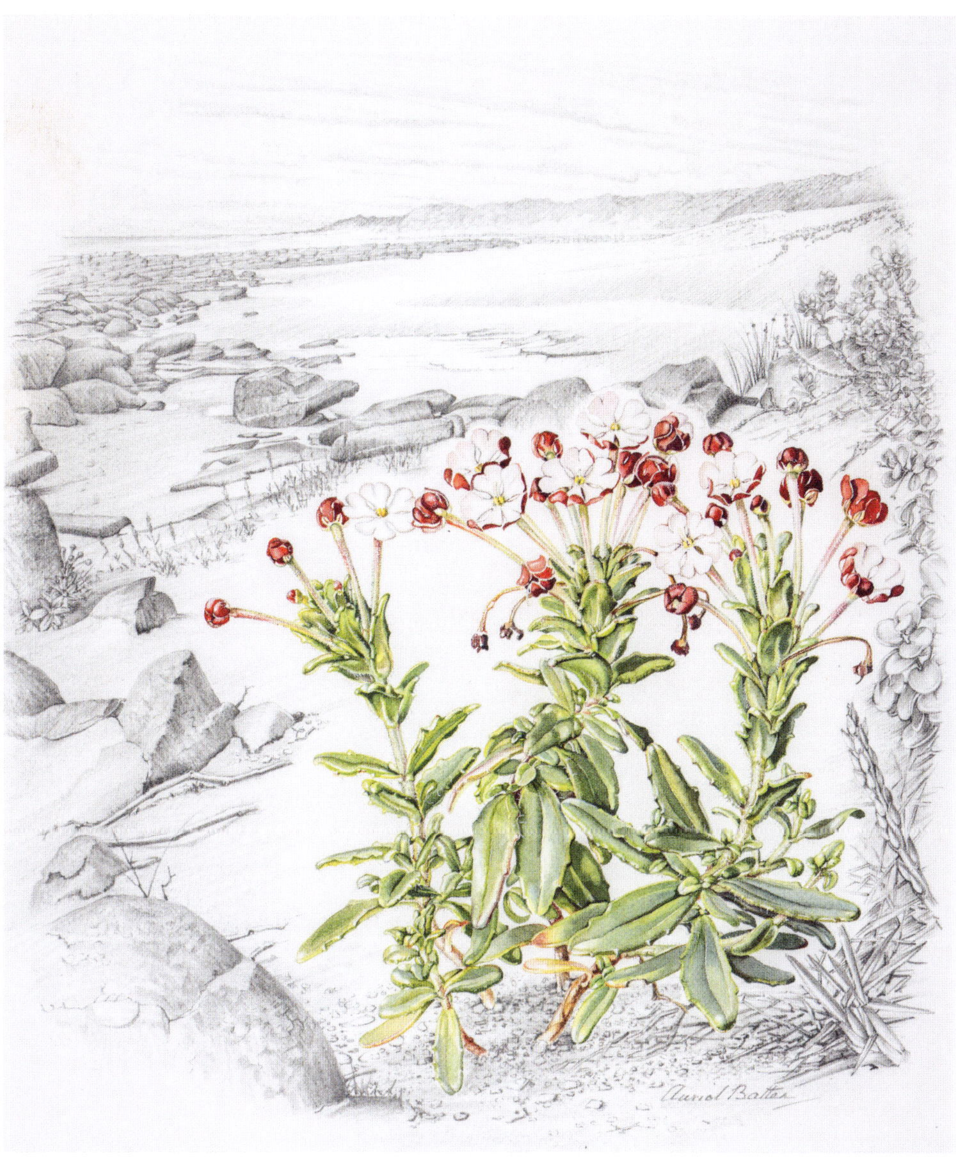

May 2025 saw the second worldwide online exhibition mounted by the ever-increasing number of botanical art societies in more than 30 countries. Cross-fertilization provides innovation and the rediscovery of old techniques such as woodcuts, nature printing or silverpoint. Numerous tuition opportunities, many online, are widening the cohort of artists, and diplomas are providing a basic standard.

Fresh concerns provide new themes in botanical painting, including climate change and habitat loss. Margaret Mee, who took 15 canoe expeditions into the Amazon basin, was one of the earliest to record disappearing flora in the 1950s. Several Brazilian artists have now trained at Kew by way of the Margaret Mee Fellowship Programme. South Africa's rich flora is reflected in the work of artists such as Auriol Batten, who set her plants against a background of the habitat in which they grow. Exhibition themes depict rarity, endemism, lost heritage or a renewed interest in regional flora.

It is at the international exhibitions that eastern and western artists share the best scientific rigour and the best aesthetical approaches in illustration, which makes for innovation and reveals the beauty and the fragility of the plant kingdom today.

Botanical art and the RHS

The artistry and skill of botanical art today is deeply influenced by the annual Royal Horticultural Society (RHS) Botanical Art and Photography Show. The collection of botanical artworks in the RHS Lindley Library is also an inspiring resource for new artists. The RHS show is held in London and is the World Cup of exhibitions, and medals gained there are prized by artists all over the world.

It's a tough assignment. Artists need to exhibit six pictures on a theme, and label them correctly. All must be of the same standard, and the exhibit must be coherent. The RHS pioneered this form of juried exhibition, with its pre-selection process followed by an invitation to show a multi-work exhibit. The rewards are a Gold, Silver Gilt, Silver or Bronze medal. Awards are also given for best picture and best exhibit. An artwork may also be purchased by the Lindley Library for its collection. Artists are rightly proud of simply getting an invitation to take part.

The RHS has long commissioned artists to provide a record of plants. William Hooker provided the earliest illustrations for publication, starting in 1806. The society's current official Orchid Artist is Deborah Lambkin, who has been in post since 2005. The Lindley Library holds an extensive collection of books for researchers interested in the history and practice of botanical art. It now holds more than 30,000 items, including works by most of the important botanical artists of their day, such as Claude Aubriet, the Bauer brothers, Augusta Withers and Georg Ehret. Today, contemporary art is the focus of the collecting strategy and is seen as complementary to the shows.

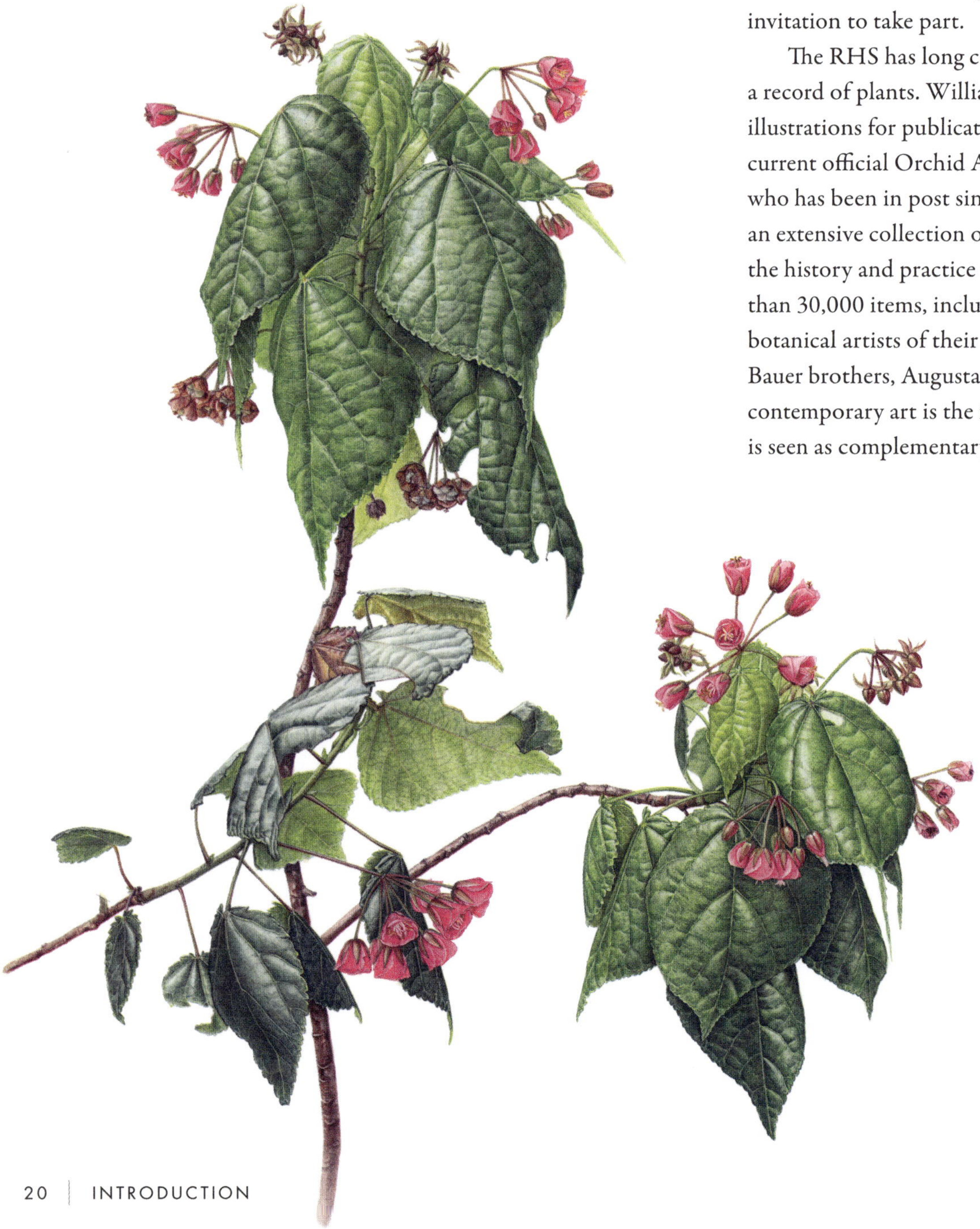

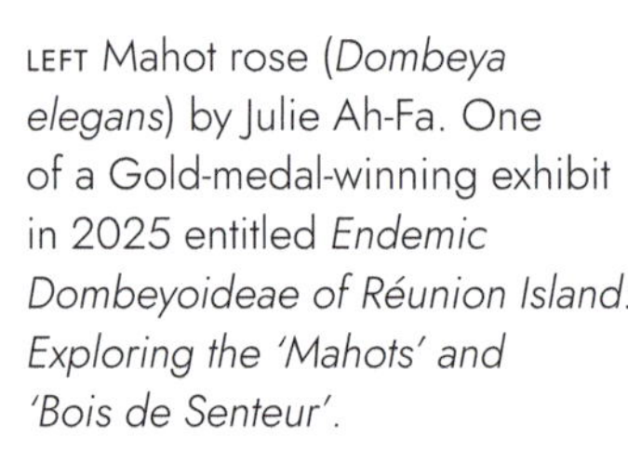

LEFT Mahot rose (*Dombeya elegans*) by Julie Ah-Fa. One of a Gold-medal-winning exhibit in 2025 entitled *Endemic Dombeyoideae of Réunion Island: Exploring the 'Mahots' and 'Bois de Senteur'*.

OPPOSITE Jersey lily (*Amaryllis belladonna*) by Lilian Snelling (1872–1972). She was one of the most important botanical artists during the first half of the 20th century, working first at Royal Botanic Garden Edinburgh and then at Kew Gardens.

Initially, widely interpreted artworks were exhibited at numerous fortnightly or annual horticultural shows. They were popular exhibits, and by the late 1920s the concept of botanical, rather than garden, art had to be refined. Today, it is defined as 'a genre of art that endeavours faithfully to depict and represent the form, colour and detail of a plant, identifiable to species or cultivar level…The best botanical illustration successfully combines scientific accuracy with visual appeal.'

Medals were awarded for artworks in the earliest shows, but these were secondary to those for plants. The first Gold Medal awarded to an artist was given to Frank Galsworthy in 1930. He was one of only five artists to receive one before 1963.

Although botanical art is now shown and judged separately, it is seen as an important contribution to horticulture. The exclusive Victoria Medal of Honour (VMH) has been awarded to only four botanical artists, including Lilian Snelling, a prodigious artist and the longest serving illustrator for *Curtis's Botanical Magazine*. The Veitch Memorial Medal (VMM) has been awarded to three patrons of botanical art: Wilfred Blunt (botanical art historian), Shirley Sherwood (collector) and Anne-Marie Evans (teacher of botanical art), and eight artists including, in 2015, Gillian Barlow, a Gold Medal winner and judge for many years.

The model for the present art show was set in the 1930s. Judging criteria were drawn up and now covered aesthetic appeal, scientific accuracy and technical skill. Entry had become more prescriptive. 'Fanciful' works and 'needlework' were out, and botanical paintings were in. After World War II, however, interest waned.

A reawakening in the 1970s produced a new crop of skilled artists who, in turn, influenced the next generation. They included Paul Furse, a naval officer and plant hunter; Mary Grierson, Kew artist and picture judge and advisor on the RHS's picture collecting; Marjory Blamey, book illustrator; and Elizabeth Cameron, artist – all Gold Medal winners.

Thus, from the late 1980s each show displayed better exhibits than the last. A refurbished Lindley Library began collecting contemporary works based on artists and their skill, rather than what they painted. The collector Shirley Sherwood was influential in both the library and in the shows as a judge.

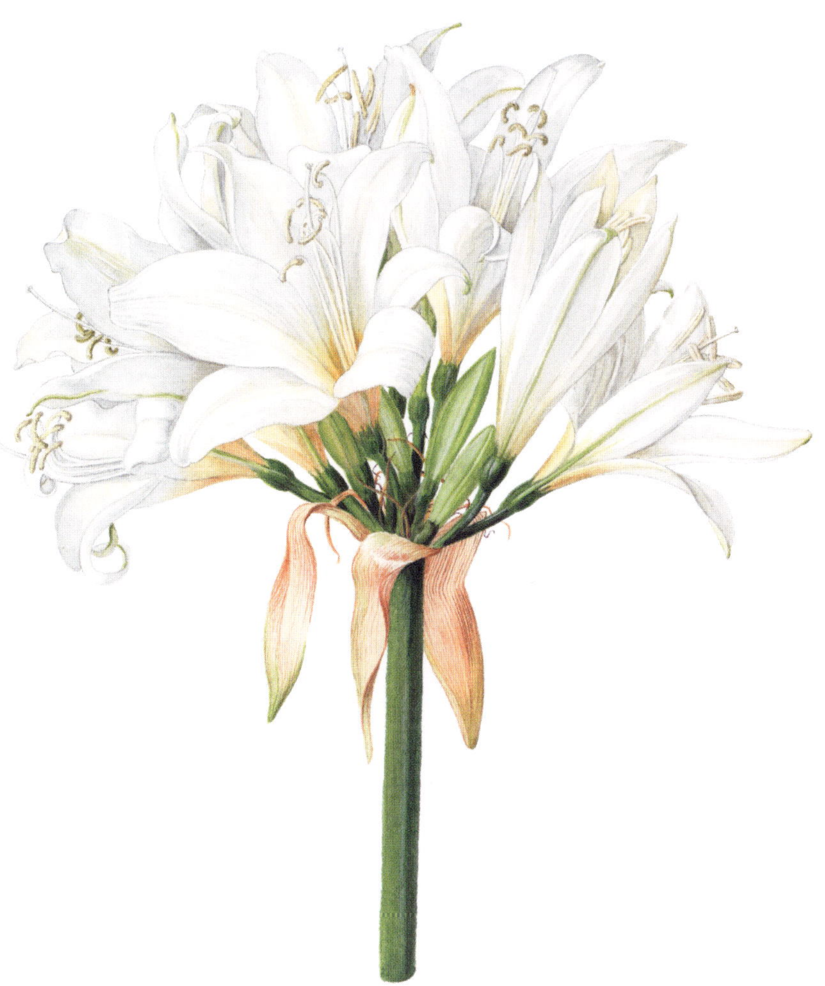

Gradually, other nationalities took part and by 2018 Gold Medals had been won by artists from 23 different countries. Many of them earned kudos from the shows to do more. Dr HyeWoo Shin found inspiration from her award-winning exhibits to become a taxonomist and teacher of botanical illustration in Korea. She has been awarded four RHS Gold Medals plus trophies for Best Botanical Art Exhibit and Judges' Special Award, followed by the Jill Smythies Award for outstanding, diagnostically relevant, published illustrations, from the Linnean Society in 2025.

From the 1990s until the mid-2000s the art shows were held in London's Westminster Halls, near the Lindley Library. Artists returned to their countries with renewed vigour, and future exhibitors went away with new ideas and high standards to aim for.

Today, it is recognized that botanical art needs to be professionally presented and made accessible to more people. The esteemed Saatchi Gallery in the heart of London is now home to the show and attracts a wider public than ever before.

Chapter one

Practical considerations

Botanical drawing and painting requires surprisingly little equipment. Anyone venturing outdoors soon learns to take just the essentials: paper, drawing implements and a small box of paints. Indoors, it need not be any more complicated.

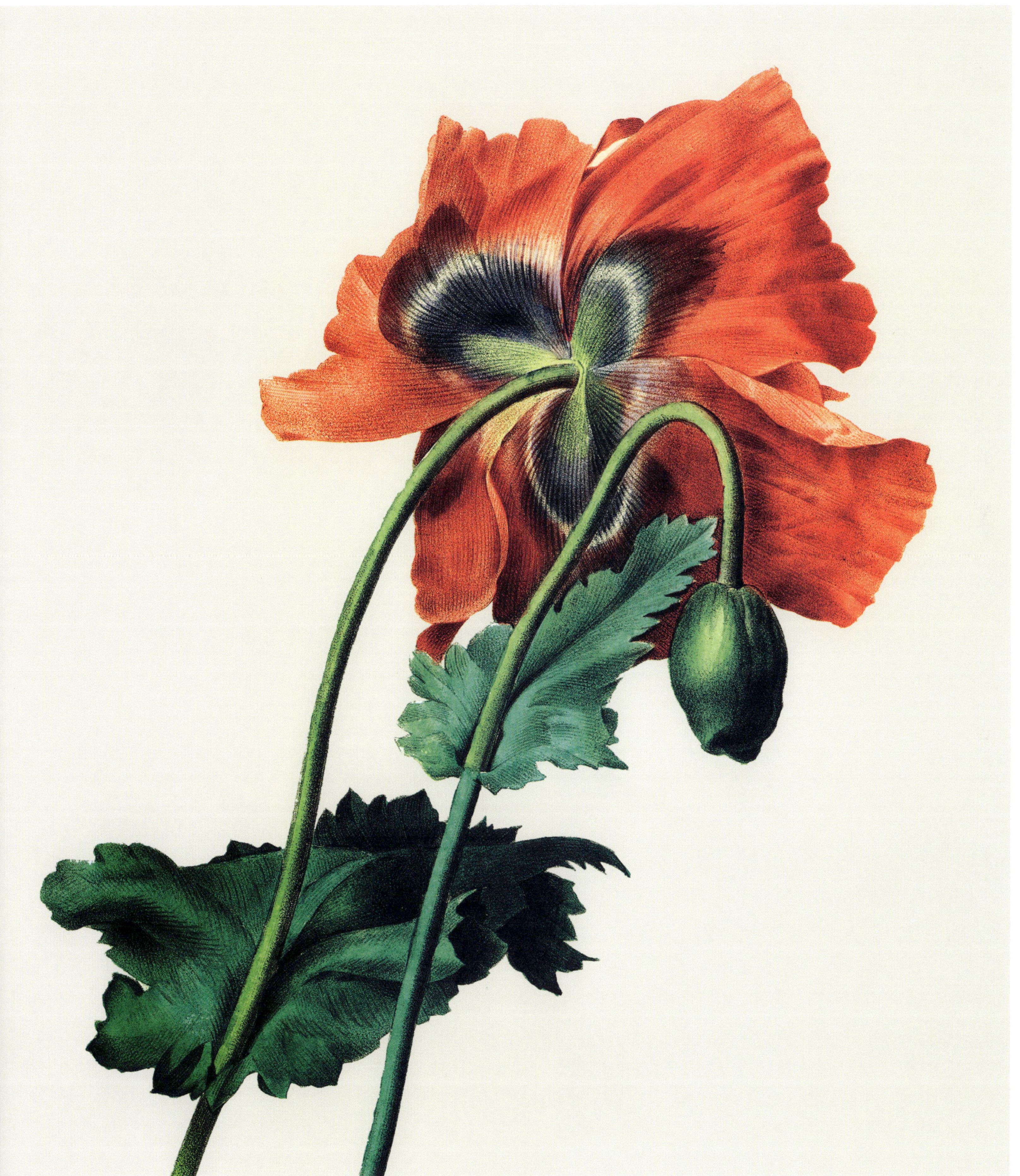

Equipment

Keeping your equipment to a bare minimum has an important advantage: you can concentrate on the thing you're drawing without constantly having to make decisions about your tools. In the spirit of adventure, only add items when you feel you want to extend your artistic range or overcome a technical problem.

A beginner artist should first 'stay' with a plant to explore its potential in any way the fancy takes them. Use the tools you have to hand to reflect and draw what it is that struck you in the first place. You may not like the result, but

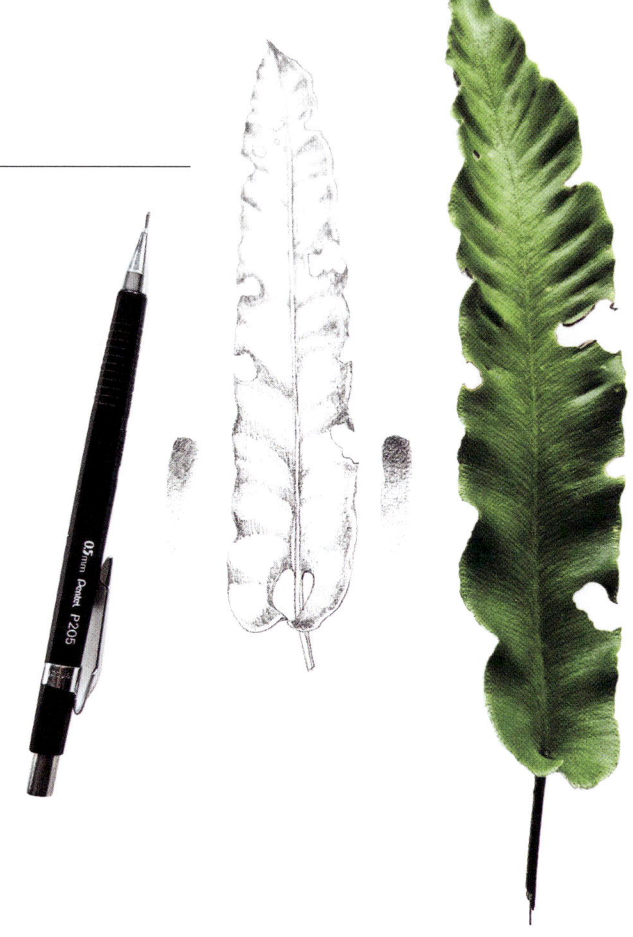

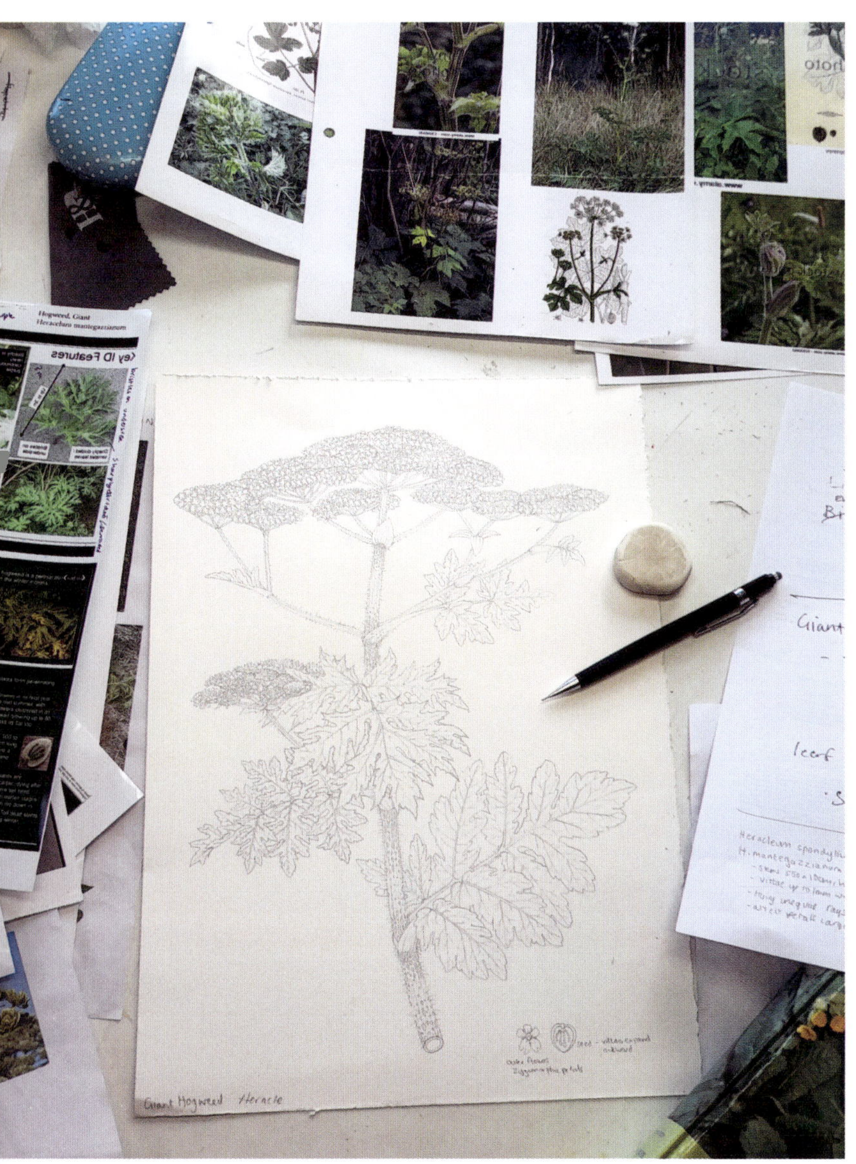

you will have found a rapport with the plant. Finding delight in compositional possibilities, or a striking colour and shape, or being charmed by the inner as well as the outer is as important as the technical process, as the Chinese discovered so many centuries ago.

Practise your pencil markings to develop eye-hand coordination and memory. Find new details and ask yourself what function they may have in the context of a plant's survival and reproduction. For, if any difference is to be made between a 'flower artist' and a 'botanical artist', it's in developing observational skills, the uncovering of small details, and the faithful rendering of them in a precisely executed painting.

LEFT Less is more: use the minimum number of tools you can to sketch and paint until you get to a stage where necessity dictates that you need to add to them.

ABOVE Study your subject and get to know it well while you sketch, in order to develop your drawing skills.

Travelling kit

On my field trips in Ethiopia, I have turned hotel rooms into studios, had kind guides hold an umbrella over my sketches to reduce the glare, and on one occasion I had to ask a boy to hold flower to stem after he over-enthusiastically cut it. My kit was kept in a small waterproof bag; paper was cut to size and pieces joined with tape if necessary. A lightweight board protected the paper and kept it flat in a waterproof bag in a rucksack. Added to this was a box of 12 basic colours in pans, rather than tubes (which can ooze after flights or dry out), and two sizes of good brushes (very carefully covered). In a school pencil case went pencils, erasers, sharpeners, hand lens and dividers, plus extras in case any were lost.

This kit was strapped on the back of donkeys and carried through rainstorms. At hotels, staff would tiptoe past my room to look at progress, always pleased to know 'their' flora was special.

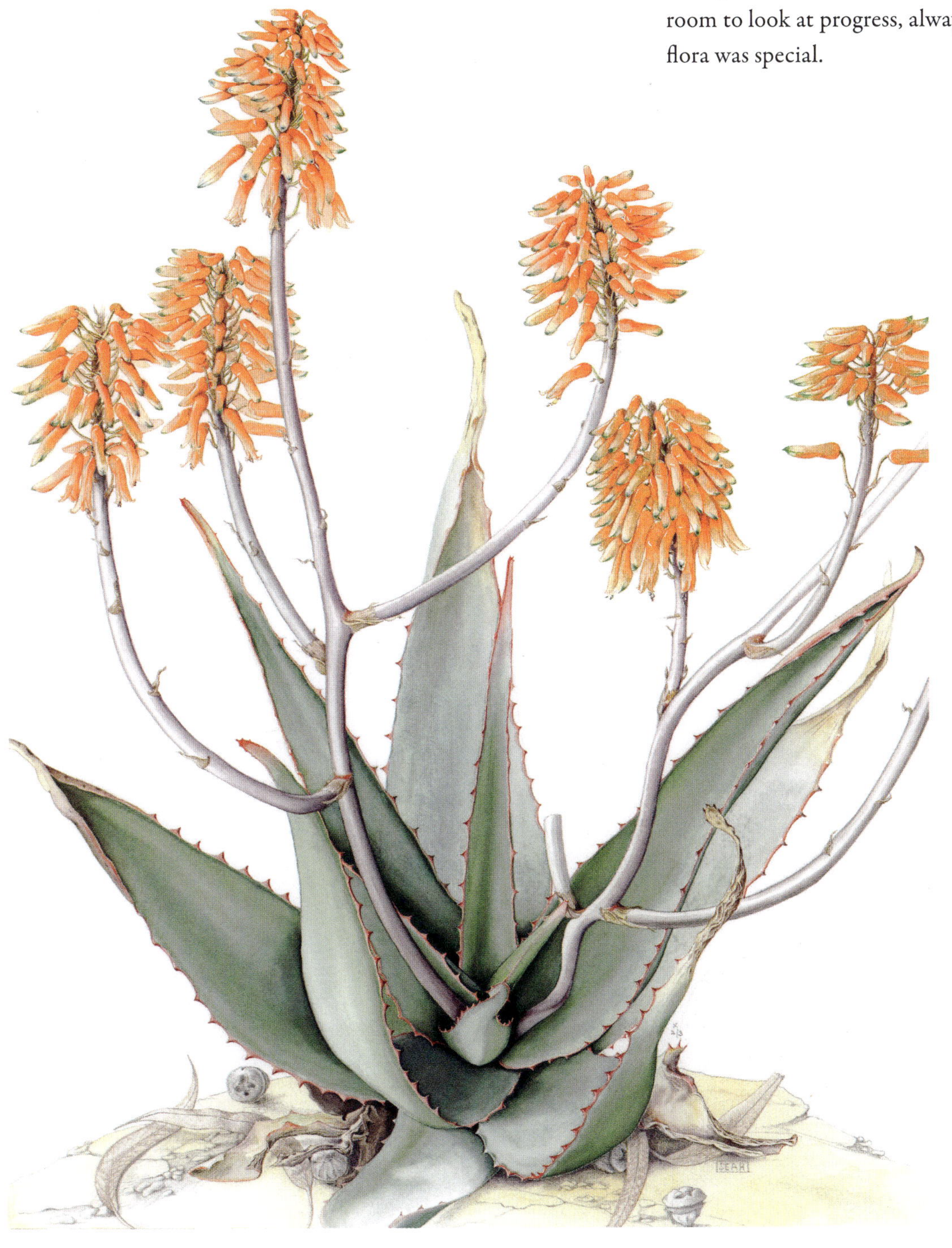

LEFT *Aloe elegans* by Sarah Howard. Succulent plants can often survive journeys and be drawn later, but it is best to draw the flowers in situ if possible.

Artists' tools

To draw and understand plants, a pencil and paper are enough to begin with. Add paints, brushes and other kit as you gain confidence and move into colour work.

Sketchbooks

This is not only a place to practise, but also somewhere to sketch as many drawings of the different parts as possible, which can be rearranged and transferred into a pleasing composition later. Sketchbooks come in many different sizes and papers. It's entirely a matter of personal choice, though an A4 size is small enough to carry outdoors and large enough to draw most plant features. When it is open, you get an A3 size for larger, or longer, plant parts. Loose sheets of paper, say A3 size, have the advantage

indoors of enabling you to draw not only larger parts, but also to keep most parts of a plant on one page so you can see all the deconstructed parts in one glance.

Test the sketching paper with your preferred drawing pencil. It should neither be so soft that fine lines are compromised, nor so smooth that your pencil struggles to get a grip and be seen. If necessary, it should also function as a painting surface.

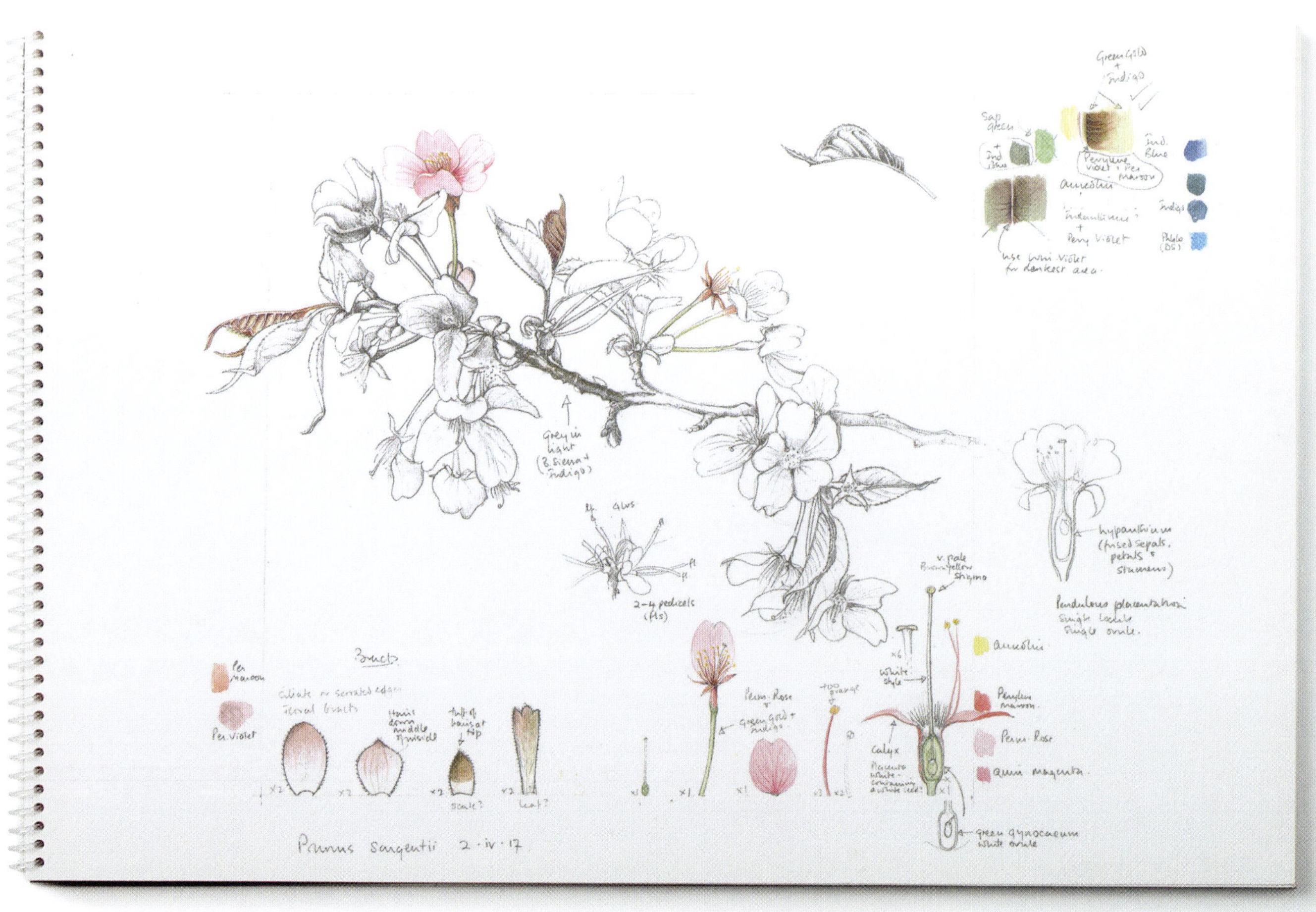

Drawing kit

Pencils (or pens) are usually used in botanical drawing with a measuring device; dividers are best, but you can get away with a ruler.

PENCILS needed are at least a hard one, which can be sharpened to a good point and provide clear lines, and a soft one, which can provide darker lines and quick shading: 2H, HB and 2B provide a good range from hard (H) to soft (B). There are different kinds of pencils, ranging from the traditional wood-encased ones to the more expensive mechanical ones, which need graphite refills.

MEASURING is important and a pair of school dividers will usually do the job. More expensive, but much easier to use, are proportional dividers; these also measure 1:1 but will quickly downscale or upscale if necessary. It means large parts can be downscaled to fit a space, or small parts upscaled so they can be seen easily.

ERASERS and SHARPENERS are necessities. A good eraser should remove graphite cleanly without tearing the paper: plastic, vinyl and polymer erasers are usually preferable to rubber. Kneadable putty erasers are good for dabbing loose graphite off a drawing. Sharpeners need to suit the pencil and achieve a good point – some sharpen better than others.

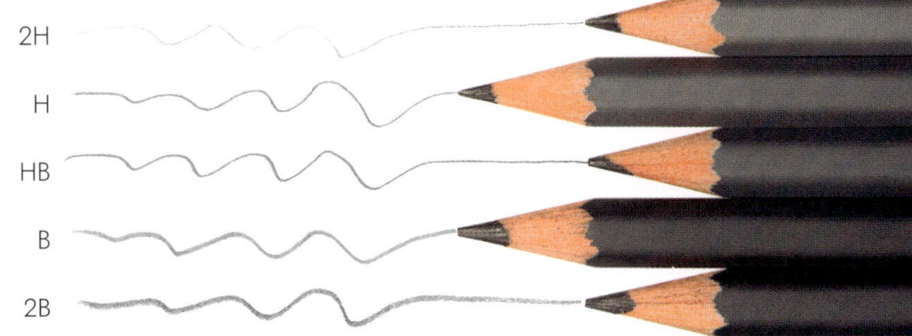

Graphite pencil grades from hard (H) to soft (B)

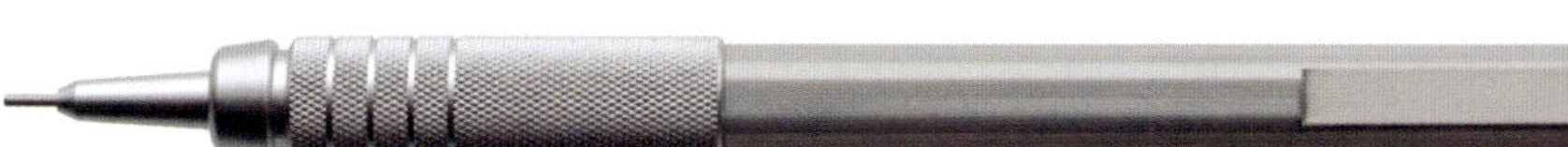

Mechanical or leaded pencil

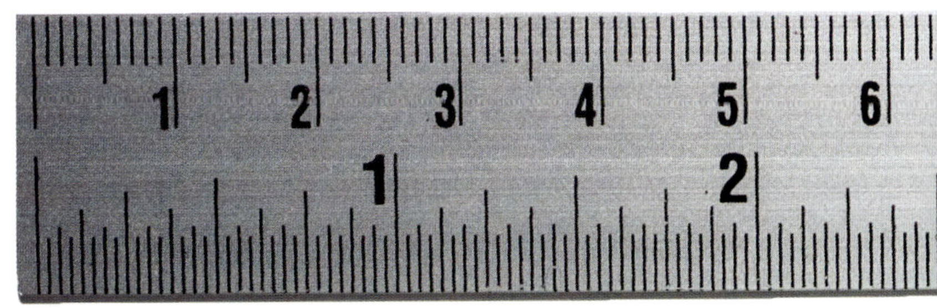

Rulers can be steel or plastic

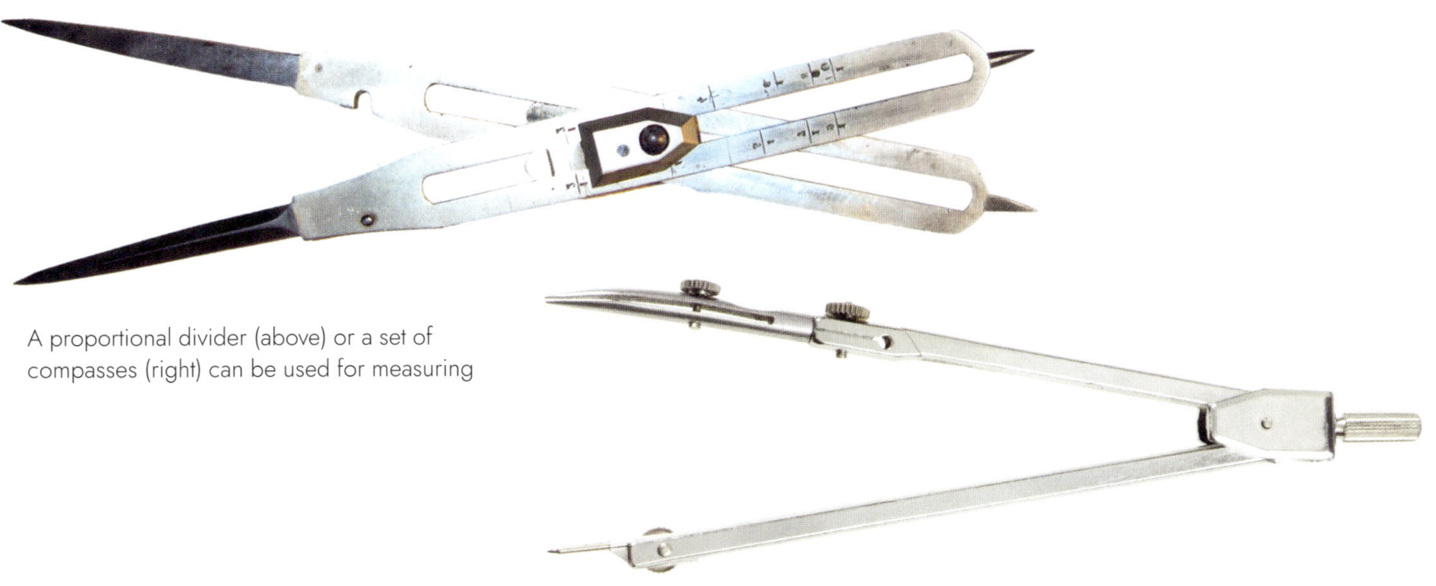

A proportional divider (above) or a set of compasses (right) can be used for measuring

Painting kit

Good-quality equipment helps an artist to paint precisely, though practice is essential if you want to master any technique – even for the experts.

Brushes

Brushes should be high quality, otherwise you will not get the result you want. They should have a good point, a 'spring' and be able to hold diluted pigment for as long as your stroke, to avoid getting a streaky appearance.

Brushes can be long- or short-handled, made from synthetic material or a natural fibre such as sable. Travel brushes that retract into a container are useful. Many botanical artists prefer shorter handles and brushes for the control they allow.

The size of brush matters. Smaller ones are capable of making tiny marks and larger ones are good for working over large areas. A compromise is to buy a medium size with a good tip and well-bodied brush. You can paint fine strokes with the tip, or use the side of the brush, loaded with pigment, for large features. Useful sizes range from 0 to 4.

Brushes can last through several paintings. When they show signs of splaying out or hairs have become loose, retire them and use them as mixing brushes.

Paints

'Artist'- or 'Professional'-quality watercolours are the usual medium, rather than oils or acrylic, for the fine delicacy that can be achieved with them. They are sold in tubes of various sizes, or in half or whole pans, and contain more pigment than 'student' quality. This is an important consideration as paper may threaten to buckle or 'scar' under several layers of pigmented wash, and paint with more pigment will give you good results more quickly.

Make a note of the paint's lightfastness, transparency, staining and granulation properties. These are all written on the tube. Earth pigments are likely to granulate (grains of pigment become visible), which might sometimes matter. Paints that stain won't be so easy to remove, once applied, and transparent colours will show through subsequent layers, but are useful for a velvety effect.

Most instruction books recommend their own palettes, but all agree that, at the very least, you should have a cold and a warm colour in each of the primaries (red, yellow and blue). With so much green in the natural world, it's wise to make a colour chart of greens by mixing a variety of blues and yellows, with a hint of red or pink to subdue the brightness. Whites and blacks are not generally used, though many artists might add a burnt sienna to mix with blue to achieve a natural grey or dark shadow, and an opaque white is useful for adding white hairs.

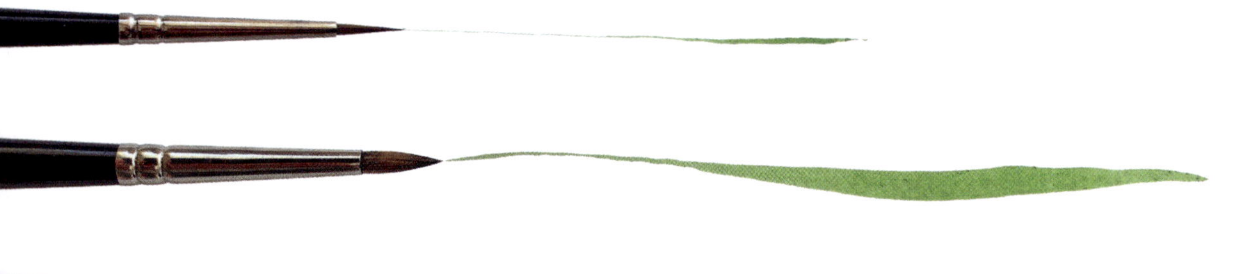

LEFT From top to bottom: size 0, miniature (short) size 4, size 4. Experiment with a few different brushes to find out what suits you.

Paper

Smooth Hot Pressed (HP) paper is used for botanical work. Manufacturers produce different qualities of absorbency, colour and weight, which are made from pulp and/or cotton – sometimes 100 per cent rag. It is sold by the sheet or in pads. Sample packs are available.

Check the weight, so the paper doesn't buckle if wet techniques are used (otherwise stretching is necessary). Absorbency is also a factor if you use lots of water. The smoothest paper might require layers of dry brush, otherwise you will sweep away a previous layer of paint with a wetter brush. Paper's whiteness varies a lot, and you might prefer an off-white rather than a purer white.

Vellum (calf skin) is a beautiful surface to work on if you are using a dry brush technique, and the background can provide its own character.

Transfer (tracedown) paper comes in different sizes and there are two types. Transparent tracing paper needs to be covered on the back with soft pencil, then drawn over the top to transfer lines to the painting paper. Wax-free transfer paper is simply placed between the drawing and the painting paper, thus cutting out a step. However, the opaque nature of transfer paper makes it difficult to see through onto the painting paper.

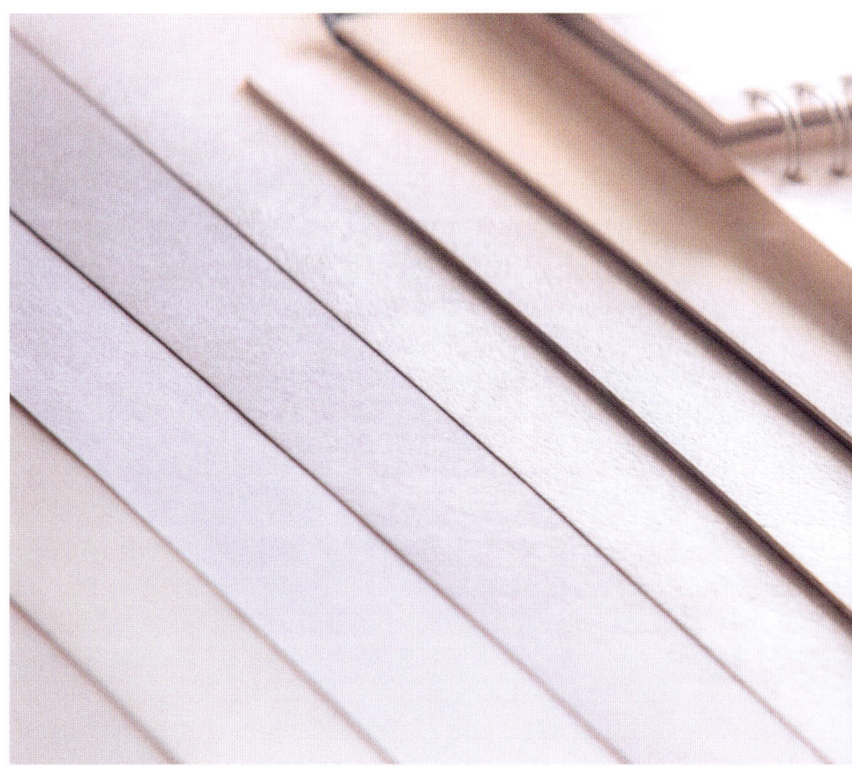

ABOVE Using good-quality paper is important in getting a good final result. Try buying sample packs to find one that works with your painting style.

OTHER KIT

Brush holders: brush cases and brush rolls are essential if travelling. A DIY solution is rolled-up corrugated card. Good brushes are usually sold with a hollow cylindrical top which can be popped back over the brush.

Paint boxes: used to store tubes, pans and half pans, or for travelling. They are made in metal and plastic, and can have mixing trays and a cavity for a small brush.

Palettes: white porcelain dishes are easily washed, show the colour mixture well and provide spaces to prevent mixtures blending. Plastic palettes or white tiles are good alternatives.

Camera: photographs are used for reference only or to provide a record of your progress. When taking reference images, put a ruler or graph paper next to the object to show scale.

Portfolio cases: these range in size from A4 to A1. The ring-binder style uses individual sleeves, while more expensive examples have zip closures. The cheaper card cases do not last long.

Mount (matt) board: used to check the composition and edges.

Masking fluid: protects white areas when swishing a brush over the top.

Where to work

Setting up a good drawing position, both indoors and outdoors, helps to achieve precise results. Get as close as possible to the subject outdoors, whilst indoors good lighting over the subject is essential.

Outdoors

Drawing outdoors is always desirable to understand a plant's habit and to achieve a lifelike result. It also provides an ecological context that might include any animal life enjoying the plant's leaves or pollinating its flowers. The disadvantages of working outside are glare, changing light conditions, getting a good aspect – and rain and insects! Pack a small bag or rucksack with the following items to cope with most circumstances:

- lightweight board

- sketchbook that opens flat

- a selection of pencils (or pens) with dividers, erasers and sharpeners

- a shade of some sort

- lightweight seat

- painting kit comprising a small box with a mixing area, selected watercolour pans or tubes, a small jar of water, and brushes that are carefully wrapped so as not to ruin the tips

- a box/plastic bag in which to put specimens for studio drawing.

A camera, sun hat, insect repellent, sunblock and the always-useful toilet paper complete the list.

THE COUNTRY CODE

The countryside code varies from place to place, but artists should be aware of the following:

It is illegal to remove endangered plants and to take any from a protected area.

If you collect plants outside of protected areas, do not deplete the population, and try not to dig up a plant.

It can be illegal to export or import plants and their seeds.

Only collect on land where you have permission.

Always close gates behind you.

BELOW Capture the emotional essence of your plant, as well as its habit, from field drawings and colour studies.

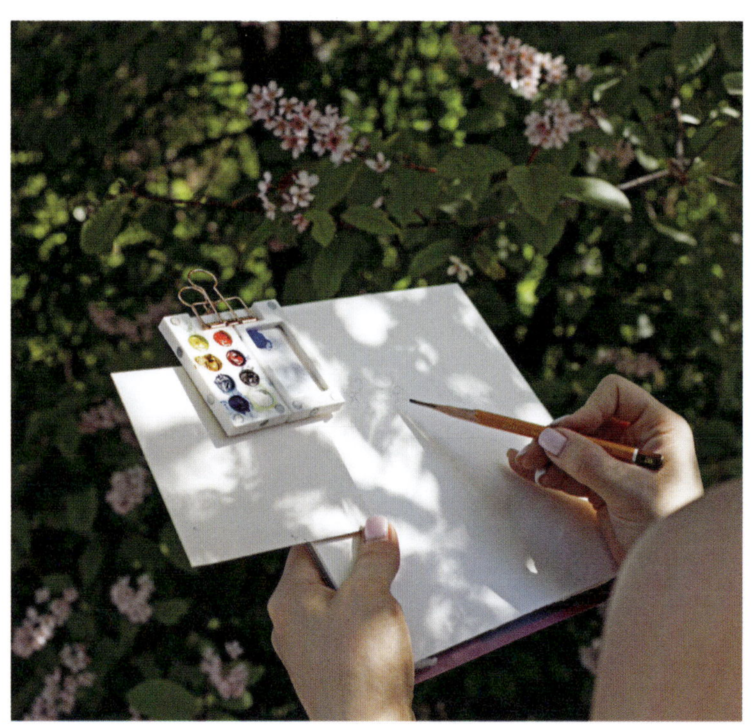

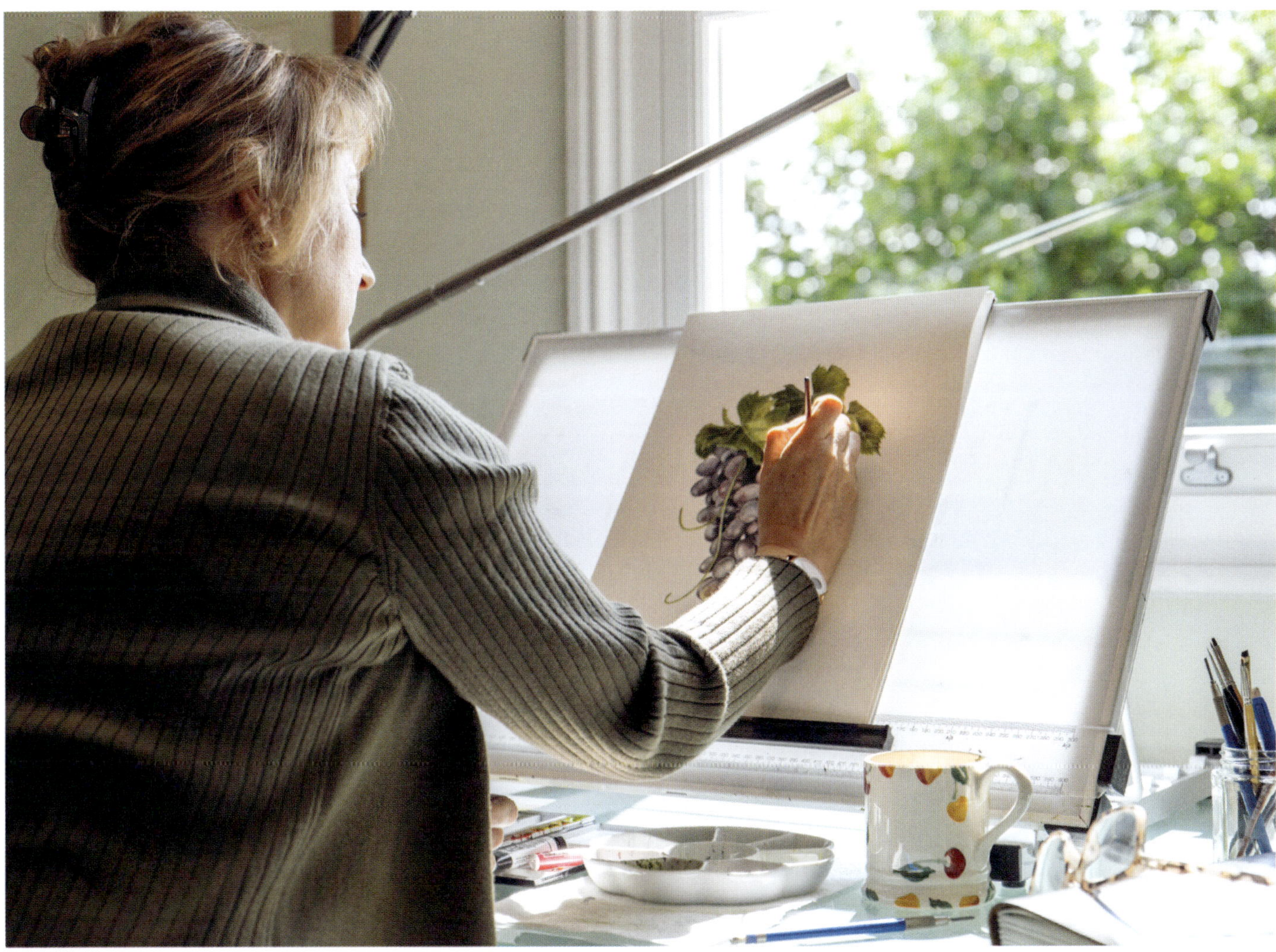

Indoors

Back indoors you will have a table, with a board (or table easel) set at a slight angle. You can position a plant beside it to provide the best aspect and lighting for drawing, and have tools handy for painting, magnification and dissection.

Choose a room that gets indirect light to avoid changes with the sun's movements. The table should be positioned so that your drawing hand does not cast a shadow over your work. If you use artificial light, invest in a lamp with an arm that can be adjusted to shine on different aspects of the plant – especially important for highlights.

Set up the plant using a vase, with plenty of string, tape and pins to keep it still. A laboratory retort (or stand) is useful to hold the plant at eye level. Florists' tubes, jam jars and bottles in which to put small specimens are also handy.

A hand lens is an inexpensive way to look at details but, better still, use a basic dissecting kit to cut open flower parts and a student stereo microscope with a zoom lens to view them. A magnifying glass may also be useful.

What to put on the page

Drawings provide the basis for the painting, but very few botanical artists draw straight onto good-quality paper. Their final paintings are based on initial drawings of all the parts needed, or that might be needed. These are made on sketching paper then transferred to good-quality paper using tracings. This method provides scope for rearranging parts, ignoring what is not needed, and being certain the outline is in the right compositional position before painting.

A comprehensive 3-D drawing, indicating shadows and highlights, also provides a good reference for painting and can make constant referral to photographs less necessary.

What to include in your sketches

1. The main feature. The final painting is likely to consist of a main flowering stem or inflorescence, a branch or whole plant, whether fruiting or in seed.

2. Explanatory parts, such as details of the flower, stem or seeds can be added in the space around the main feature. Draw as many as you can, because you never know what might be useful in the composition.

3. Add notes as reminders, or to suggest an aspect to be careful about or that has not worked for you – no one else need see it. You can also include a date and place of collection — at a later time you'll be able to see how well you've progressed.

4. Scale; if you decide to draw at larger than, or less than, full size, make a note – for example, x 2, x 3 and so on, or use a scale bar with the actual length marked out and placed against the enlarged feature.

5. Margins can help keep a drawing within the boundary and not wandering off the page by mistake. They are also useful places to add notes and colour matches.

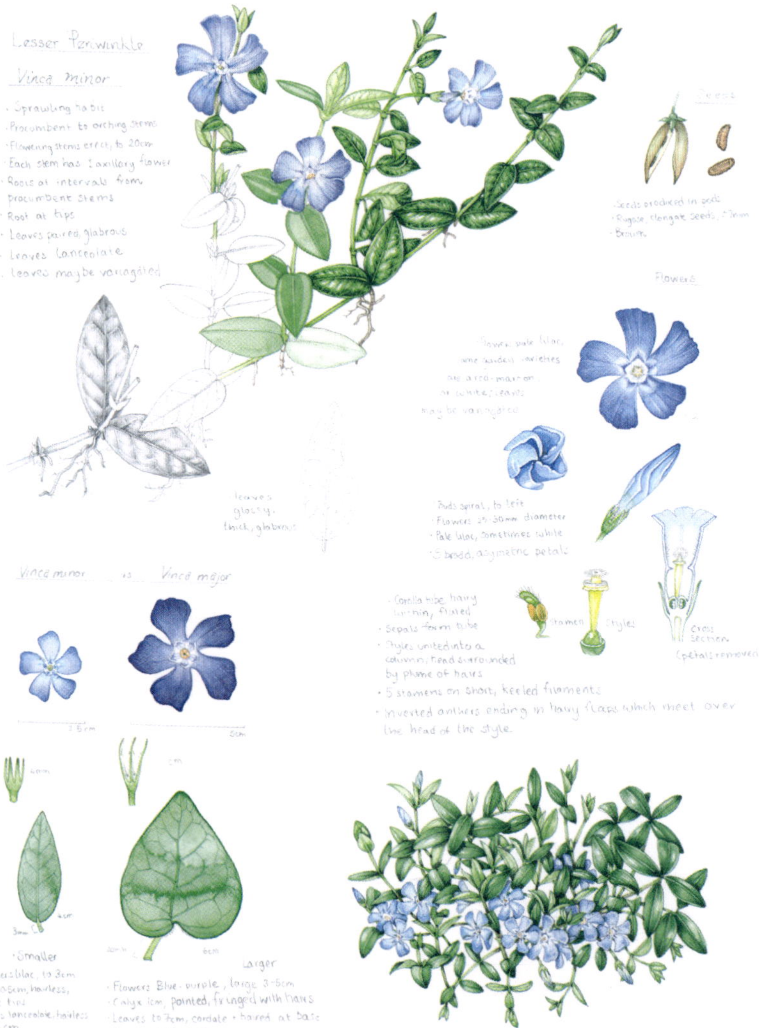

LEFT Lizzie Harper's sketchbook page for the lesser periwinkle (*Vinca minor*) includes sketches of different aspects of the plant, as well as comparisons with the closely related greater periwinkle (*Vinca major*).

RIGHT Understanding colour, from mixing the right shades (top) to building up colour using small brush strokes (bottom), is key to creating an accurate painting.

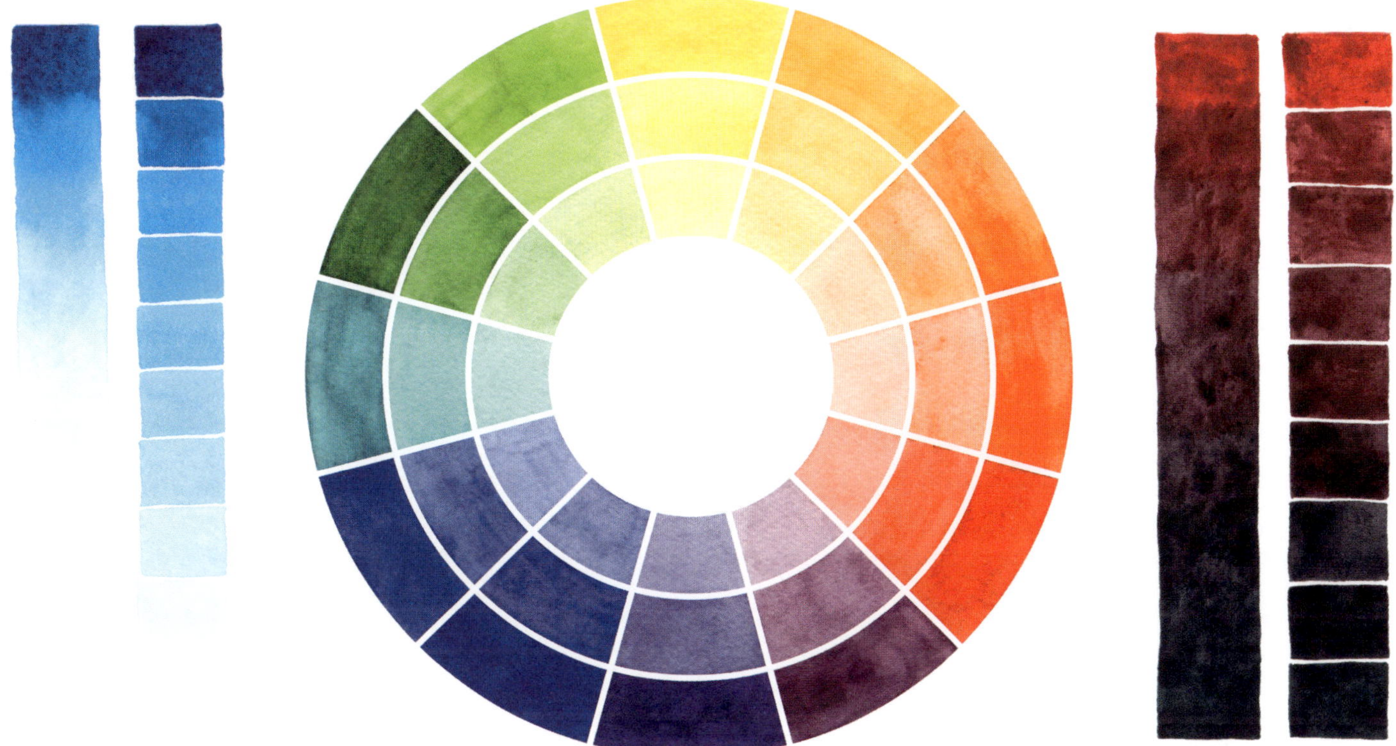

Colour matching

This is the fun bit for many people, though you should not stint on perfecting your sketches. Colour is an unreliable diagnostic feature, and becomes even more so when you're matching it in full sunlight, or in dubious indoor lighting where bulbs and background can significantly alter a colour matched outside. Decide what your starting point is and try to stick with it.

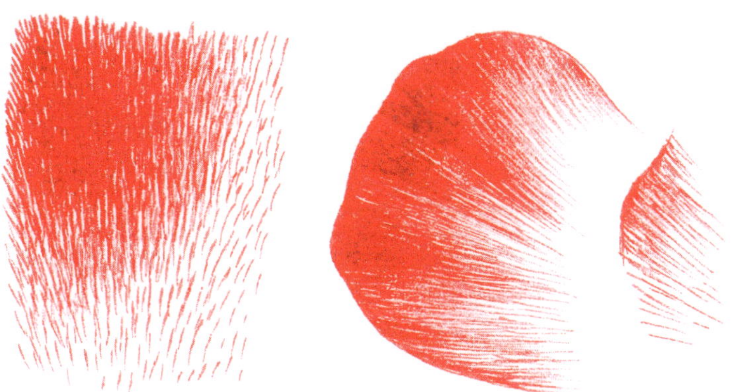

1. Make your colour blobs on the sketching paper, or use a spare piece of the paper on which you intend to paint. This will give you the best idea of how the colour will look when it dries. Colour match every aspect of the plant from petals to leaves and stem. Try to limit the basic colours to prevent complicating your palette and to minimize the risk of not being able to replicate the mix.

2. Note mixes for a shadow colour in each case.

3. Experiment with applying the paint, how you might insert veins, deal with spiny structures and whether the paper surface is suitable for your technique. This is also the time to decide whether to use a dry or a wet technique, and what paper to use. You might want to achieve a highlight by lifting off paint, so you won't want to use a staining paint or paper that buckles under too much water; or you might want to try scratching lines for hairs and so you don't want your paper to tear.

Composition

Composing is a matter of intuitive taste. Sometimes tweaking the orientation of a leaf or leaving out a flower or leaf, because it gets in the way, makes all the difference. Always ensure that botanical integrity is maintained. As a guide, make thumbnail drawings as soon as you can. Look at other paintings with similar subjects to get ideas and be clear what you want to communicate. Here are a few elements that should be considered in order to create a pleasing composition:

- Overall shape: for example, oval, square and so on.

- Orientation of lines: for example, where do they face in relation to the margins, or how do they relate to each other within the painting?

- Focal point: where is it in the painting? Is it achieved by tone, colour or size?

- Balance. This often conforms to the idea of the golden section, golden mean or Fibonacci ratio – approximately 5:3 (see page 49). It is unwise to be too rigid about the theory, as it is usually achieved intuitively. Even more approximate is to apply the 'rule of thirds', that is, to place your focal point off-centre; or point a slender leaf a third of the way down a margin, rather than in the middle.

- Balance can also be achieved by 'grounding' the plant to avoid it looking top heavy – for example, with leaves or another flower, or filling awkward negative spaces.

The composition can make or break an otherwise beautifully painted flower. Copy your drawings onto tracing paper, cut them out and play with each component to find a pleasing design. Tips of leaves pointing to another feature can close gaps and negative spaces can also make a statement. Adding a dried leaf or scattered seeds can be a playful way to balance the whole.

Allow wiggle room at every stage, such as leaving a decision on where to place a cut and what angle it should be to match a nearby feature to the end.

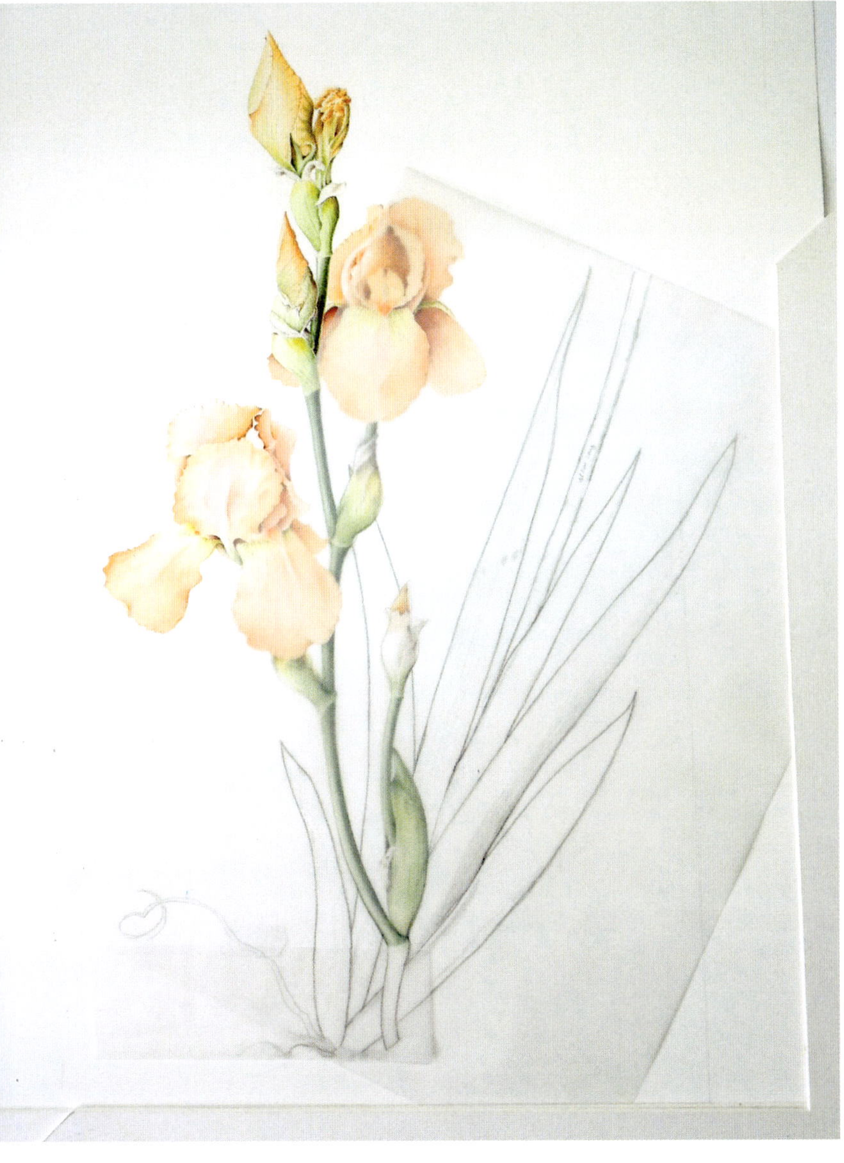

LEFT A tracing of a fan of leaves is used to work out their most appropriate position. To create a guide for repositioning the leaves later, make two lightly pencilled Xs on the paper and the tracing paper.

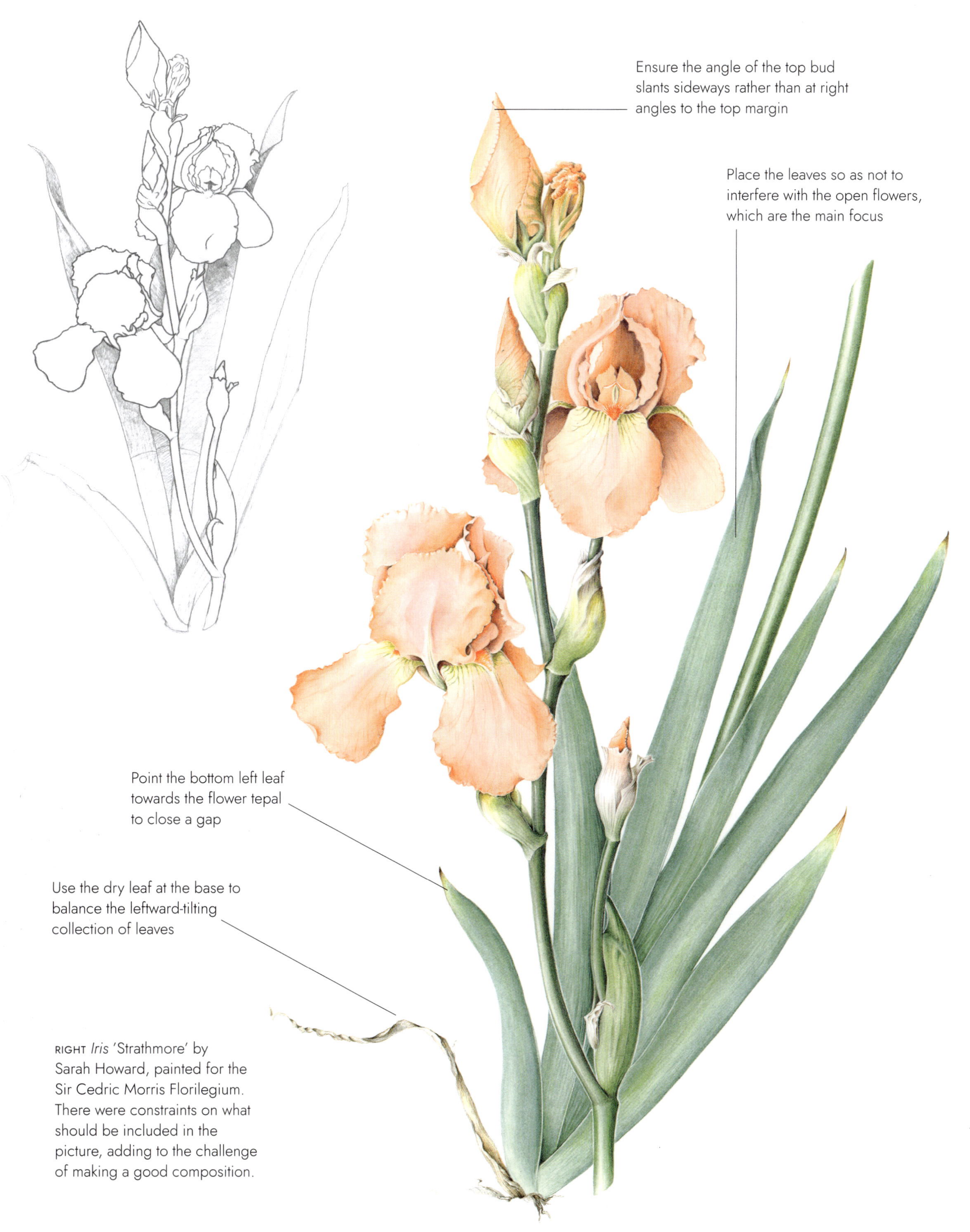

RIGHT *Iris* 'Strathmore' by Sarah Howard, painted for the Sir Cedric Morris Florilegium. There were constraints on what should be included in the picture, adding to the challenge of making a good composition.

Basic watercolour techniques

Two basic painting techniques are used to paint flowers: a watery one and a dry one, with many approaches in between.

Every botanical painting tutor has their own method, described in many excellent 'how to' books. Consult several of them for, in practice, you will need different techniques for different purposes, and everyone achieves what they want in different ways. It is common to use both wet and dry techniques in the same painting.

The application of watercolour is usually pale and dilute at first, working towards an intense saturation. Many artists choose to work in this way from the darkest areas of the form to the lightest, leaving highlights to last.

Wet techniques

Wet techniques involve wetting or moistening the surface of the paper first and then applying watery pigment in strokes. The stroke should bleed subtly into the surrounding paper without leaving watermarks or streaks, depending on how wet the brush is. Always have a clean, spare brush to moisten the paper or to smooth out an edge. It takes practice!

Washes; some artists prepare their painting surface with a wash of dilute paint. This can helpfully alter the paper surface in some cases and provide shapes to follow, or can be used as a final application.

'**Puddling**' requires a lot of watery strong colour, which is pushed into a marked-out shape. This is useful for structures with a complicated edge, such as a serrated or spiny leaf.

Glazing is done by stroking dilute colour lightly over a moistened shape. It is applied in several layers, always after the previous layer is thoroughly dry. It enables you to control the colour process, working from dilute to saturated. It is useful for large, smooth areas needing tonal gradation.

BELOW Christmas cactus (*Schlumbergera* cv.) by Gloria Newlan. It has the appearance of a loose, watery style, but is deceptively accurate. Beginning with a few layers of colour washes, painting continues with stronger colours gradually defining detail and shape. 'Peripheral' colours are added in further washes. The brush is large for most botanical artists — a size 6 — and used to blend the edges with each wash, so as not to leave a streak. A tiny size 0 is used for the detail.

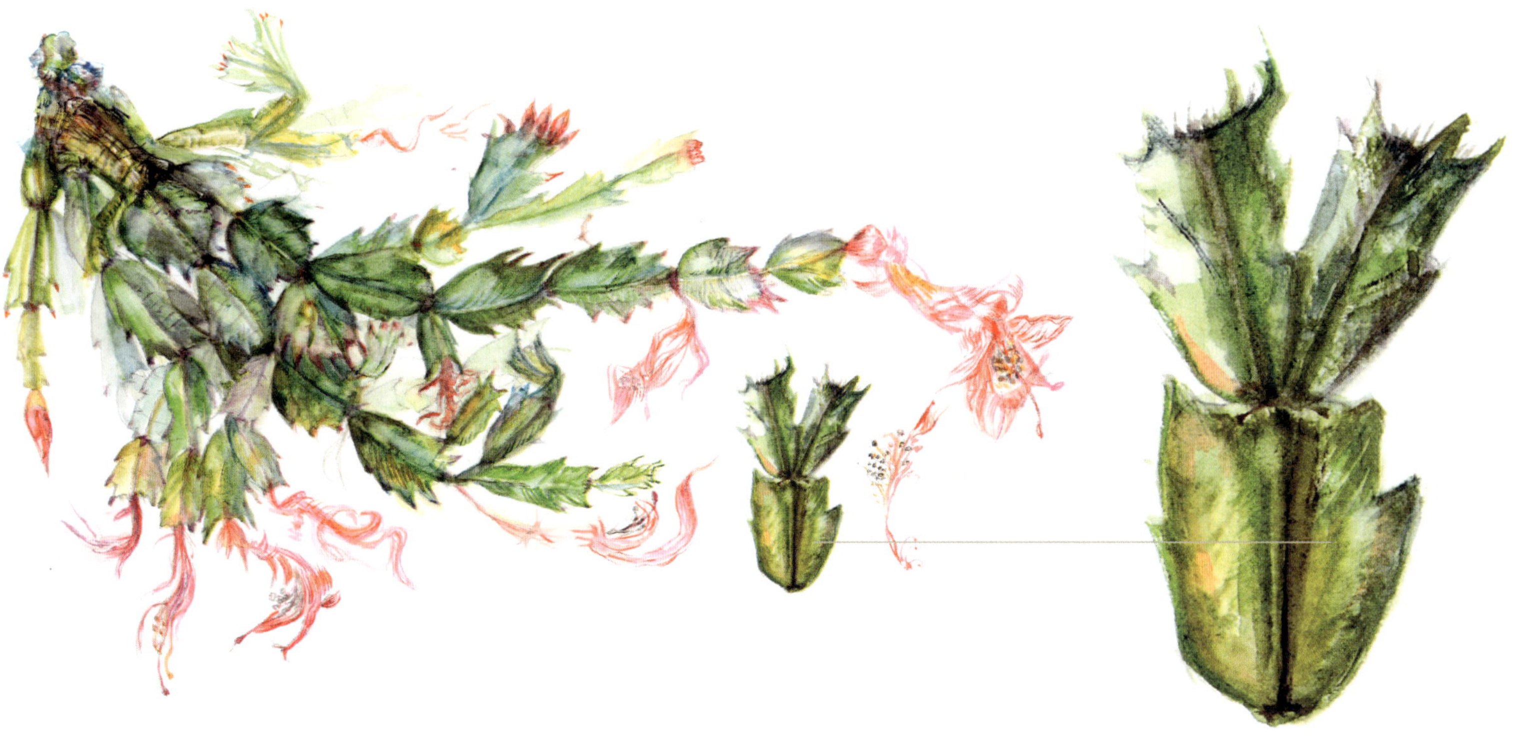

Dry brush

A dry technique is like drawing with a paint brush. It provides absolute control, avoiding rivers of watery pigment with a life of their own.

The brush should be dry enough to apply tiny markings, even stippling. Dry brush is essential for dealing with tiny structures or spaces after the basic washes have been applied. It's also a good way of tidying up edges, applying surface colour on small details and emphasizing dark shadows behind pale-coloured stamens.

Dry brush is often not as dry as is suggested, but always uses minimal water. A little moisture on the brush or on the paper, often both, might be needed to make the paint sink into the paper.

Feathering is essentially a dry process. The brush is loaded with the pigment then very lightly 'feathered' over the form, so only the surface of the paper is covered. It could be used to cover a highlight quickly, or to apply another colour over a velvety feature.

Stippling — the application of tiny dots of colour — is slow but meditative. It can result in tiny white spaces on the paper, which has the effect of 'clouding' a subject, so some artists apply a wash colour in advance or afterwards.

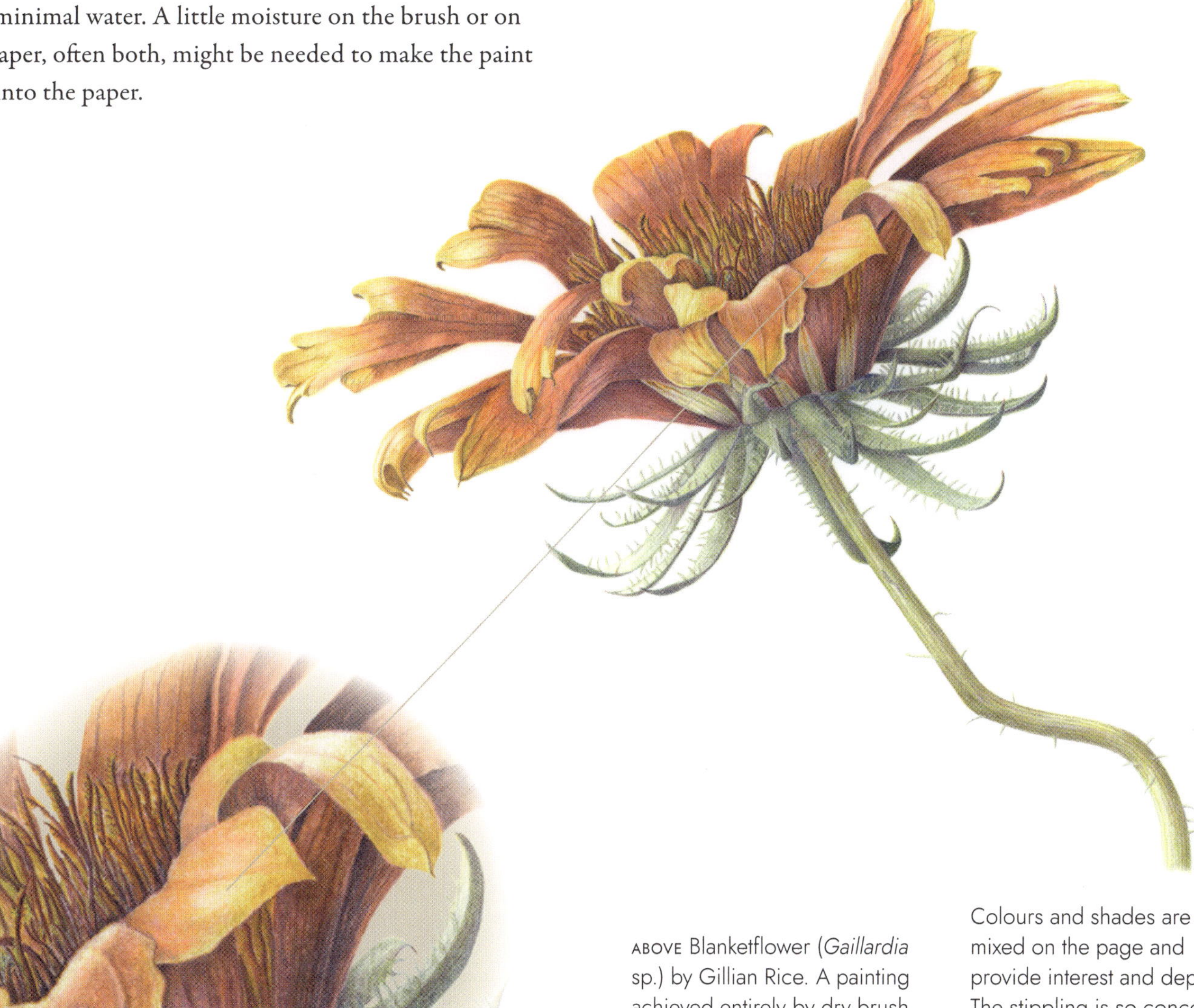

ABOVE Blanketflower (*Gaillardia* sp.) by Gillian Rice. A painting achieved entirely by dry brush with a stippling technique. The basic palette colours are applied and matched as work progresses, dilute and random at first, then more concentrated later.

Colours and shades are mixed on the page and provide interest and depth. The stippling is so concentrated, no wash is used. Gillian's tools are Winsor & Newton Series 7 Miniature size 0 to 2, or Rosemary Series 323, size 2/0. For very fine detail, Interlon 1023, size 3/0.

Further techniques

Colour mixing

Colour mixing in watercolour is often trial and error. There are many variables, such as the quantity of pigment added from each colour, the amount of water, and the type of paper.

1 | The best way to mix a colour is to mix a wash of the lightest colour and then slowly add in tiny amounts of the second colour, testing at each point. If the colour becomes the approximate shade but too dark, add more water to dilute the pigment. Adding more of the dark colour will make the mix darker while adding more of the lighter colour will return the wash to a paler shade.

2 | Colours interact in interesting ways. Experiment by testing mixes on small sections of paper and creating charts to see what colours might do.

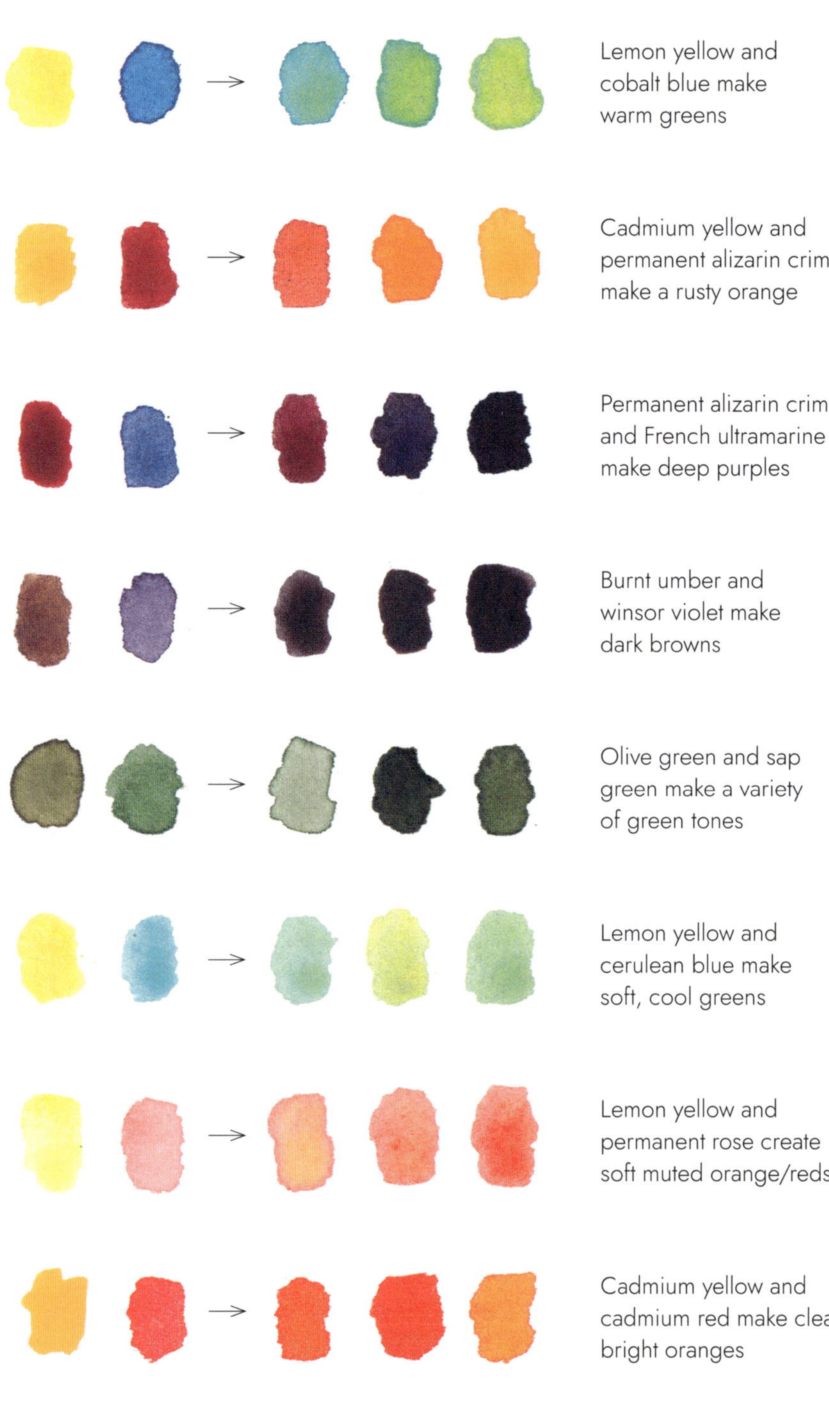

Lemon yellow and cobalt blue make warm greens

Cadmium yellow and permanent alizarin crimson make a rusty orange

Permanent alizarin crimson and French ultramarine make deep purples

Burnt umber and winsor violet make dark browns

Olive green and sap green make a variety of green tones

Lemon yellow and cerulean blue make soft, cool greens

Lemon yellow and permanent rose create soft muted orange/reds

Cadmium yellow and cadmium red make clear, bright oranges

French ultramarine and winsor violet create clear, bright violet

Building colour in layers

Building up layers of colour is a key technique in watercolour painting, adding a new wash as each dries and fades. The more layers used, the more vibrant the colour becomes. The red apple here demonstrates how dark yet vibrant colour can be achieved.

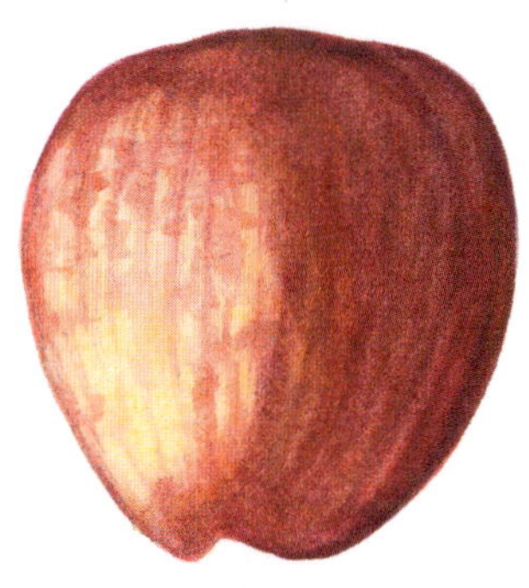

4 | Apply a wash of cadmium yellow over the whole of the dark red side of the apple, this will brighten the colour. Use a permanent rose wash for the pink pattern areas on the light side of the apple.

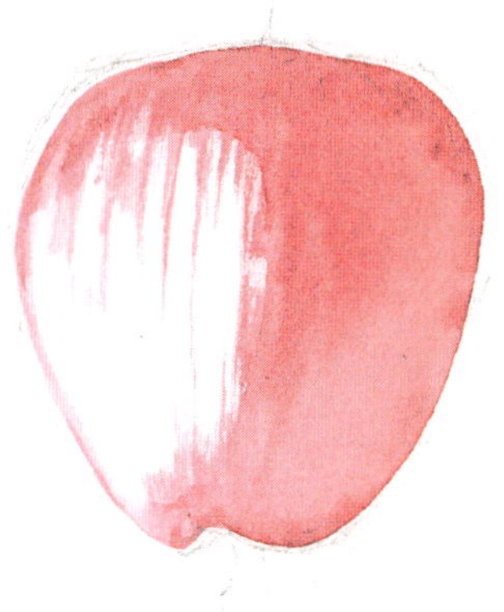

1 | Draw the apple outline in pencil. Mix permanent alizarin crimson and cadmium red to create a deep red wash. Use a wet-on-wet technique, putting clean water onto the page and then adding the colour, allowing it to spread over the apple, using broad brush stokes to follow its shape. Use a small brush to sharpen up the edges all the way around.

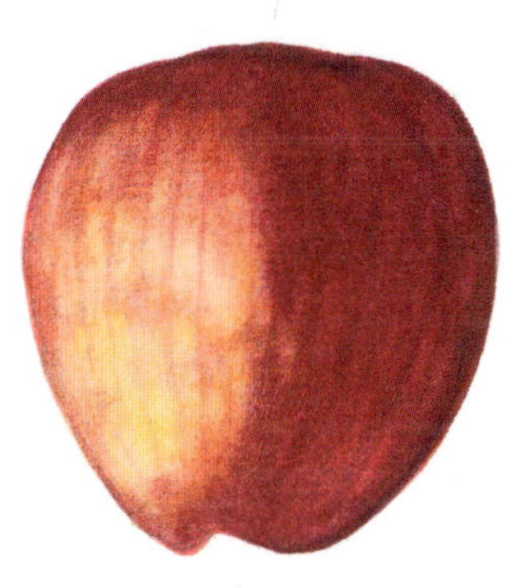

5 | Use a wash of permanent alizarin crimson on the dark red side of the apple, following the form and defining areas. Apply a lemon yellow wash to the yellow section of the apple and use this to smooth the colour and pattern so the skin looks more rounded.

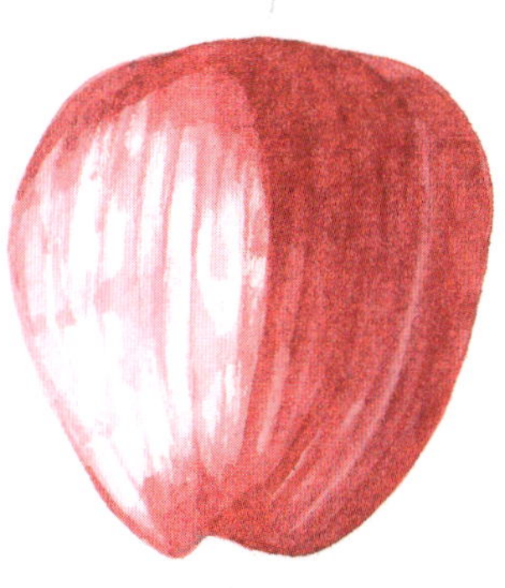

2 | Rub out the pencil lines. Following the curve of the apple, add another layer of the initial wash to the dark parts. Add some of the 'blocky' pattern to the lighter side of the apple using a watery wash of the same colour.

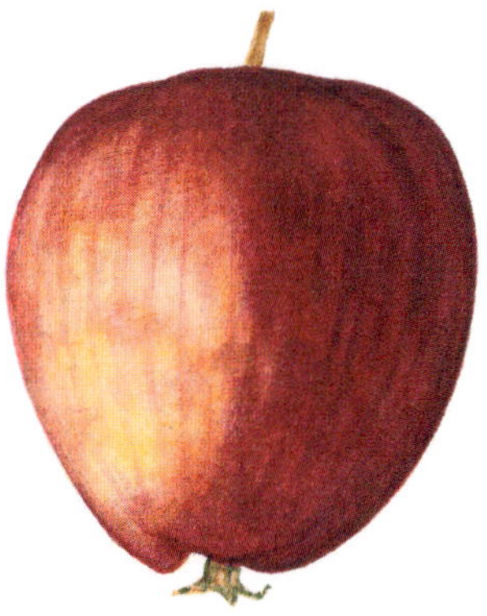

6 | Apply another permanent alizarin crimson wash to the darker side of the apple, especially the darkest areas. Use raw umber to paint the stem and raw umber and sap green for the sepals at the bottom.

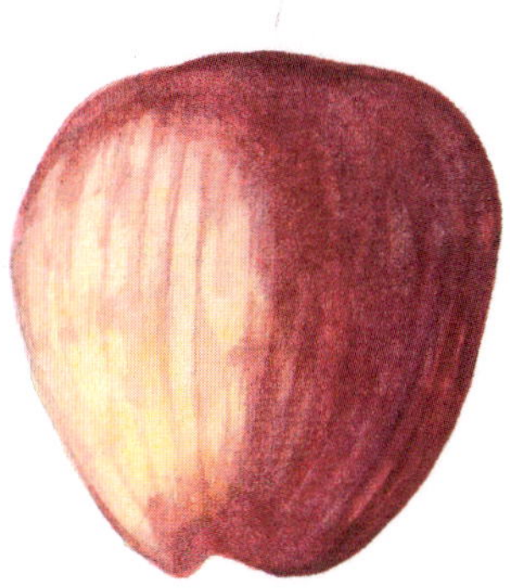

3 | Mix winsor violet into the wash and use this on the darkest parts of the apple. Mix lemon yellow and sap green to make a greeny yellow and add this to the light side of the apple.

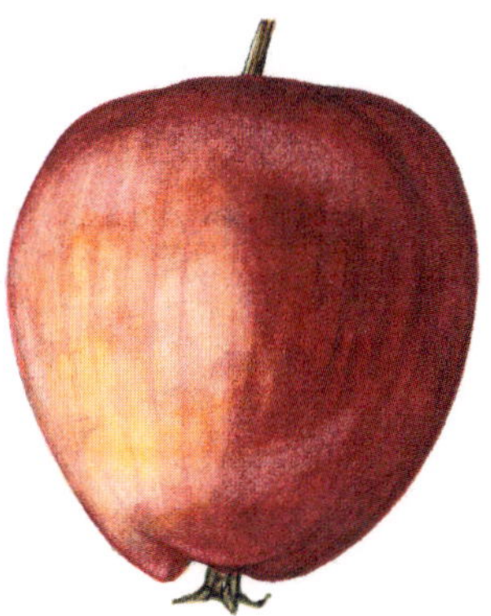

7 | Apply clean water to the very light areas of the apple, then use a tissue to lift off the colour by pressing it hard on the paper. Soften the resulting edges with your brush. If too much paint is removed it can be reapplied and will still look lighter. Apply further details to the stem and sepals using raw umber and sap green and, finally, a tiny bit of Payne's grey.

Painting dark and shiny objects

This is the seed of pod mahogany (*Afzelia quanzensis*), a tropical tree with orange arils on the seeds. This technique can be used for any dark and shiny surface.

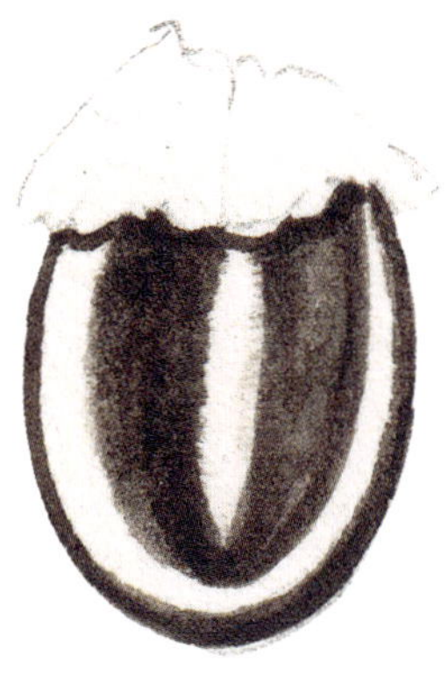

1 | Draw the seed and mark out the lighter areas where the paper needs to remain white. Mix burnt sienna, burnt umber and Payne's grey into a dark wash. By using these three colours a really dark colour can be achieved without using black, which is considered quite a dead colour and best not to use if at all possible. Apply the first wash.

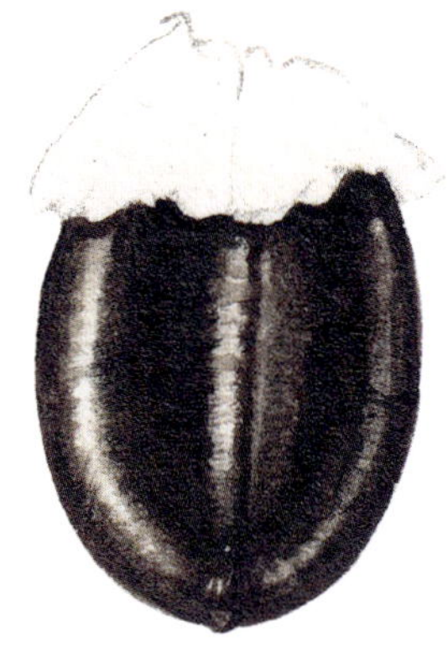

4 | Add more Payne's grey to the wash to make it darker and use this to darken some areas of the seed surface.

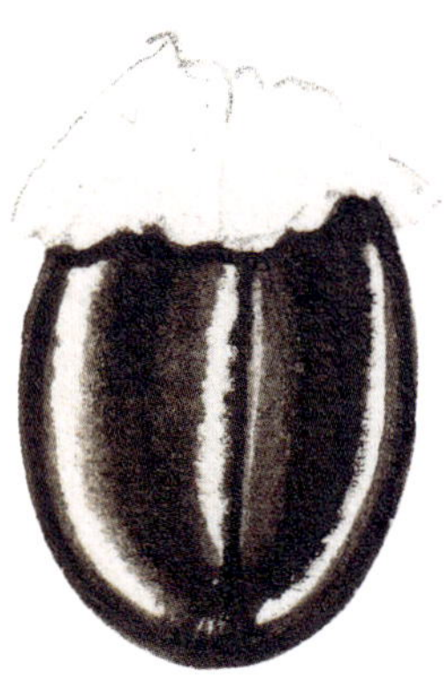

2 | Use a second wash of the same colour, darken some areas further and, using a small brush, feather the colour using very small lines that follow the form of the object into the white areas. This takes away the hard edge of the dark line.

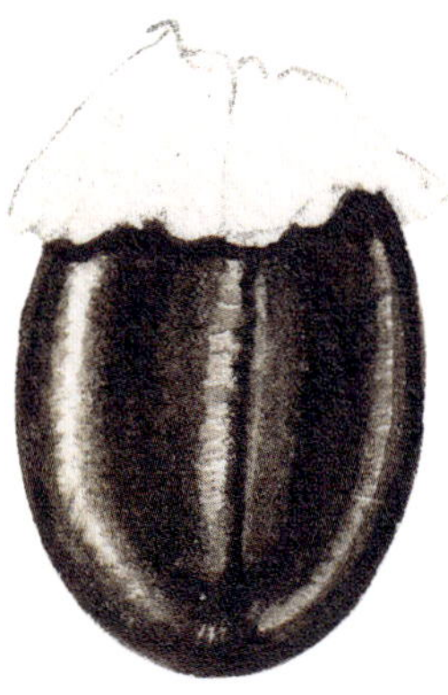

5 | Apply a cadmium yellow wash to the aril, leaving lighter areas white. Mix cadmium red into the wash and apply this to darker areas, leaving a white edge around the aril.

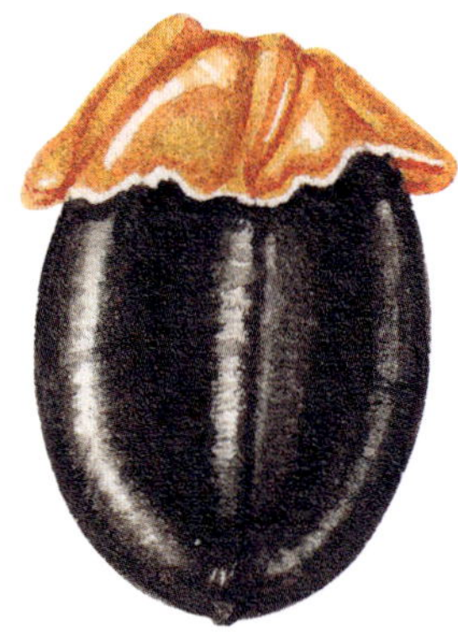

3 | Still with a small brush, use clean water to feather the colour into the light patch on the darker side of the seed to create a three-dimensional shape by making one side darker and one side lighter. Further feather the edges on the light side, following the slight curve of the surface.

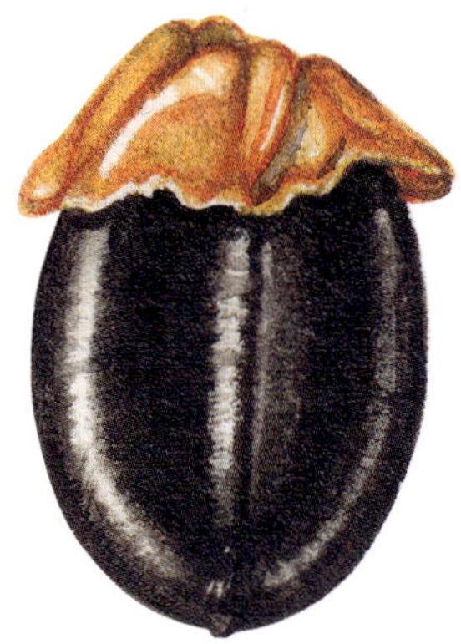

6 | Mix a shadow wash of Payne's grey and cadmium red and apply this to areas of shadow on the aril. Finally, apply cadmium yellow to the white line on the aril on the darker side of the seed to show that one side is darker.

Creating form using light and dark

Rendering a three-dimensional object on a two-dimensional surface involves using light and dark areas to show the form. As it is almost impossible with watercolour to go back and make something lighter, it is vital to be mindful of areas of light right from the start and to keep these areas with little or no paint.

Leaves quite often have bumpy surfaces, with many changes of light and dark. Only one colour – sap green – is used here, applied in layers to build up the areas of darkness.

This project is by no means a finished leaf, and many further washes can be added. But these first five steps show the general process of creating form using light and dark areas.

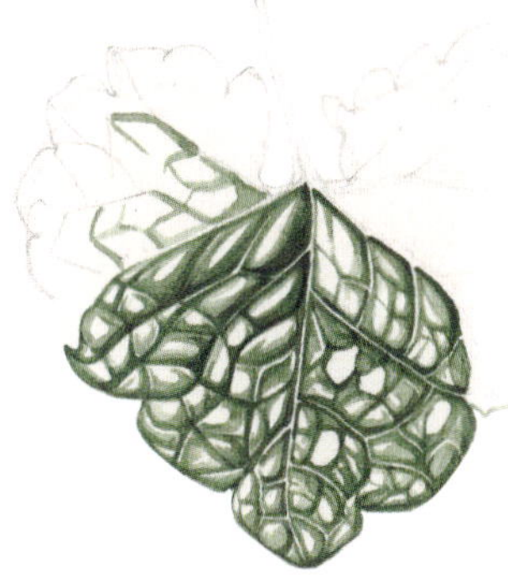

3 | Use the next application of sap green to clarify the darker areas of each bump, applying it to the far right-hand side of the bumps. Also use this wash to outline the edges of the leaf and make these lines clear and solid.

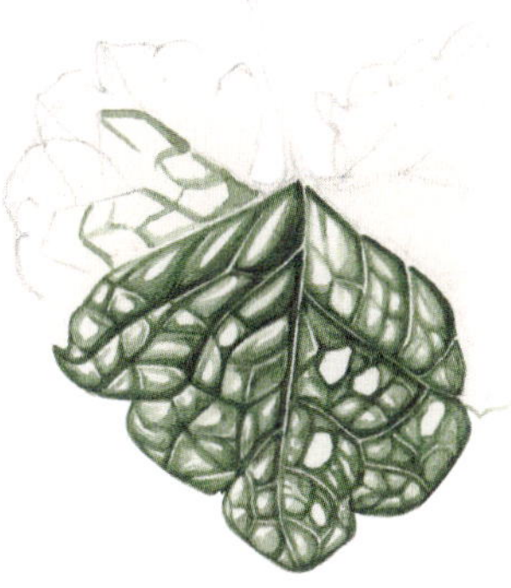

4 | Using clean water, gently pull colour from the darker areas into the lighter areas, make any harsh lines softer and blend areas into one another. Using a small brush and clean water, work on each bump separately and smooth the colour. There should still be areas of white left on each bump.

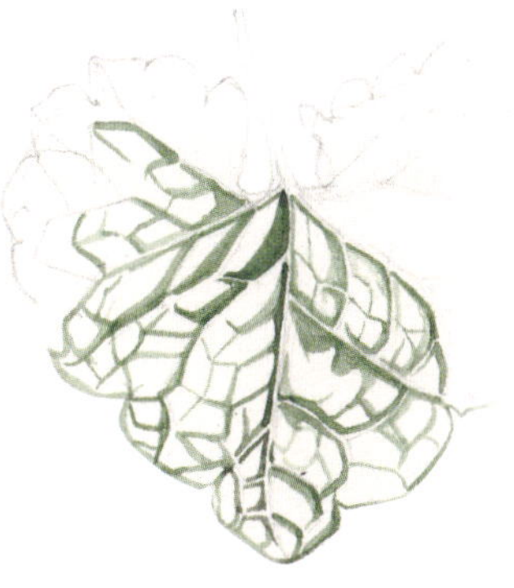

1 | Draw your leaf, showing the areas where bumps are created by the veins running over the surface. Mix a watery wash of sap green and apply this to the darker side of each bump, leaving large areas of each bump white — the light is coming from the left-hand side, so the right-hand side is the darker side of each bump. This should leave the veins white. This first application of paint maps out the painting to follow.

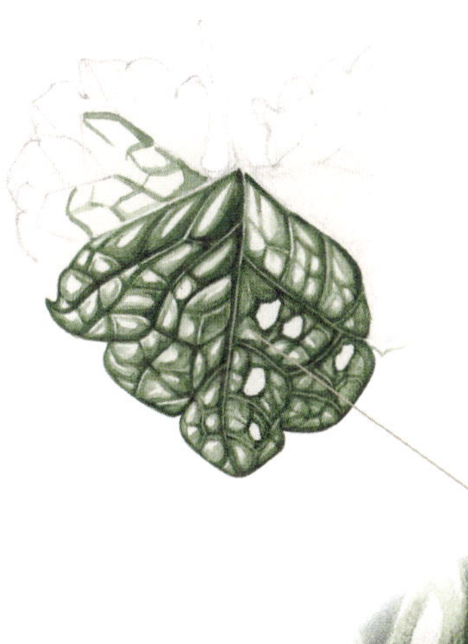

5 | Finally, apply a light wash of sap green to areas of the leaf where the bumps are darker and have no light areas, and also to some of the veins.

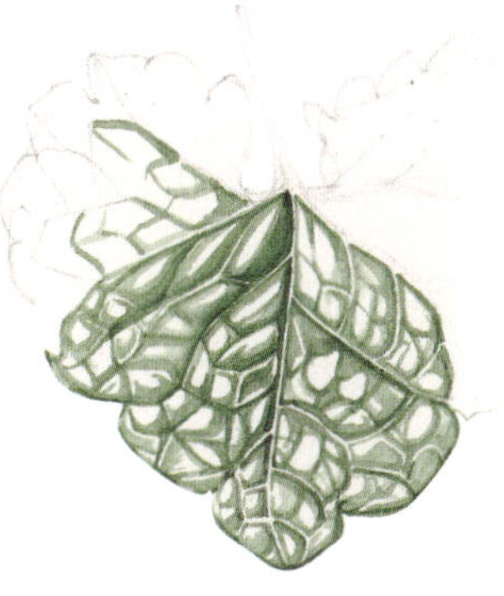

2 | Apply a second layer of sap green, still leaving an area of white clear on the middle/left-hand side. Add another wash to the areas in shadow.

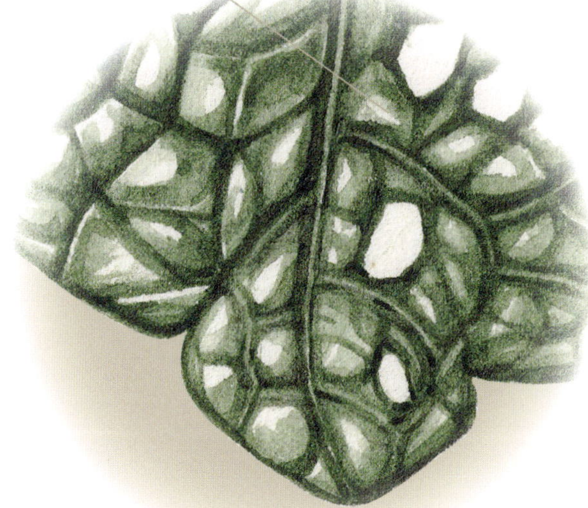

Chapter two

Botany essentials

Botanical artists are mostly concerned with what they see, but they need to understand the structure of a plant, as well as its classification and ecology. The reasons why plants look the way they do can often be explained by other aspects of botanical enquiry involving the study of function, chemistry and molecular biology.

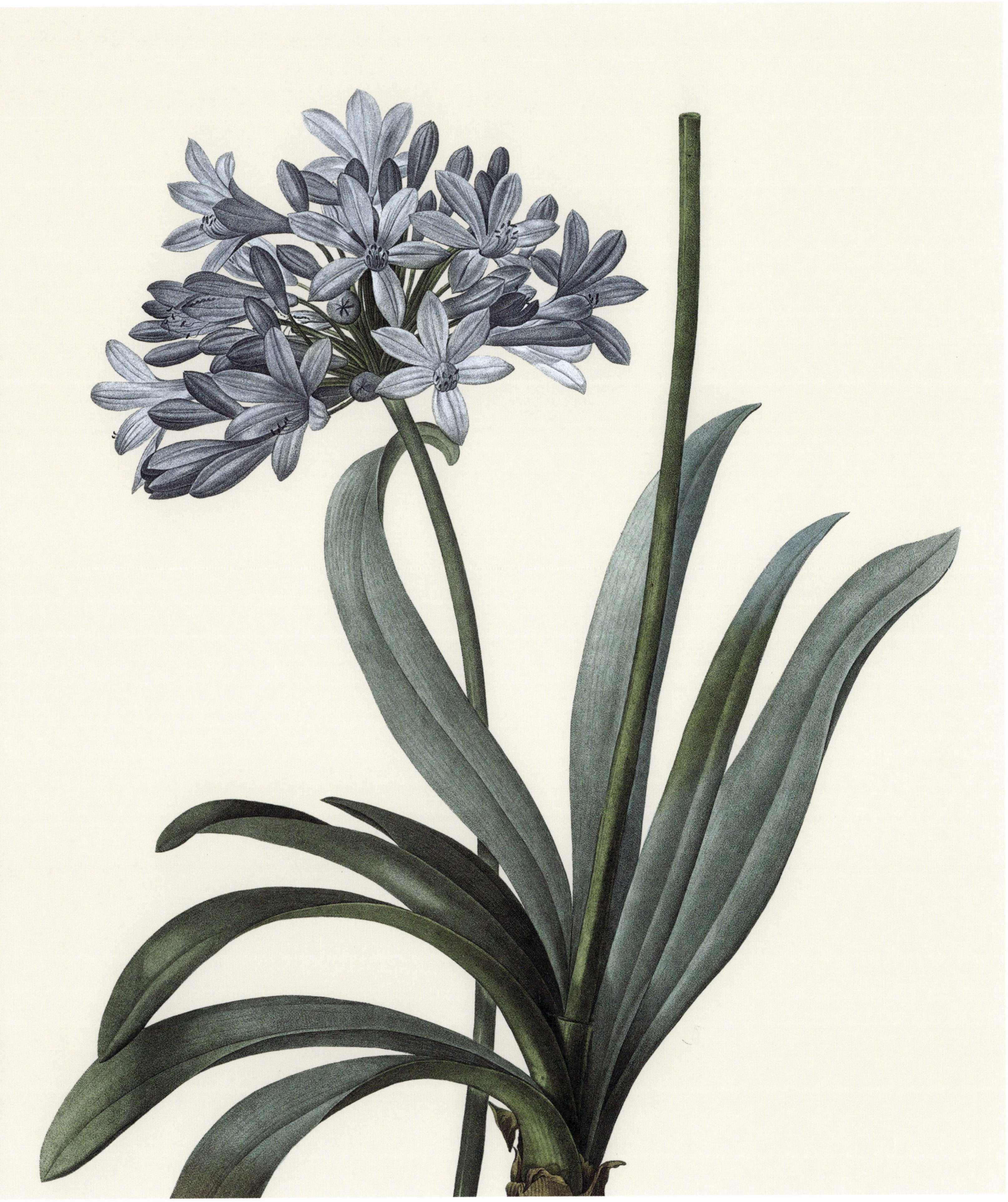

What is a plant?

There are nearly 400,000 species of plant on Earth, most of them flowering, though some are not yet discovered or named. They come in an infinite variety of forms, which enables them to occupy every possible ecological niche.

One definition is that plants share all, or some, of the following characteristics:

- With the exception of a few wholly parasitic species, plants make their own food by using sunlight to photosynthesize carbohydrates from water and carbon dioxide, using a green pigment called chlorophyll.

- They are multicellular, and the cell walls contain cellulose, giving the plants rigidity and strength.

- Although they cannot lift themselves up and move about like animals, their roots, shoots and leaves can explore the environment around them in search of food, light and water.

PHOTOSYNTHESIS

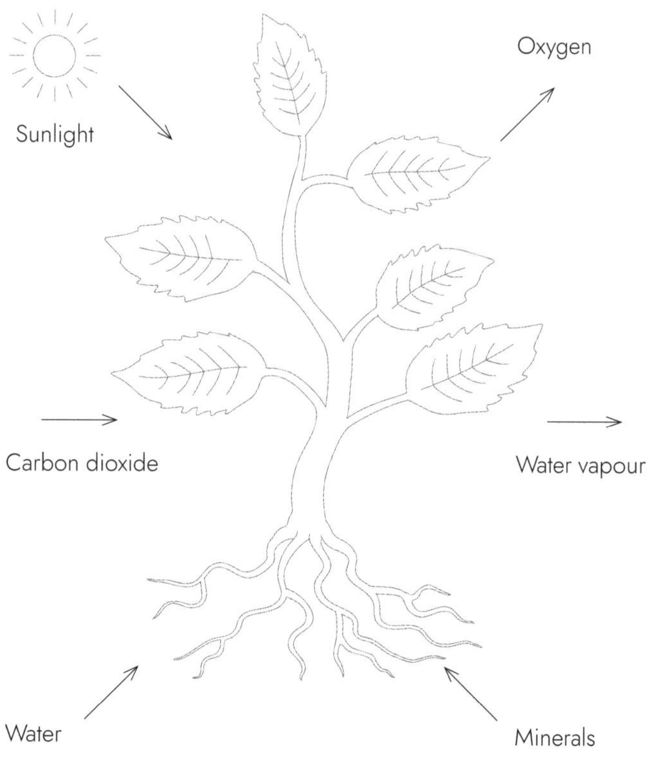

Fungi are attractive to paint and used to be categorized with plants, but are now a separate group, because they are biochemically quite distinct from plants and derive their food from other organisms that are either living or already dead.

Life on Earth is dependent on the plant kingdom (the scientific name for which is Plantae). The food chain begins with the carbohydrate energy stored within plant organs. A by-product of carbohydrate production is oxygen, which is expelled from leaves and sustains animal life.

Basic structure

To fulfil its basic requirements a plant has a root system, which keeps the plant fixed to the spot and absorbs water and nutrients from the soil; and a stem, on which the above-ground features are attached. Essentially, these are leaves, flowers and fruits.

Plants manufacture food by using the green chlorophyll in their leaves to absorb the energy from sunlight, essential to fusing carbon dioxide with water. Leaves have pores on their surface (stomata), which enable carbon dioxide to enter and oxygen to leave (the opposite to gaseous exchange in animals). A vascular system, seen on old leaf scars and leaf veins, carries water and sugars around the plant.

Flowers are also attached to the stem. Petals are really modified leaves. Showy ones attract animal pollinators, whereas petals of wind-pollinated flowers may be reduced or absent. Inside the flowers are the male and female parts (the androecium and gynoecium respectively).

LEFT Plants derive their energy from the photosynthetic process. It enables them to grow and gives leaves their green colour.

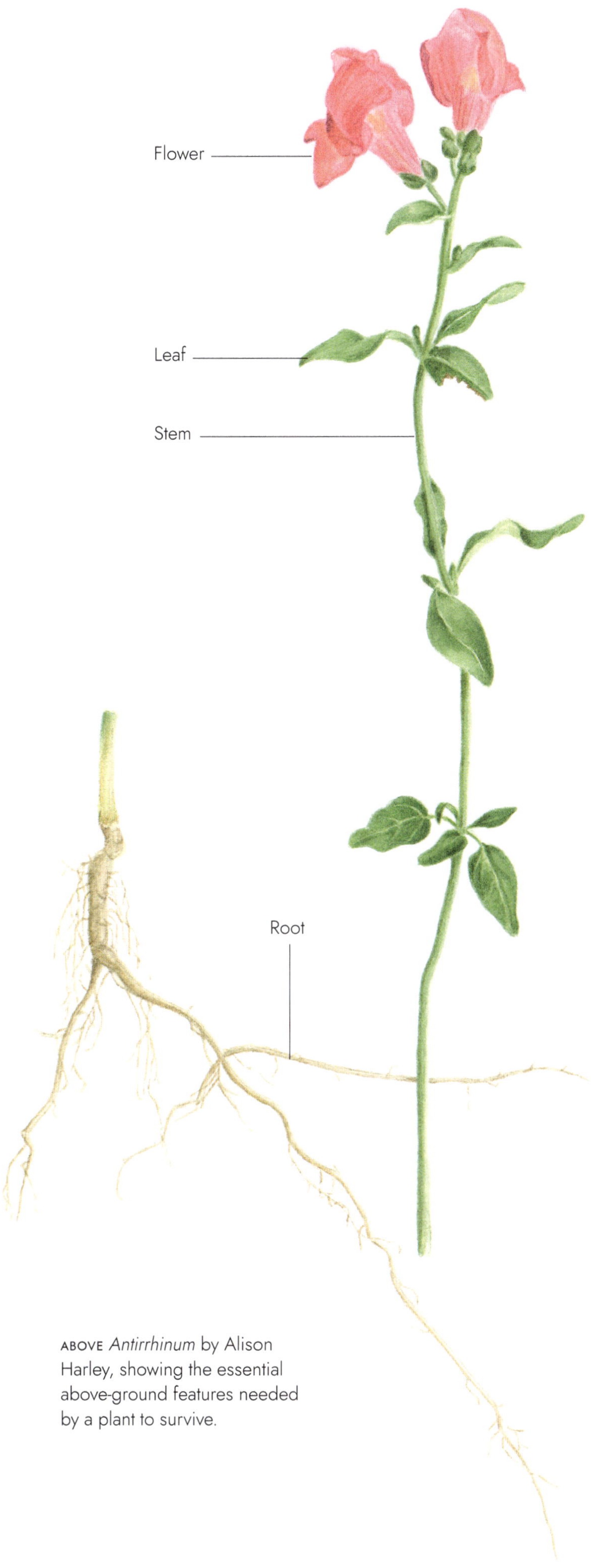

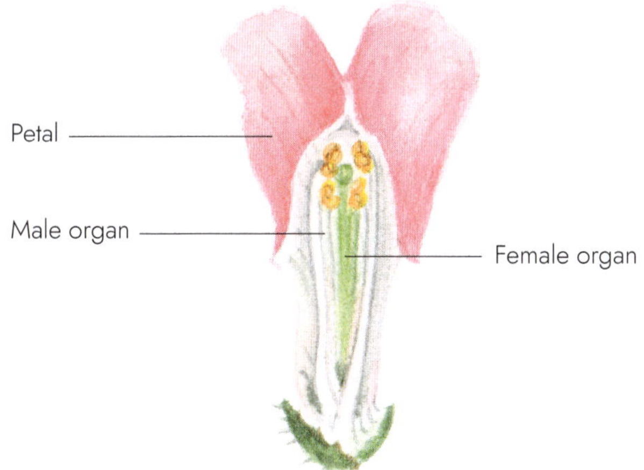

ABOVE *Antirrhinum* by Alison Harley, showing the essential above-ground features needed by a plant to survive.

Everything about a plant's structure is designed to maximize its chances of survival and reproduction. In practice, an artist needs to look out for an infinite variety of modifications to the basic structure. This is what makes the art of botanical illustration interesting, and a good botanical painting will show finely tuned observational skills.

Some roots exist above ground level; not all plants have male and female parts in the same flower – they may be differentiated on the same plant, or in different plants; some plants do not have a stem as such, or the stem might be flattened and leaf-like (see page 36, Gloria Newlan's *Schlumbergera*). A few plants are not green because they are parasitic and get their food from other plants, or their food might be supplemented by animal proteins, as with some carnivorous plants.

MIXING DIFFERENT GREENS

- Make swatches of greens by mixing both warm and cool yellows and blues and make a note of what you used.

- Include in your swatches diluted as well as concentrated forms of the greens.

- Purples help to achieve the deepest tones or shadow greens.

- Reduce an artificial looking, or over vibrant, green by adding tiny amounts of red or pink.

- Keep a separate mixing palette for green to avoid muddying purer colours used for other parts.

- Vary the tones and shades over a large area to provide vibrancy to a feature.

Life cycle

Every plant goes through a cycle of generation, maturation and decay. Most plants reproduce sexually, where an exchange of genes produces variety in their offspring. After the ovule (female) is fertilized with pollen (male), the flower decays, whilst the fruit containing seeds matures until it is ready for dispersal. When conditions are right, germination begins and the first seed leaves (cotyledons) appear above ground. Seedlings grow into mature plants ready for flowering.

Cross-pollination is the transferring of genetic material to a flower on another plant, resulting in more varied offspring. Some plants have ways to prevent self-pollination, such as releasing pollen before the female organs on the same plant are fully ripe.

However, many plants keep their options open and can self-pollinate in adverse conditions. This keeps the gene pool stable and overcomes the problem of not having the right pollinators available, as might happen with invasive plants. Garden pea flowers transfer pollen from upturned stamens to their own stigmas before the flowers open, and when they do it's likely that 'spent' anthers will be found inside. Tomatoes have pistils surrounded by joined up (connate) stamens, enabling pollen to drop directly onto the stigma inside, often with the help of vibrations from bees.

Regeneration is also possible without the transfer of pollen with bulbs, runners or tubers capable of producing new plants. The new plants are genetically the same as their parents (this kind of reproduction is called asexual or vegetative).

An artist needs to decide at what point to 'freeze' the cyclical process, whether to illustrate the whole cycle, and whether to include atypical forms.

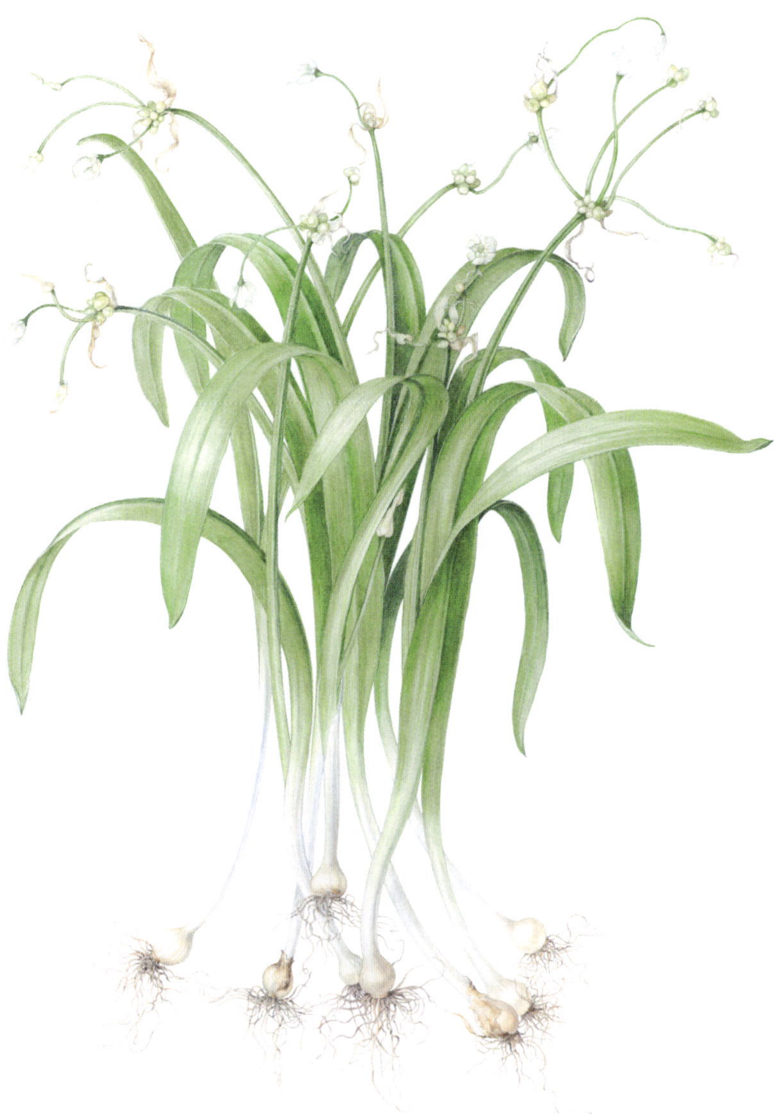

ABOVE RIGHT A detail of few-flowered garlic (*Allium paradoxum*) by Janet Dyer. It spreads easily from bulbils that drop to the ground after developing in the floral parts.

LOOK OUT FOR UNUSUAL POLLINATION METHODS

- Keep an open mind by putting yourself in the place of the plant and how it might achieve cross-pollination, or avoid it.

- Dissect several flowers at different stages from when they are in bud to when they decay.

- In each flower, establish how many stamens there are, how they are arranged, where they are placed, and their lengths.

- Do the same with the style, taking note of any changes in its appearance as it develops.

- You may be surprised to find anthers that fall off easily, styles that grow through a ring of stamens, stamens that shrivel before the style is functional, and styles that display hairs at certain points in their development.

Plant habits and architecture

Habit refers to the general appearance of a plant, involving size, orientation and growth forms, all of which can be modified by the environment. Architecture strictly refers to the three-dimensional shape or organization of the plant resulting from its genetic predisposition. It includes branching patterns, the all-important placement of leaves (phyllotaxis) and inflorescence shapes.

The terms are often loosely applied and can apply to particular groups of plant.

BELOW Ivy (*Hedera helix*) by Fiona Ward has a habit that tends to droop with gravity as it scrambles on its host structure.

RIGHT Red hot poker (*Kniphofia foliosa*) detail by Sarah Howard. The architecture of its conical shaped inflorescences and upright stems comes from its genetic make-up.

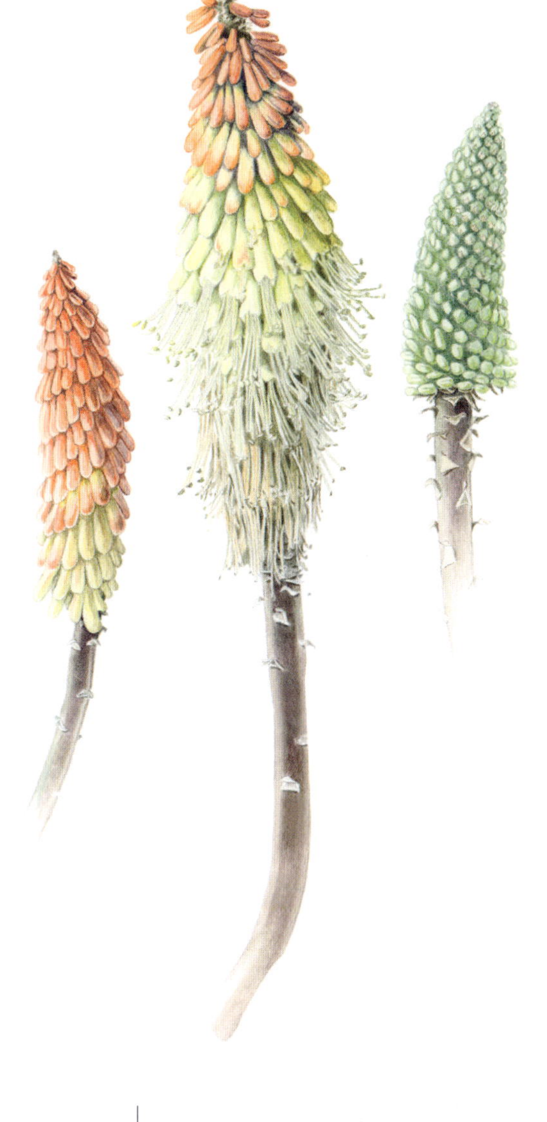

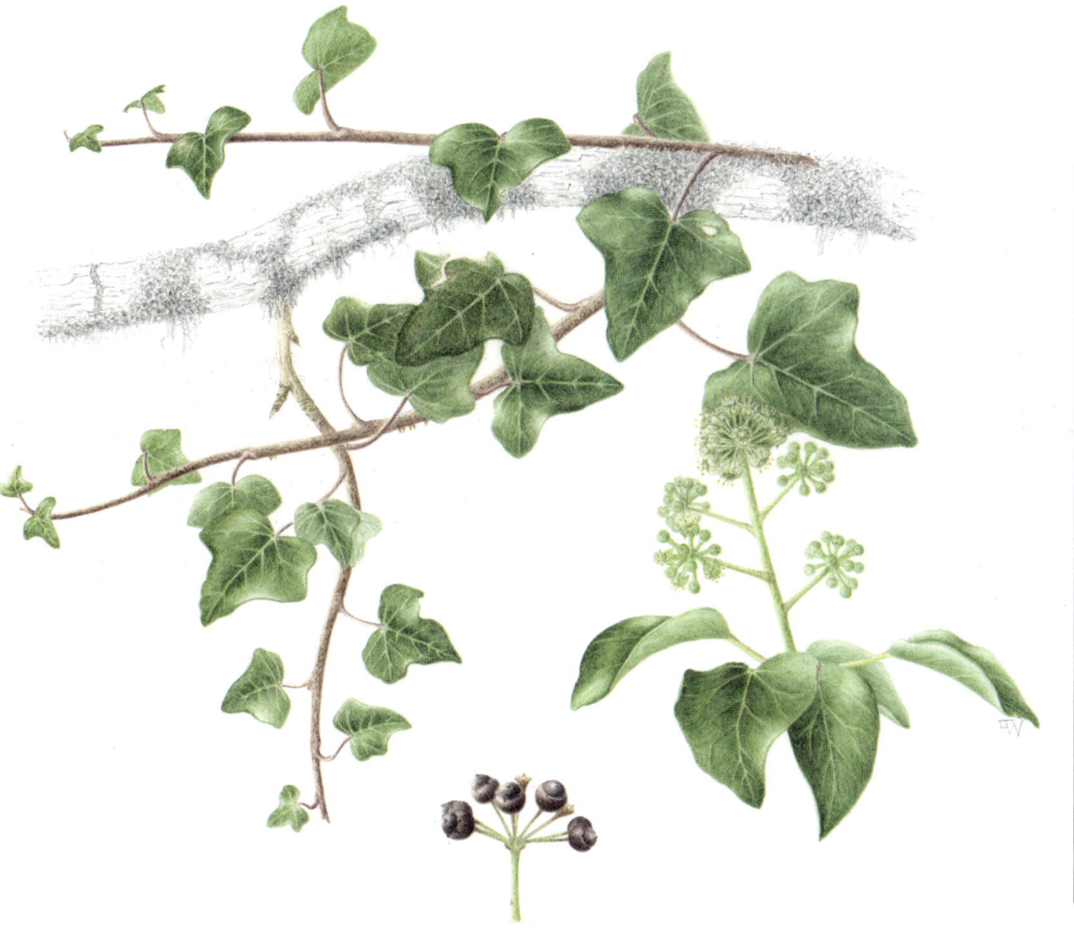

WHAT TO LOOK FOR

- Trees may be erect, fastigiate or weeping.

- Shrub foliage may be dense or sparse.

- Small, non-woody herbs can be upright, prostrate or creeping.

- Ground orchid flower heads in Europe may be described as pyramidal, rounded, lax or dense.

If any of these features are important in identifying the plant, it is best to show the feature to its best advantage or provide a scale, if a tree or shrub is being painted.

Growth forms

Evolutionary adaptation means diversification and new growth forms. For example, the yellow common ragwort (*Senecio jacobaea*) found in European fields, which is a bane to horses, is very different from the tree-like *Senecio* on African mountains. Conversely, many plants may look similar in a specialist environment such as a desert, but belong to quite different families. The fleshy, spiny Old World *Euphorbia* species, with their fluted stems and tiny leaves look like New World cacti with similar traits (convergent evolution).

Growth form descriptors include trees, shrubs, annuals, perennials, epiphytes, geophytes, succulents and aquatics.

It is sometimes desirable to give context to a painting of a flowering tree branch by including the whole tree, the bark and a scale bar. Artists should also be aware of any differences between a wild plant and a cultivated plant.

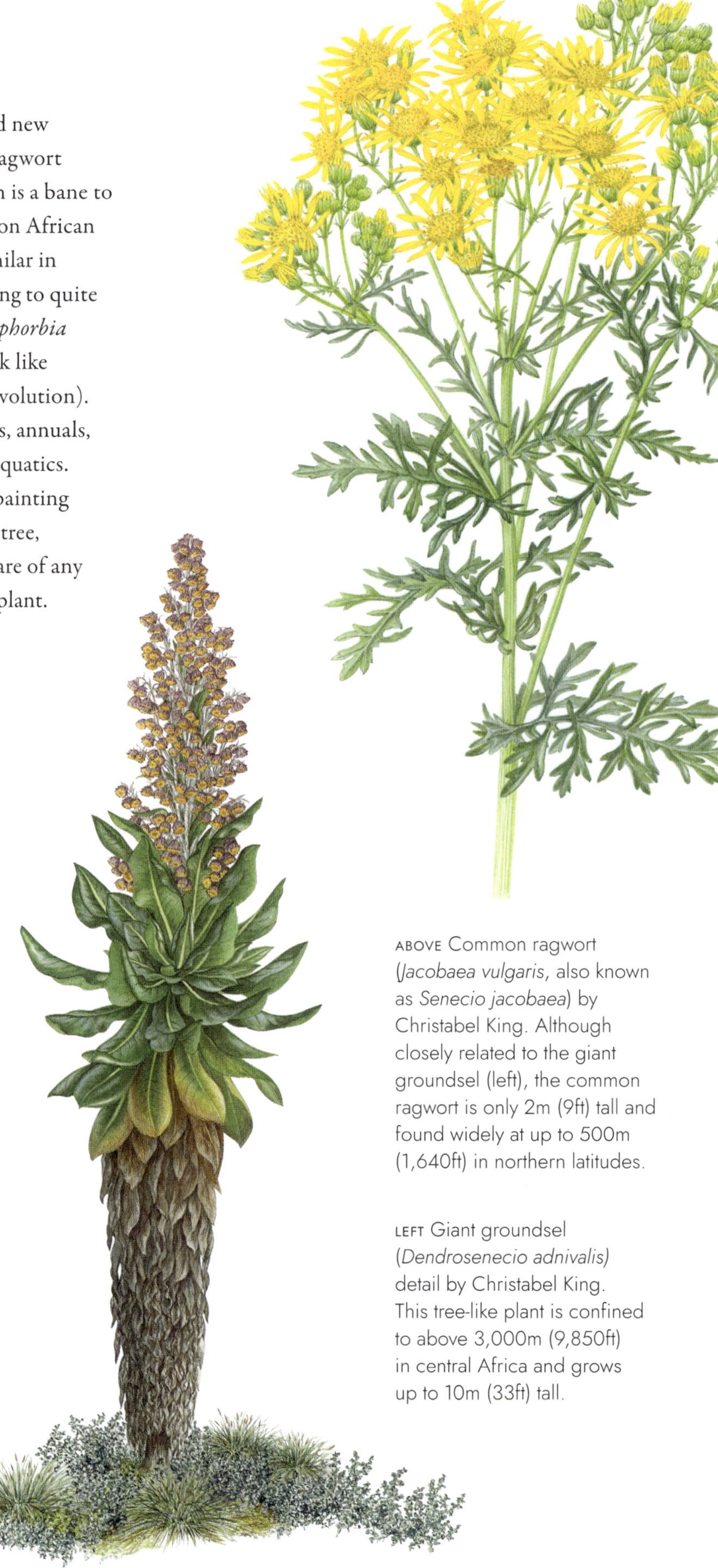

ABOVE Common ragwort (*Jacobaea vulgaris*, also known as *Senecio jacobaea*) by Christabel King. Although closely related to the giant groundsel (left), the common ragwort is only 2m (9ft) tall and found widely at up to 500m (1,640ft) in northern latitudes.

LEFT Giant groundsel (*Dendrosenecio adnivalis*) detail by Christabel King. This tree-like plant is confined to above 3,000m (9,850ft) in central Africa and grows up to 10m (33ft) tall.

PROVIDING ECOLOGICAL CONTEXT

- Set a colourful flowering plant against a pencil drawing of the landscape.

- Provide a few other plants alongside, or around, your chosen plant.

- Paint in an insect or caterpillar on the flower or leaf, ensuring they are the correct combination and the right size.

- Set the base of the stem in a substrate, such as gravel or stones.

- For a forest setting, make a background of hanging plants and vertical structures.

Patterns in plants

For the mathematically inclined, patterning in plants is a fascinating subject. It includes symmetry and Fibonacci spirals and sequences. This pattern was described in India more than 2,000 years ago, but was not studied in Europe until the 13th century in Italy.

Fibonacci sequences follow a pattern by which each number is the sum of the previous two numbers. Thus, one finds the sequence 1, 1, 2, 3, 5, 8, 13, 21 and so on. It's immediately obvious that the number of some floral parts (for example, monocot floral parts in multiples of 3, dicots often in multiples of 5) relate to this pattern. Fibonacci spirals can often be discerned in the placement of leaves on a stem and the spirals on a pineapple or a pine cone. The areole arrangement on most cacti follows a Fibonacci spiral pattern.

On a flat plane, a spiral pattern can be seen in the centre of composite plants, such as sunflowers, or the arrangement of *Echeveria* leaves.

The Fibonacci sequence is closely linked to the golden ratio, which artists have used for centuries to create the most pleasing placement of objects, roughly one-third from the top, bottom or sides.

Tessellation

Tessellation is a pattern of shapes (or 'tiles') without overlaps or gaps. It covers a surface and may or may not involve Fibonacci sequencing. The pattern on a jackfruit is tessellate, as is that on a pineapple, and the latter's tiles are also arranged in a Fibonacci spiral. The shape of the 'tile' varies, though the polyhedron pattern (as in a honeycomb) is the most efficient. A fritillary (*Fritillaria meleagris*) has a tessellated pattern like a chequerboard. *Vanda* orchids are famous for their tessellate patterns on the petals, and one is even named *Vanda tessellata*.

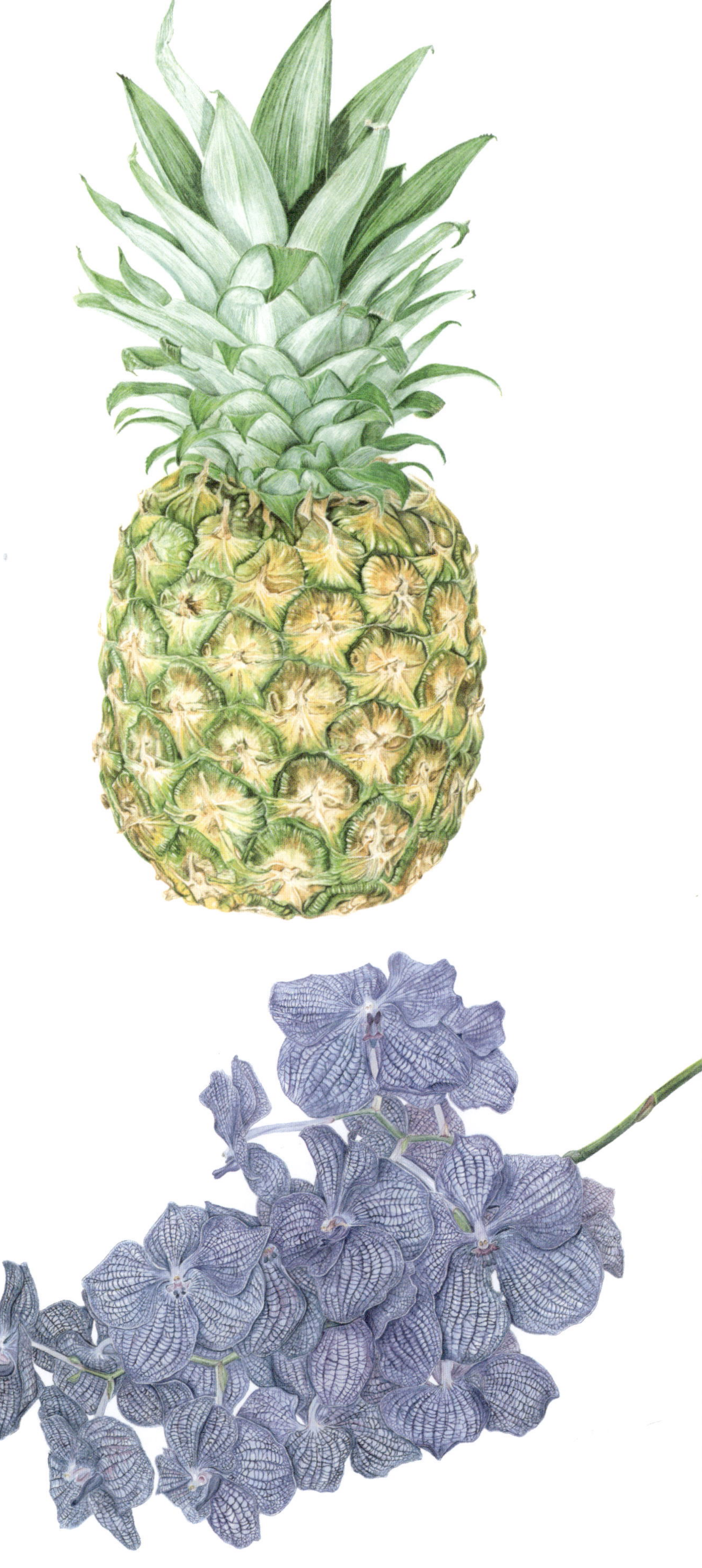

ABOVE RIGHT Pineapple (*Ananas comosus*) by Jean Harlow. It has eight spirals one way and thirteen spirals in the other direction, both Fibonacci numbers.

RIGHT Orchid (*Vanda* 'Blue Magic') by William D. Phillips. Note the rectangular 'tiles' that follow the veining pattern of this orchid.

Evolution and diversification

It has taken more than 470 million years for plants to evolve from single-cell algae and spore-bearing organisms, to sexually reproducing flowering plants. Genetic variation from one generation to the next enables plants to survive (or not) in ever-changing conditions.

Key morphological shifts over time provide an evolutionary timeline:

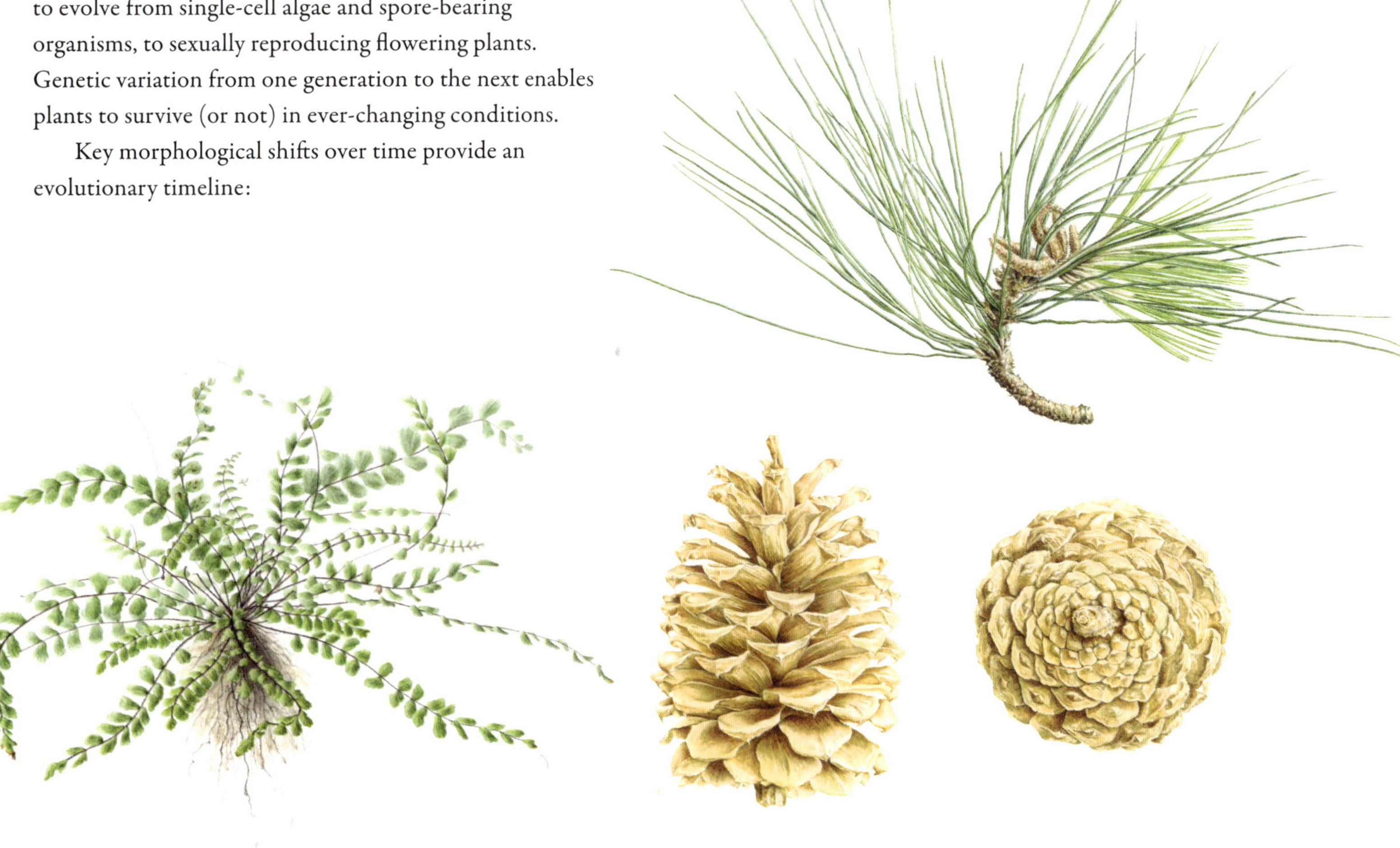

455 million years ago

420 million years ago

The appearance of a vascular system and larger green leaves in the first spore-bearing plants such as ferns meant nutrients and water could move around the structure.

ABOVE Spleenwort (*Asplenium* sp.) by Mary Ellen Taylor. The leaves are much larger than their evolutionary ancestors, the mosses and liverworts.

Conifers are representative of the gymnosperms, which were among the first seed-producing vascular plants. Their seeds are unenclosed, and typically sit on the surface of cone scales

ABOVE Pine (*Pinus jeffreyi*) by Nicola Macartney. The first seeds, often held in cones, appeared in this group of plants.

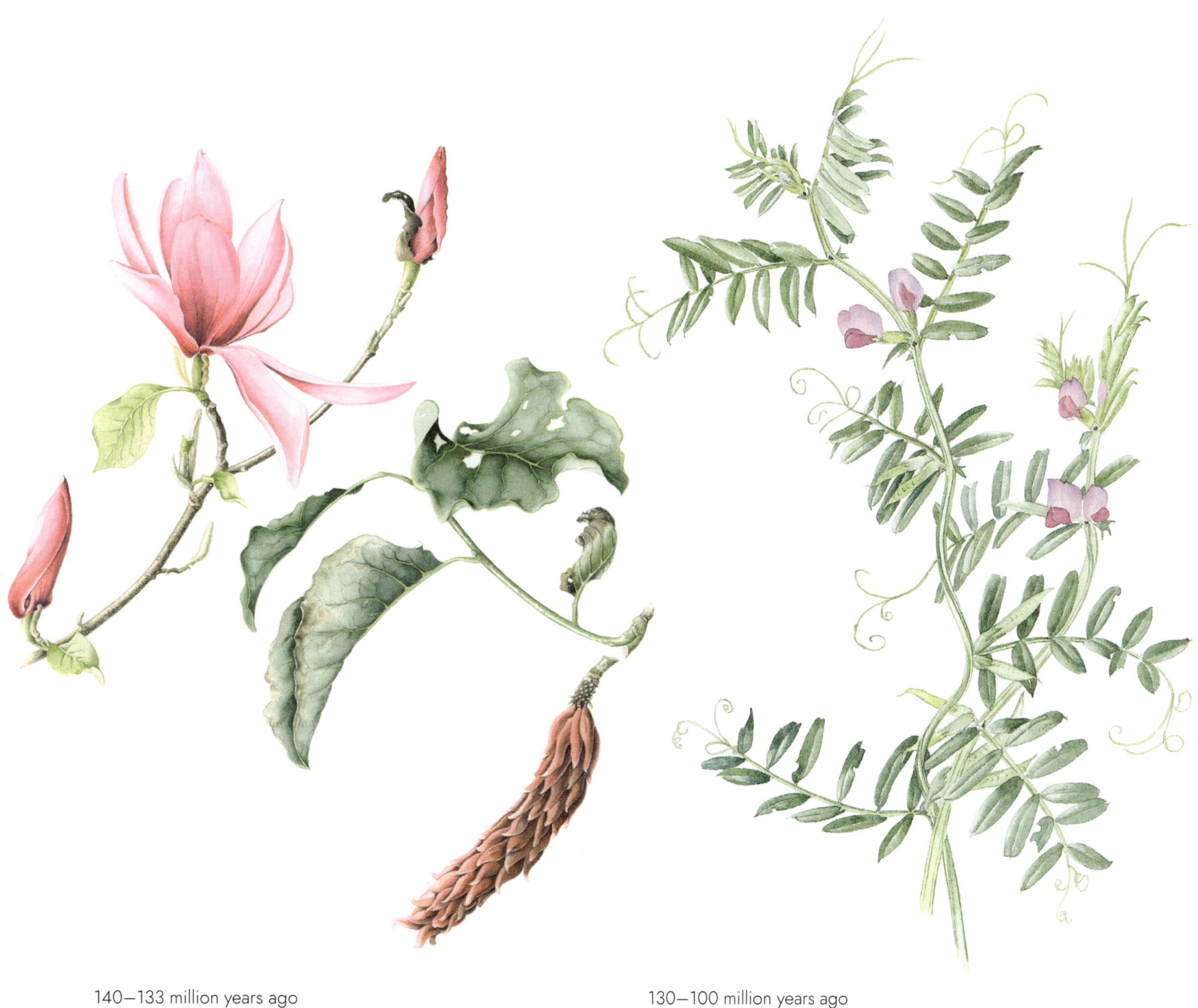

By the early Cretacean period, the angiosperms evolved, which have seeds that are enclosed in an ovary. Insect pollination and reproduction methods evolved together to produce a diverse array of forms. Angiosperms comprise 94 per cent of plant species. Water lilies and magnolias are similar to the earliest flowering plants.

ABOVE *Magnolia* 'Galaxy' by Sarah Howard. The earliest flowers appeared at the end of the dinosaur age and evolved with organs to protect the seeds.

Most angiosperms (70 per cent) are eudicots (formerly known as dicotyledons), a diverse group that can be compared with the other important group, the monocots – the grasses, lilies and orchids, which split early from the main line of angiosperms.

ABOVE Common vetch (*Vicia sativa*) by Annie Morris. This eudicot has features such as bilaterally symmetric flowers and compound leaves with several pairs of leaflets terminating in a branching tendril.

Taxonomy and nomenclature

Classification

From Aristotle onwards, various authors have tried to bring order to the number of plants by classifying them according to a set of visible traits. Linnaeus developed a hierarchical system, the highest level being the kingdom Plantae, and the lowest being an individual plant, or taxon (species, subspecies, variety or forma). The main levels to know are Family and Genus. Taxonomists assign each Species to a Genus, which has its place in a Family, and so on. A large Family or Genus might be subdivided. Family names end with -aceae.

The latest approach, however, uses genetic studies. This has resulted in some shuffling of the original groupings, and there is less emphasis on hierarchy. A value of this method is that it's not subjective and it reveals better evolutionary connections and divergences.

In practice, field guide authors usually adapt traditional classification systems to make it as easy as possible for non-professionals to identify a plant. And it must be remembered that most classification systems are a human construct: plants look the same however we classify them! The value for artists is that a classification system helps to decide what features to highlight, so that the image can be readily identified by a professional.

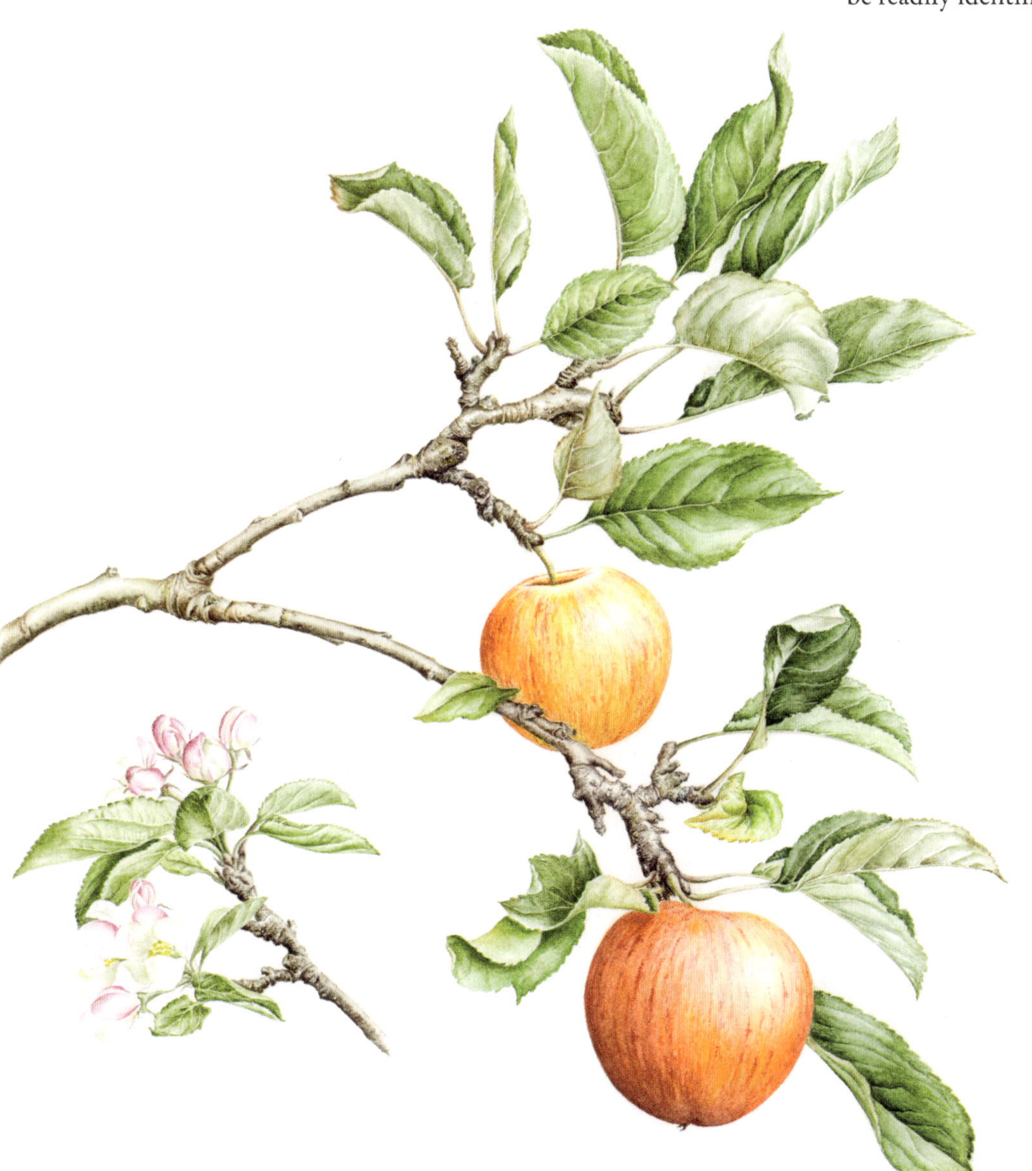

LEFT Apple (*Malus domestica* 'James Grieve') by Nicola Macartney. This plant belongs to the genus *Malus* (along with wild apples), which is in the *Rosaceae* (rose) family, and is ultimately an angiosperm — a flowering plant. Its name, '*domestica*', refers to the domesticated apple of an orchard.

Below are the hierarchical levels to which an apple belongs, the most specific at the bottom and the most general at the top.

SCIENTIFIC CLASSIFICATION

Kingdom:	Plantae
Clade:	Tracheophytes
Clade:	Angiosperms
Clade:	Eudicots
Clade:	Rosids
Order:	Rosales
Family:	*Rosaceae*
Genus:	*Malus*
Species:	*Malus domestica*

Naming of plants

The binomial system, also invented by Linnaeus, provides two Latinized words to every species (shortened to sp. or spp. plural), making each one unique. The first word in the name is the genus name, and it is always written with an initial capital letter. The second is the species name or epithet, which might honour a person, describe its look, or where it is found, and always begins with a lower-case letter. Species might be divided into one or more subspecies (shortened to ssp. or subsp.), varieties or forms.

The binomial system provides a common language for people to use across the world. It avoids confusion when the same common name is used for different plants: 'Black-eyed Susan' is *Rudbeckia hirta* in North America but the name also refers to *Thunbergia alata*, a beautiful East African vine. Similarly, the word 'geranium' is used by botanists for the genus *Geranium* (or 'cranesbill'), but also commonly for members of the *Pelargonium* genus. Both belong to the *Geraniaceae* family.

The names may provide clues as to what the plant looks like. *Prunus spinosa* (blackthorn or sloe) refers to its spiny appearance. *Prunus* is the Latin word for plum, a fruit that gives its name to the genus *Prunus*. *Trillium sulcatum* (or wake-robin in the USA) belongs to a genus which has three leafy bracts. The species name refers to the furrowed sepals (*sulcus* is Latin for 'furrow').

BELOW Trillium (*Trillium sulcatum*) by Lyn Campbell. Linnaeus established the *Trillium* genus and placed it in the lily family. Now the preferred family is *Melanthiaceae*.

Illustrated field guides and regional floras are often ordered by family, with ferns and gymnosperms at the beginning, monocots at the end, and angiosperms in the middle.

Other guides may offer a simpler approach, classifying by colour (though this is usually an unreliable indicator), flower features or habitat. Today, apps offer image comparisons, but use these with caution. Always check against an illustrated written description — there's a reason why botanical illustrators still have a job!

However information is arranged, the plant's description remains similar among guides, though certain details may be added or omitted. An illustrated glossary of terms is often included. Owning several guides with different approaches is helpful for cross-checking and finding essential information for your drawing.

Identification keys in a field guide provide choices which lead to a plant's identity. The answer at one step determines the next until the correct species is reached. They only apply to the specific flora in question. A key to the genus *Rhododendron*, in an American guide, will be different from a Chinese one as the species present will be different in each place.

An index will usually include species name, genus and family, so you can find your plant at any of these levels.

Chapter three

Roots and stems

The root and stem comprise the main axis of a plant. A typical root system is an anchoring device below ground that receives water and nutrients. A typical stem is above ground and provides the framework for the shoots system and flowers. The vascular (vein) system runs throughout the plant, carrying water, nutrients and sugars manufactured by photosynthesis.

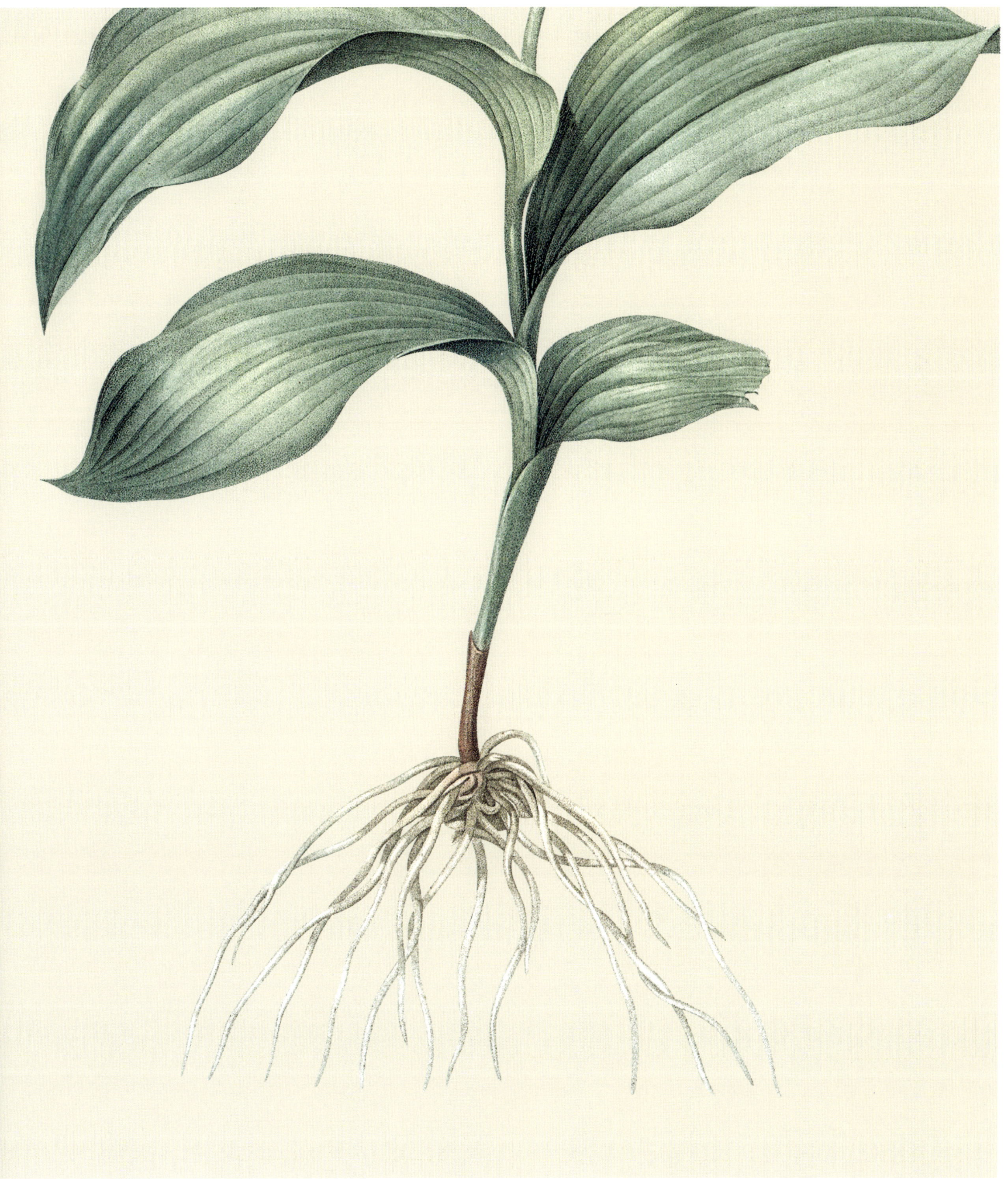

Root systems

There are two basic types of root systems: 'primary' and 'adventitious'. Primary roots are the first to emerge from a seed, and become the main or taproot; whereas adventitious roots are those that emerge from other parts of the plant.

PARTS OF ROOTS AND STEMS

WHAT TO LOOK FOR

- What is the arrangement, density and shape of the root system? Roots are distinctive and reflect the plant's biology.

- Note details such as hairs and nodules.

- Distinguish between a tuberous swollen root, and a tuberous underground stem. True roots do not carry stem features such as nodes and leaf scars.

- Distinguish, if necessary, between a stem or a root by making a cross section to see the arrangement of vascular bundles — they are in the centre of the roots. In stems, they lie on the periphery or are, in monocotyledons, scattered.

- Distinguish, if necessary, between the primary root, lateral roots, contractile roots, tuberous roots, and so on.

- What is the role of any roots not in their usual underground position?

- How do parasitic plants attach to their host using modified roots?

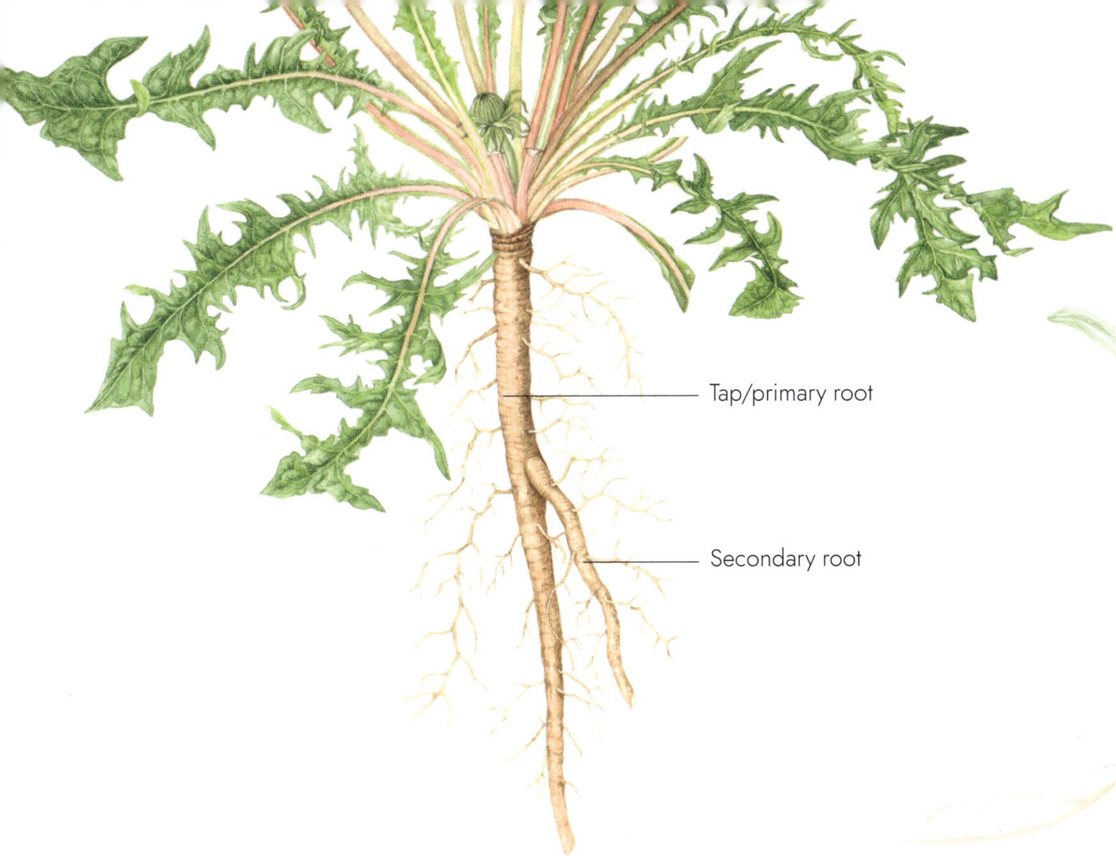

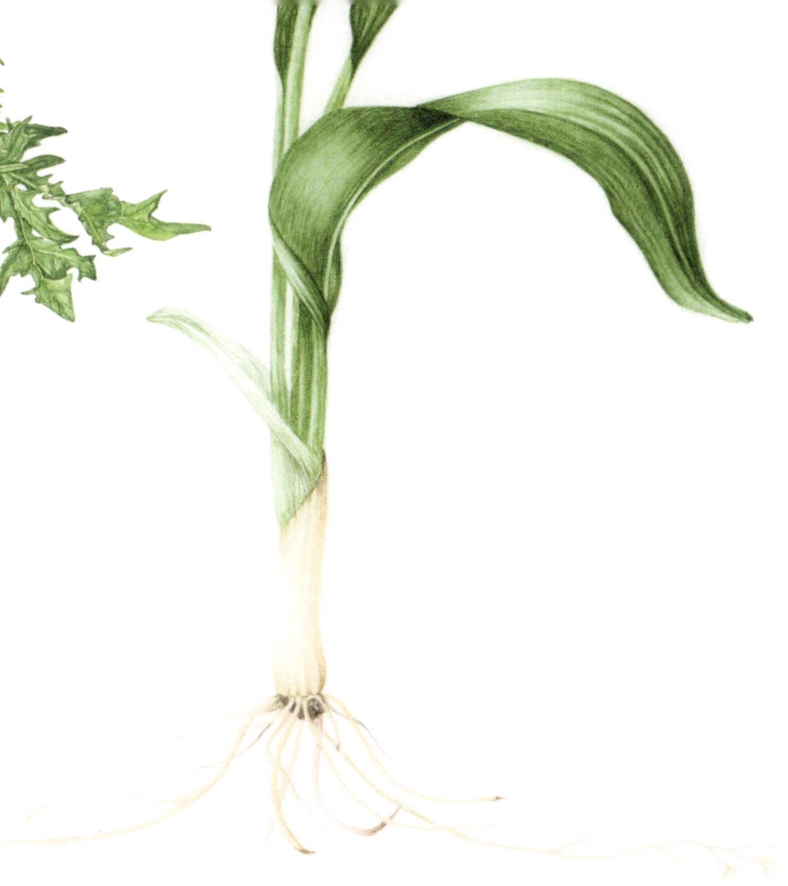

A primary root system (above left), consisting of main (tap) root and fibrous-looking lateral branching, is derived from the seedling radicle, and is associated with dicotyledons.

An adventitious root system (above right) arises from the stem and is typical of monocotyledons such as grasses. Adventitious roots can also appear from the leaves and stems of dicotyledons.

TYPES OF ADVENTITIOUS ROOTS

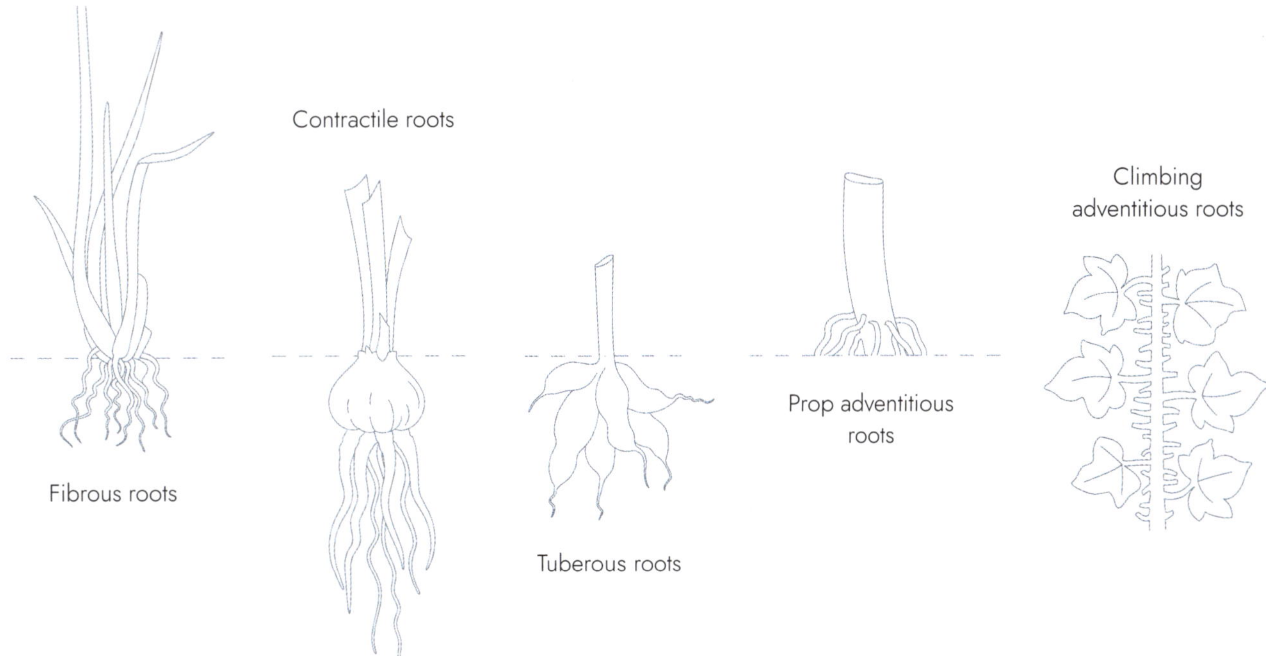

Terminology

'Adventitious' literally means 'not found in its usual place'. This word is used widely, although these roots are sometimes known as 'fibrous' or 'nodal' roots. Some people use it to refer to the short-lived primary root system of grasses and other monocotyledons. Roots that form at the nodes (growing points) of stems, above or below ground, are 'adventitious' but not necessarily out of place. Some people prefer the term 'nodal roots' in these cases.

Nodal roots, above or below ground, are widespread in dicotyledons. When they occur above ground, they aid asexual reproduction, balance a plant, help it to climb, or simply to get water from moisture-laden air. Below ground, they arise at the nodes of rhizomes and corms.

Morphological features

The root crown is either the point at ground level where the root system changes into the stem, or the point where annual shoots arising from a perennial root system decay (for example, a rhubarb 'crown').

Root nodules are small, spherical, cylindrical or branching structures on roots, which contain nitrogen-fixing bacteria. They are typical of leguminous plants and easily seen on *Vicia faba* (broad or fava bean) roots.

A root cap is a protective covering on the tips of roots, obvious on aerial roots (for example, *Pandanus nobilis*).

Contractile roots shorten in length to keep a plant at a particular level underground, by responding to new growth or environmental conditions. They are distinct from the surrounding roots because they are vertical and more wrinkled, for example, those under hyacinth bulbs, *Gladiolus* corms, or on rhizomes of skunk cabbage (*Lysichiton americanus*).

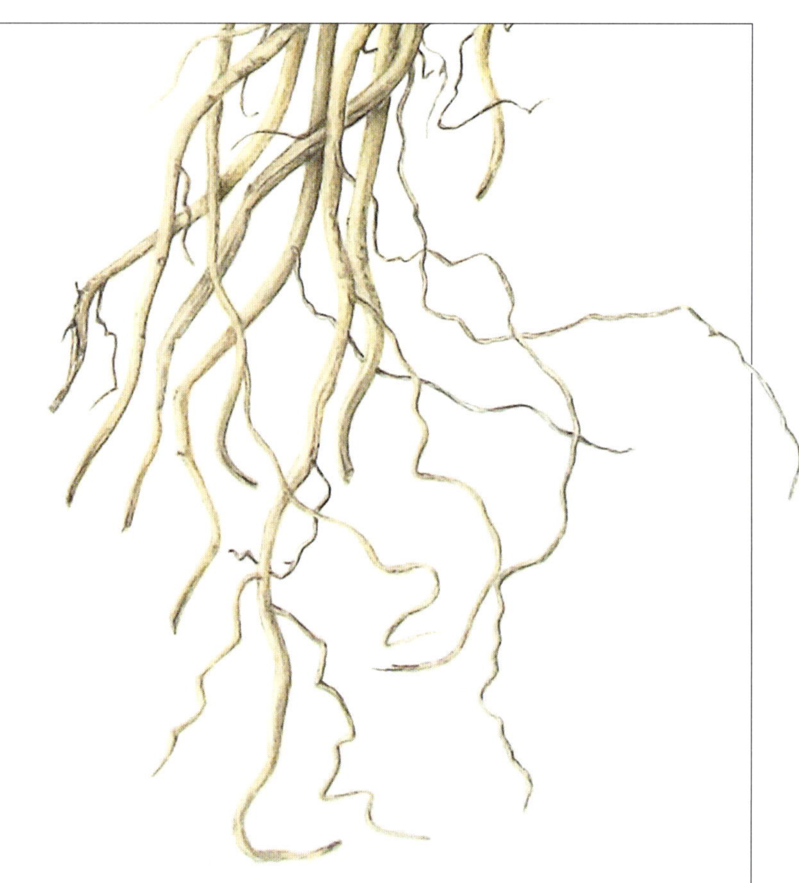

TIPS ON PAINTING A CLUSTER OF ROOTS

- Draw roots lightly first, noting bulges, then mark the overlaps.

- Mix at least three intensities of colour to develop the idea of volume in the root system.

- Use the lightest shade to paint the outlines of all the roots and their overlaps.

- Cover the spaces underneath with the next darkest colour, without losing your lines.

- Use the darkest shade to cover the darkest, and more limited, recesses of the system. In this order, you'll add intensity by painting over an existing shade.

- Finish by indicating shadows from overlaps, bending, the volume of each root, and the overall volume of the system by darkening the side in shadow.

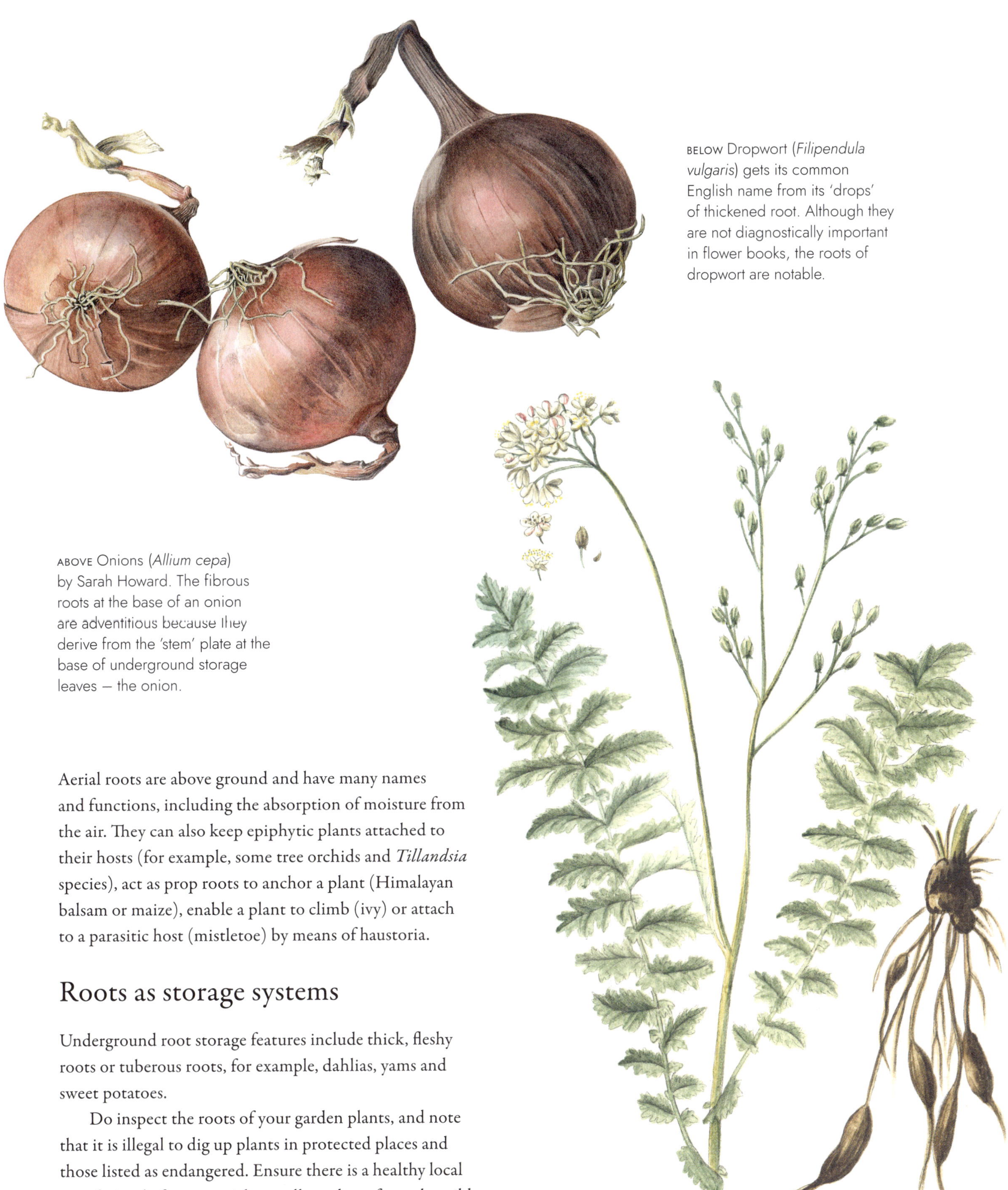

Aerial roots are above ground and have many names and functions, including the absorption of moisture from the air. They can also keep epiphytic plants attached to their hosts (for example, some tree orchids and *Tillandsia* species), act as prop roots to anchor a plant (Himalayan balsam or maize), enable a plant to climb (ivy) or attach to a parasitic host (mistletoe) by means of haustoria.

Roots as storage systems

Underground root storage features include thick, fleshy roots or tuberous roots, for example, dahlias, yams and sweet potatoes.

Do inspect the roots of your garden plants, and note that it is illegal to dig up plants in protected places and those listed as endangered. Ensure there is a healthy local population before you pick or collect plants from the wild.

Orchid root systems

Orchids have two types of root system: roots that are relatively 'normal' and grow downwards into the ground, where they find nutrients and water like most other plants; and others that are aerial and derive moisture and nutrients from the air. The aerial roots are covered in spongy cells called velamen, which absorb water from the humid conditions in which they grow. These roots are also used to anchor the orchid to trees.

From another perspective, monopodial orchids only have one root system at the base of the stem, either in the ground or above ground, whereas sympodial orchids, may have several rooting points – at the nodes of lateral growing stems.

LEFT Moth orchid (*Phalaenopsis* sp.) by Sarah Howard, is an example of a monopodial plant with its root system at the stem base. Paint the off-white, spongy cells of velamen using a variety of blue-greys if the roots are dry, and green-greys if they have absorbed water.

Stems

The main function of a plant's stem is to provide support for the leaves, flowers and buds. It holds the plant upright and helps it to grow towards the light.

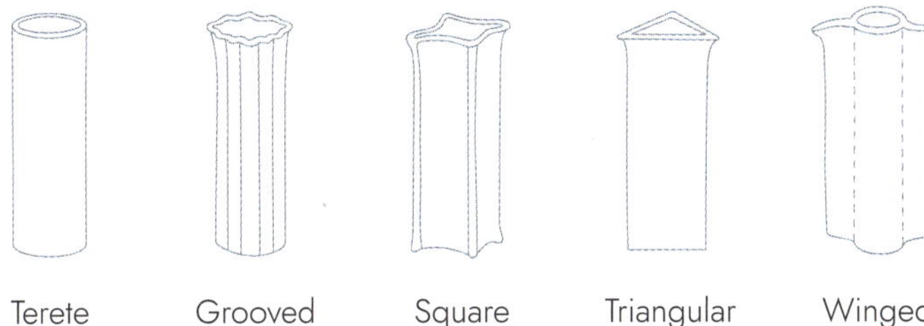

WHAT TO LOOK FOR

- Are the growth forms upright/ascending, prostrate, stoloniferous, multi-stemmed, branching, or with no visible stems except for flowering stems?

- Are the stems woody or herbaceous (usually green)?

- What is the width like? Monocotyledon stems are one width, while dicotyledons tend to taper.

- What are the essential features such as nodes (where leaves and buds are inserted, called 'phyllotaxis') like, and the internodes that connect them?

- When the transverse shape is seen in cross-section, is it square, round, oval, winged, hollow or solid?

- Does the external shape which reflects the cross-section have both grooves and wings?

- What surface features does the stem have? For example, does it have colour, hairs, prickles, tendrils or leaf scars?

- Does the plant have swollen or flattened stems indicating succulency and/or modification into a leaf?

- Does the stem have obvious openings that enable an exchange of gases, called stomata, or lenticels, which are raised and slit, as on woody stems?

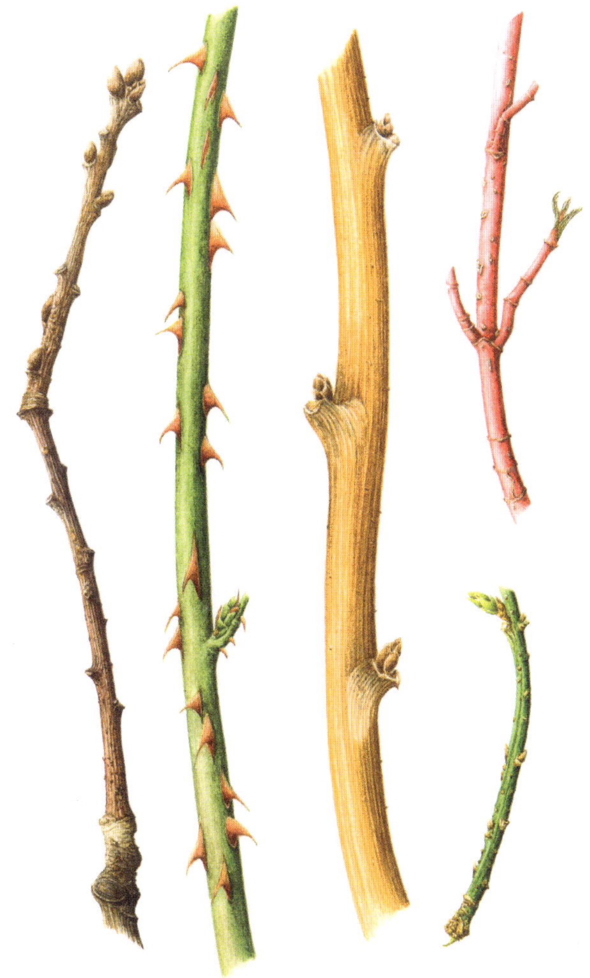

ABOVE *Winter Twigs 2* by Leigh Ann Gale. Note the lenticels on the twigs.

Leaf scars

Scars on stems mark the place where leaves have fallen off.
They are regular in shape and location and are an important
clue as to how leaves are arranged on a stem (phyllotaxis).
Any spiralling may indicate a Fibonacci pattern.

RIGHT Horse chestnut
(*Aesculus hippocastanum*)
detail by Linda Pitkin. The leaf
scars indicate an opposite
leaf pattern.

BELOW RIGHT Common stinging
nettle (*Urtica dioica*) by Leigh
Ann Gale. It is covered with
hollow hairs, containing formic
acid, which irritate when the
tips are broken off.

Terminology

Spines, prickles, thorns, and hooks are often used
interchangeably. They derive from either modified
leaves or stems, or from surface (epidermis) tissue, and
are important for identifying wild roses or African 'Acacia'.
All features could be found on the same plant. Note whether
they are dense, point up or down, or are found only on
young, low-growing plants.

Spines are sharp-pointed, hardened structures that
originate from the vascular part of a plant, whether
modified leaf, stipule, root or branch. 'Acacia' spines
are stipular, with always a pair, often with a leaf
bud between them.

Thorns are short, reduced branches and
contain vascular bundles, for example, sloe.

Prickles are sharp outgrowths of the
epidermis – the outer cell wall – and can be
pulled off easily. They do not contain vascular
bundles, for example, rose prickles.

Acicles are needle-like, also derived from the epidermis.
They can appear randomly on the surface, for example,
burnet rose (*Rosa spinosissima*).

Hooks are thin projections with a bent tip forming
a hook or double hook like an anvil.

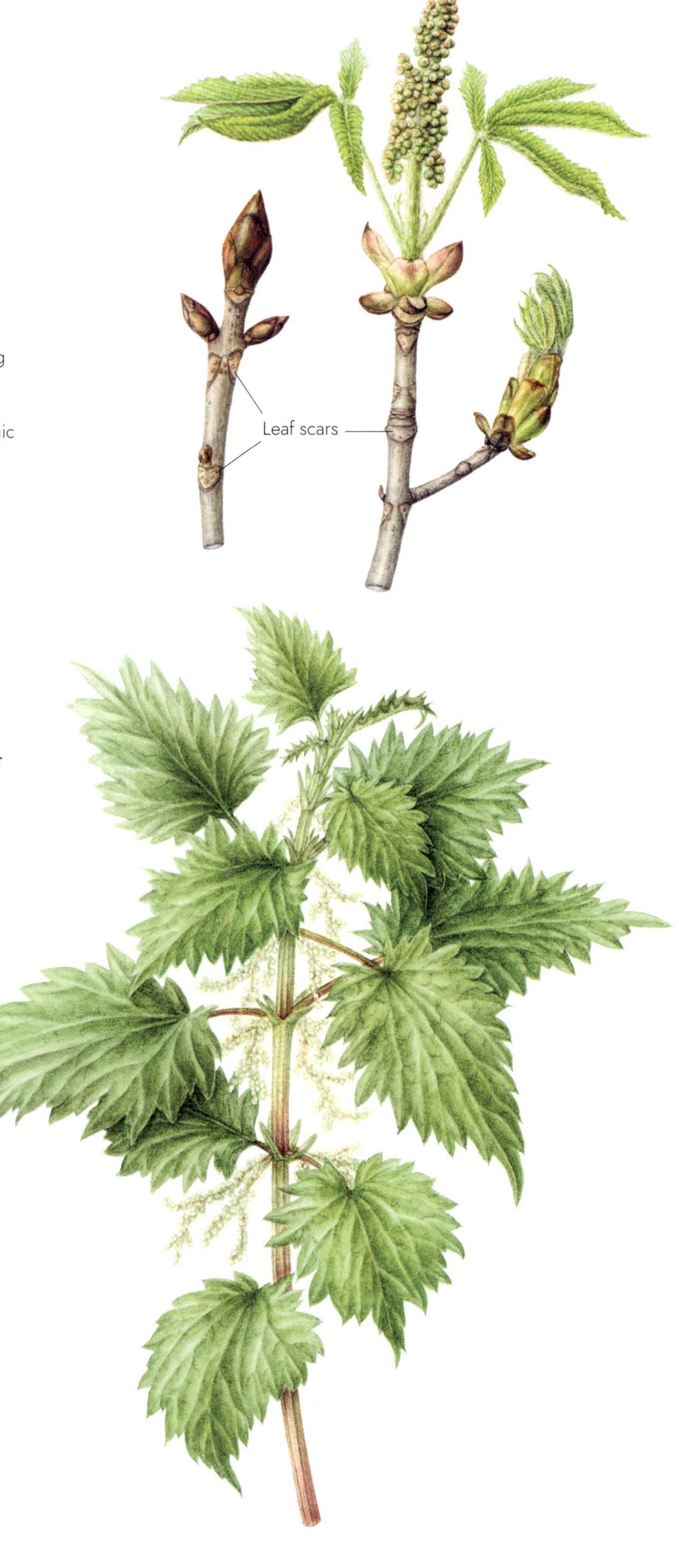

Other features on stems

Indumentum (or indumen) refers to hairs, which can occur anywhere on a plant (see page 84). Take note of what kind of hairs they are, the direction in which they grow, and where they occur. Chickweed (*Stellaria media*), for example, has a line of hairs running down only one side of its stem.

Pulvinus literally means a swollen joint and is most commonly associated with leaf attachments. On a stem, nodal swellings provide a flexible mechanism for the stem to droop without breaking in dry weather.

Wings are the flanges that form along the stem's length. They can be continuous between nodes (*Lathyrus*), discontinuous (Welted Thistle *Carduus crispus*), or spiral along the stem. Stem wings are a diagnostic feature for thistles (*Cirsium* and *Carduus*) and peas (*Lathyrus*) but their function is unknown.

Cladophylls (or phylloclades) are flattened stems that function like leaves and can be seen in *Schlumbergera* (Christmas cactus), for example. They are recognized by visible leaf scars or scale leaves which arise from them.

Tendrils are long slender threads which arise from stems, flowers or leaves as a climbing aid. It is often difficult to tell what part of the plant, precisely, the tendril represents. On stems they might either be modified shoots arising from a leaf axil or the end of a shoot. Sometimes they bear leaves. They are important features in *Passiflora*, *Vitaceae* and *Cucurbitaceae*.

Stolons and runners are often used interchangeably. Stolons are elongated horizontal shoots above ground that form small plants at the nodes. Runners may give rise to more runners at a node – for example, strawberry plants.

Rhizomes grow in a horizontal direction underground, adding node upon node. They can be fleshy like a ginger, or not, like grasses. Look out for leaf scars and adventitious roots at the nodes.

Corms are short and swollen and grow vertically underground, their roots sit immediately below the corm – for example, crocus.

Tubers are swellings at the base of the stem, rather than on a root, and are distinguished from root tubers by having nodes and leaf scars – for example, potatoes.

Dense needle-like acicles with prickles on a rose stem can be very challenging to paint. It is important to note the shape and direction of the prickles as you draw them, and decide whether an impression, rather than preciseness, is acceptable.

- If there are only a few prickles on a large stem, or you're dealing with larger features such as spines or thorns, you could try using masking fluid. It can be an inexact method, but after removing the masking fluid, dry brush around the features.

- An opaque white, such as titanium white, or gouache, tinted with the colour of the acicle, can be used to paint thin features in after establishing the darker background of the stem. On the light side, you could use the stem colour to form the outline.

- If there are only a few prickles on a fiddly, narrow stem, you could try dry brushing around the prickles, having drawn them in carefully first.

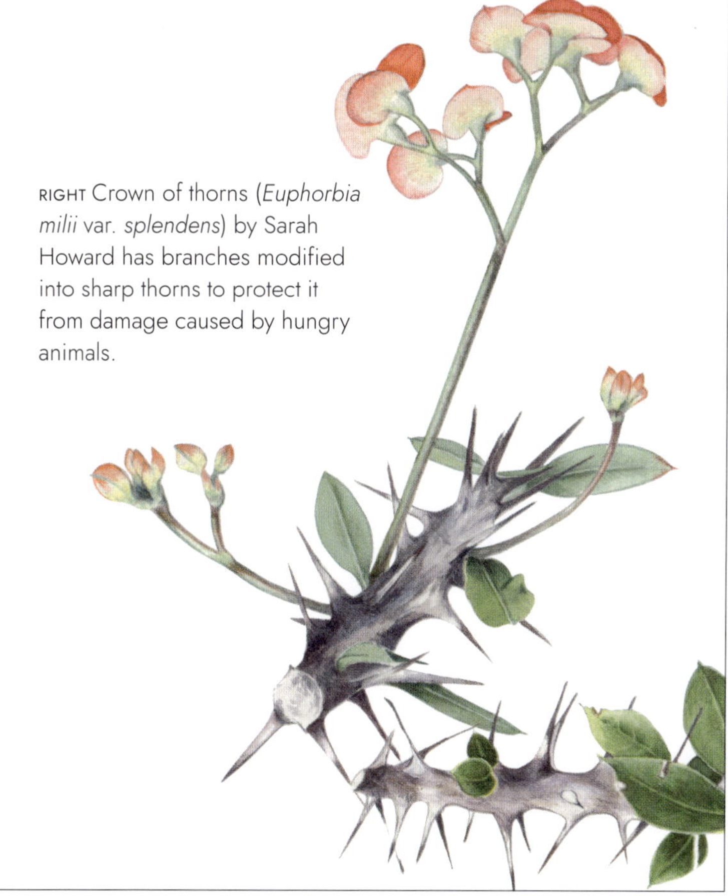

RIGHT Crown of thorns (*Euphorbia milii* var. *splendens*) by Sarah Howard has branches modified into sharp thorns to protect it from damage caused by hungry animals.

Painting a dandelion

Dandelions have long taproots, which are difficult to remove. Although they are considered a pesky weed by some, the roots are interesting to draw and paint, with their tapering shape and fibrous secondary roots. This feature helps to balance the rosette above of bumpy-surfaced leaves with toothed edges.

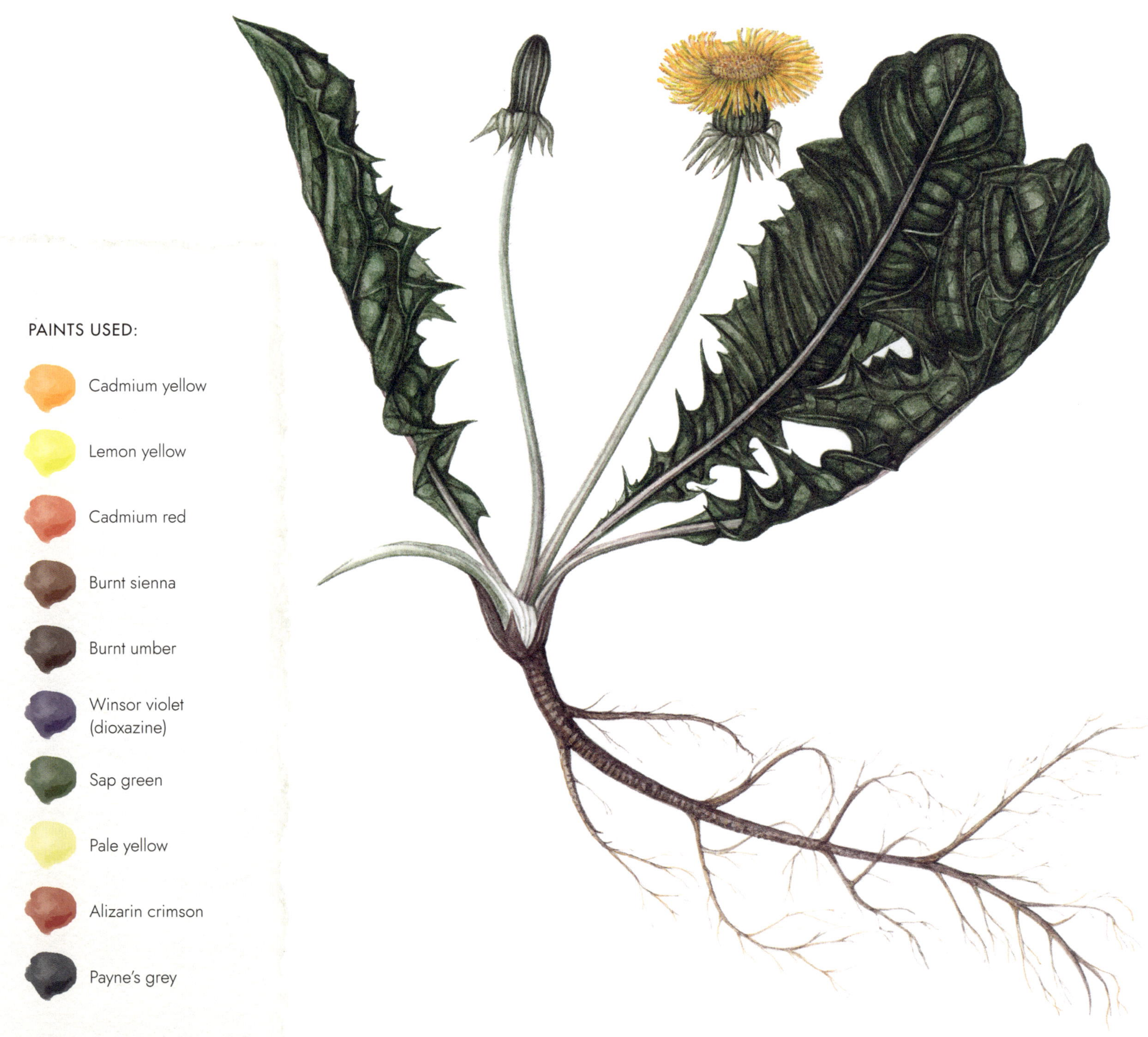

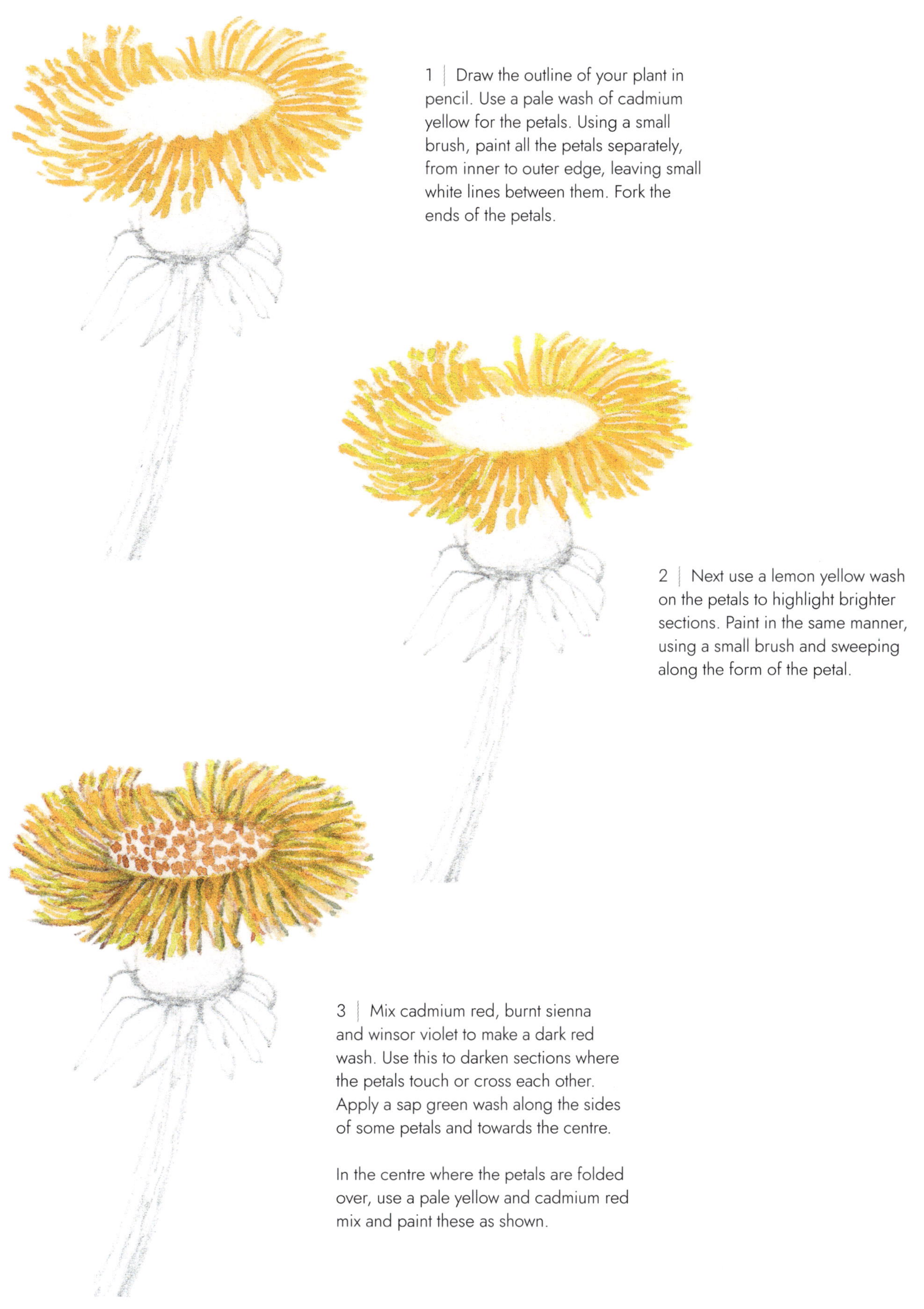

1 | Draw the outline of your plant in pencil. Use a pale wash of cadmium yellow for the petals. Using a small brush, paint all the petals separately, from inner to outer edge, leaving small white lines between them. Fork the ends of the petals.

2 | Next use a lemon yellow wash on the petals to highlight brighter sections. Paint in the same manner, using a small brush and sweeping along the form of the petal.

3 | Mix cadmium red, burnt sienna and winsor violet to make a dark red wash. Use this to darken sections where the petals touch or cross each other. Apply a sap green wash along the sides of some petals and towards the centre.

In the centre where the petals are folded over, use a pale yellow and cadmium red mix and paint these as shown.

4 | Use a lemon yellow to add colour to the flower centre where the petals are folded. Apply the dark red wash from step 3 to add details with a small brush. Use a sap green wash under the flower head, on the bracts and on the bud as a first wash.

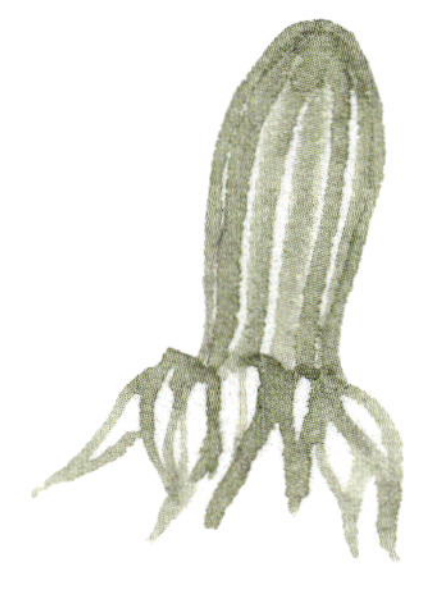

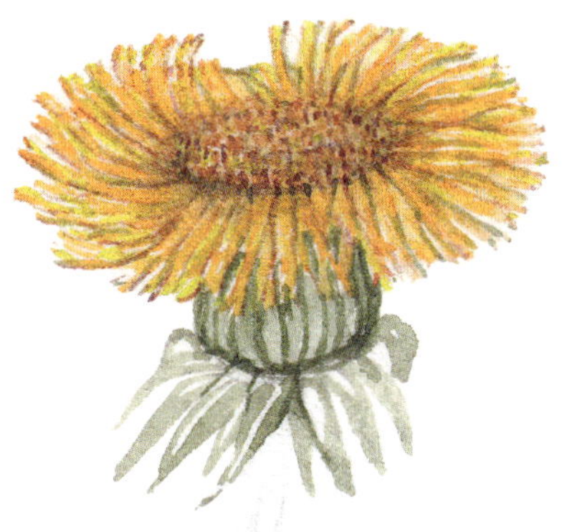

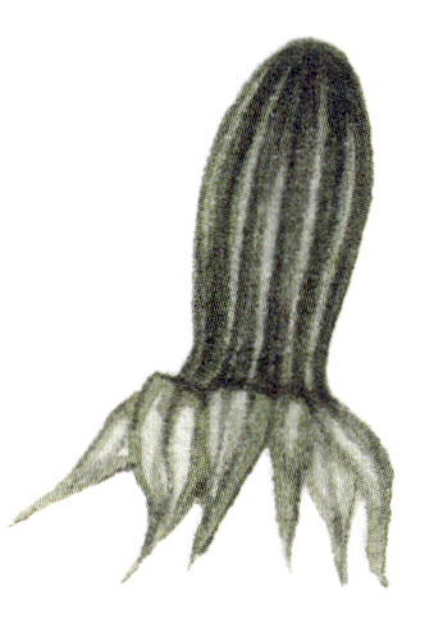

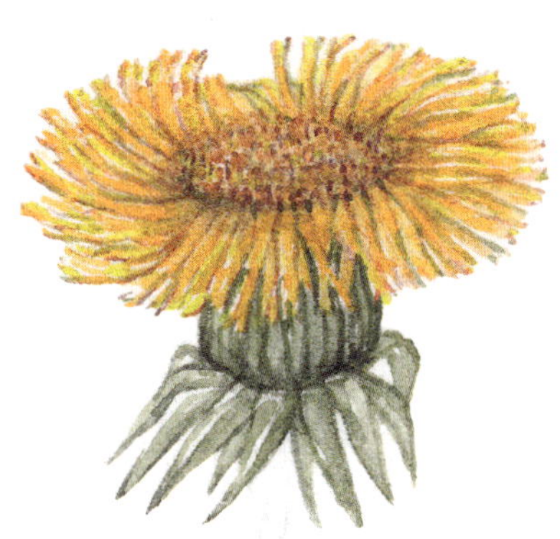

5 | Mix a wash with sap green and a touch of winsor violet, and use this to add shadows and darker areas into the bracts and bud.

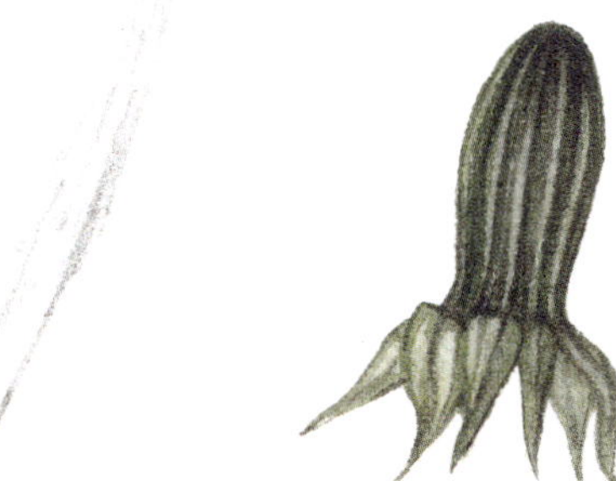

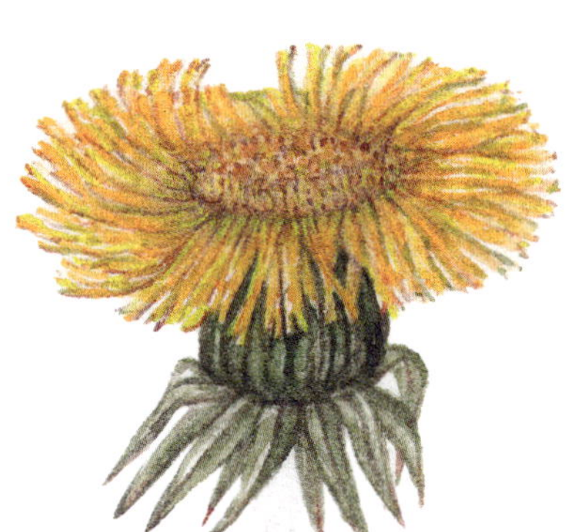

6 | Use a cadmium red wash and a very small brush to add colour to the tips of the bracts. Add further dark areas to the bracts to show the form and the vein down the centre.

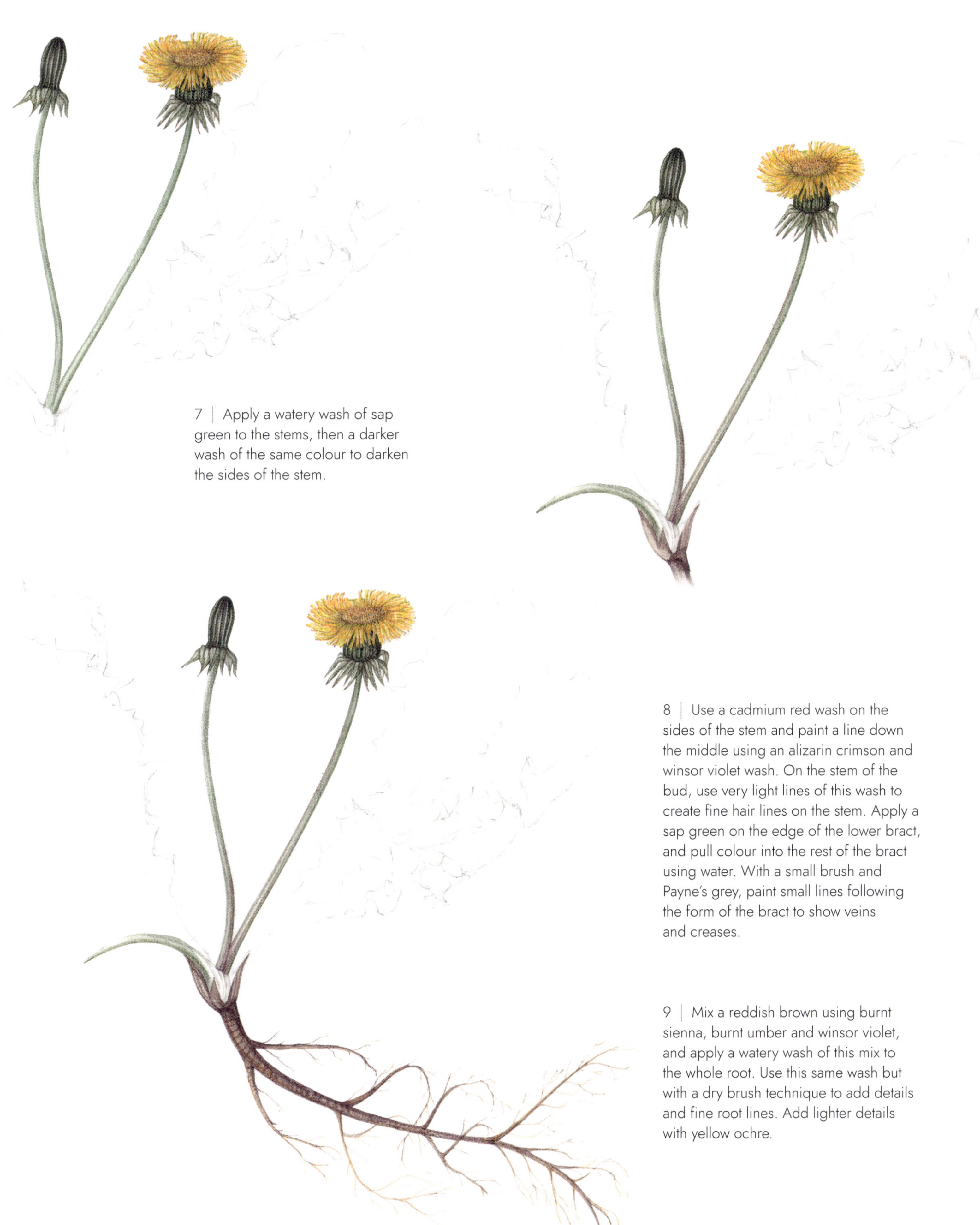

7 | Apply a watery wash of sap green to the stems, then a darker wash of the same colour to darken the sides of the stem.

8 | Use a cadmium red wash on the sides of the stem and paint a line down the middle using an alizarin crimson and winsor violet wash. On the stem of the bud, use very light lines of this wash to create fine hair lines on the stem. Apply a sap green on the edge of the lower bract, and pull colour into the rest of the bract using water. With a small brush and Payne's grey, paint small lines following the form of the bract to show veins and creases.

9 | Mix a reddish brown using burnt sienna, burnt umber and winsor violet, and apply a watery wash of this mix to the whole root. Use this same wash but with a dry brush technique to add details and fine root lines. Add lighter details with yellow ochre.

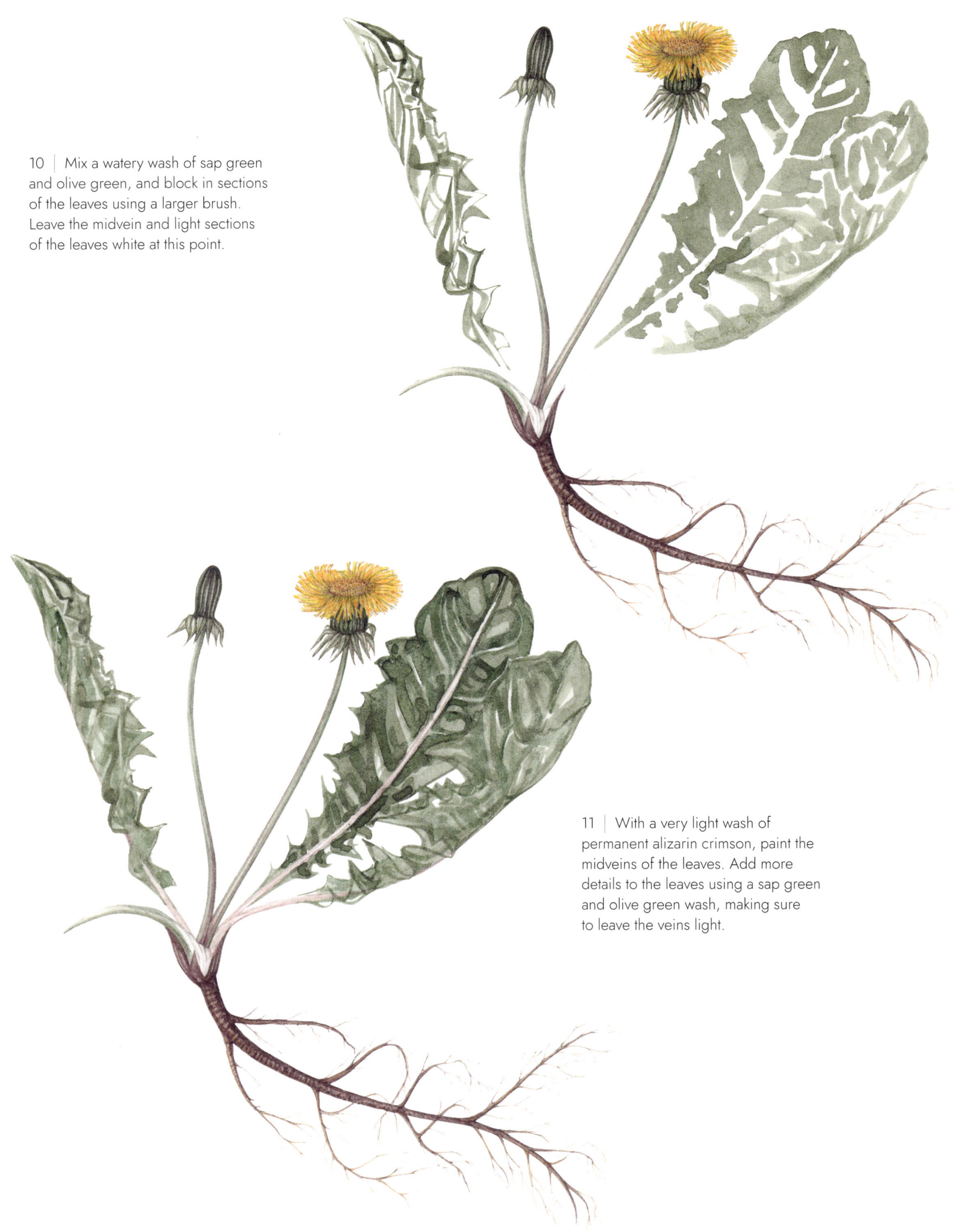

10 | Mix a watery wash of sap green and olive green, and block in sections of the leaves using a larger brush. Leave the midvein and light sections of the leaves white at this point.

11 | With a very light wash of permanent alizarin crimson, paint the midveins of the leaves. Add more details to the leaves using a sap green and olive green wash, making sure to leave the veins light.

12 | Use the green wash from step 11 and a very small brush to add details to the leaves, leaving areas of light to show the form and bumps. Add a Payne's grey strip to the midveins on one side. Add a further touch of alizarin crimson wash to the veins on the other side.

13 | Add further details to the leaves with the green wash. Add dark lines down the edges of the stems.

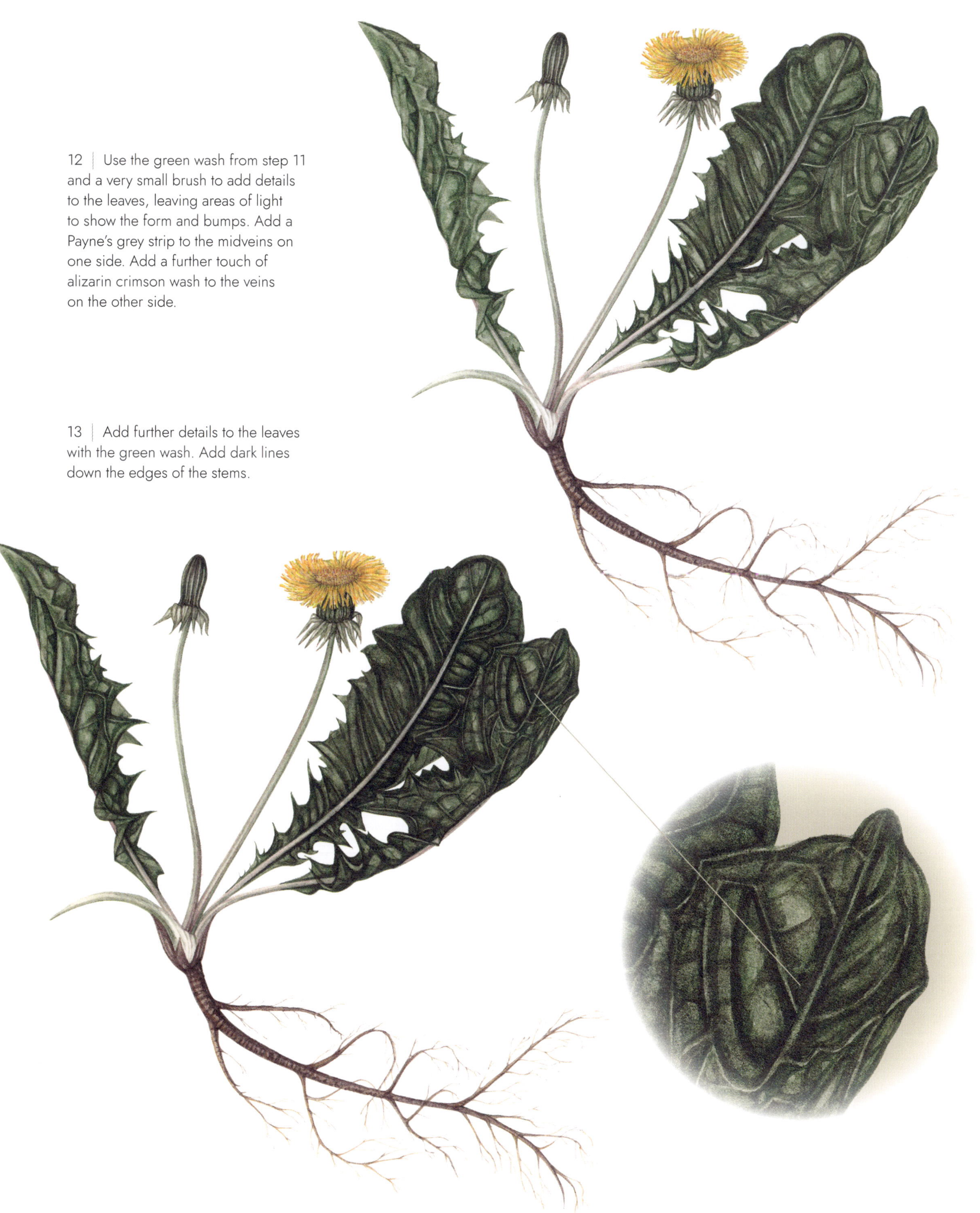

Amaryllis *(Hippeastrum cv.)*

by Julia Trickey

In this bold portrait entitled *Vintage Amaryllis*, Julia Trickey takes us straight to the roots of a spent amaryllis bulb in a wonderful study of earthy colours, showing what can be done with a feature as unloved as roots.

Technique and composition

Julia used white paper and her preferred wet-on-wet technique, but she conditioned the paper first to make it look like browning vintage paper to suit the old bulb. Rather than making the subject lift out of the page, there is a wonderful sense of the bulb's connection with the earth.

An amaryllis is normally appreciated for its large, brightly coloured, showy flowers. It belongs to the *Hippeastrum* genus from South America, in the Amaryllidaceae family.

Here, there is not a bright primary colour in sight. Instead, Julia focuses on the architecture of the root system where both adventitious fibrous and contractile roots are seen. It is a virtuoso piece and one can imagine what fun she had picking her way through the tangle to ensure each root has its place — the thick, fleshy roots on the outside and the fibrous ones on the inside. Peering into the tangle, it is possible to discern volume in the whole mass with the crisscrossing of smaller roots.

Emotional appeal

Emotionally, it is a picture that speaks of freedom as well as order. The larger roots lie in an orderly circle, whilst some small roots make a break for freedom, rather like a mischievous puppy that escapes from its leash.

But the picture would not have the effect it does without the stately green shoots that fade out above the bulb, and the daughter bulb peeping beneath. The green relieves the pressure of neutral tones but does not dominate. Loose debris floating around the main feature balances the whole composition and Julia is careful to place them, and the direction of the roots, in the most satisfying position for the viewer.

Painting roots takes time and patience. In a voluminous subject, such as a bulb and its mass of roots, with many thicker roots, Julia worked hard on the shadow areas to suggest three-dimensionality at all levels.

The rewards of technical excellence and careful composition awaken us to the mysteries of a bulbous root system, rarely appreciated, yet full of interest.

RIGHT A spent amaryllis bulb continues to fascinate with its tangle of roots. Size: 47 x 50cm (18 1/2 x 20in), watercolour on Fabriano Artistico paper.

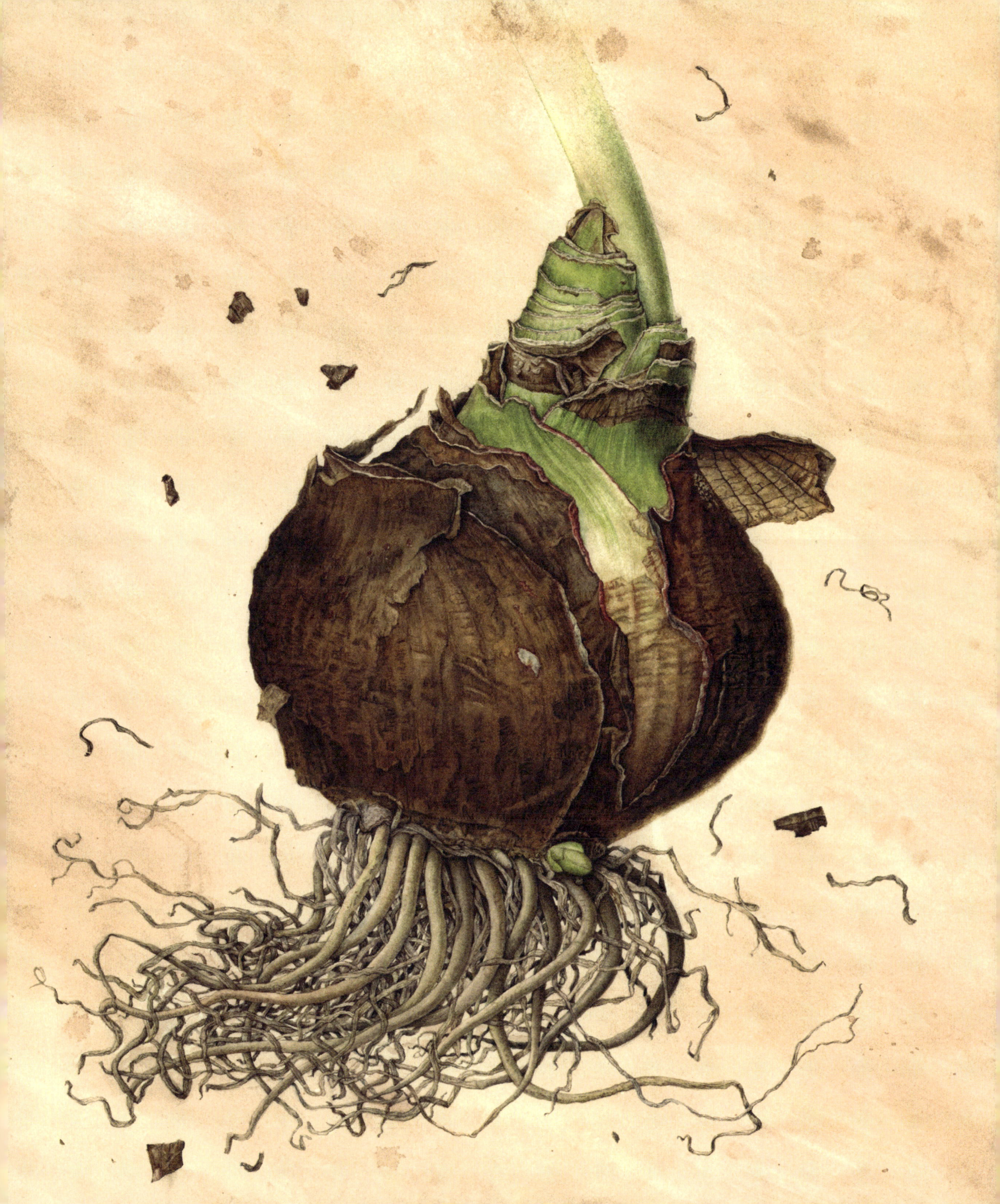

Chapter four

Leaves

Leaves are the main organ for producing a plant's energy – a kind of food production factory. Leaves also act as 'lungs', enabling gases to pass in and out of pores on the leaf's surface. They are attached to the stem and are connected by vessels carrying water and nutrients to other parts of the plant.

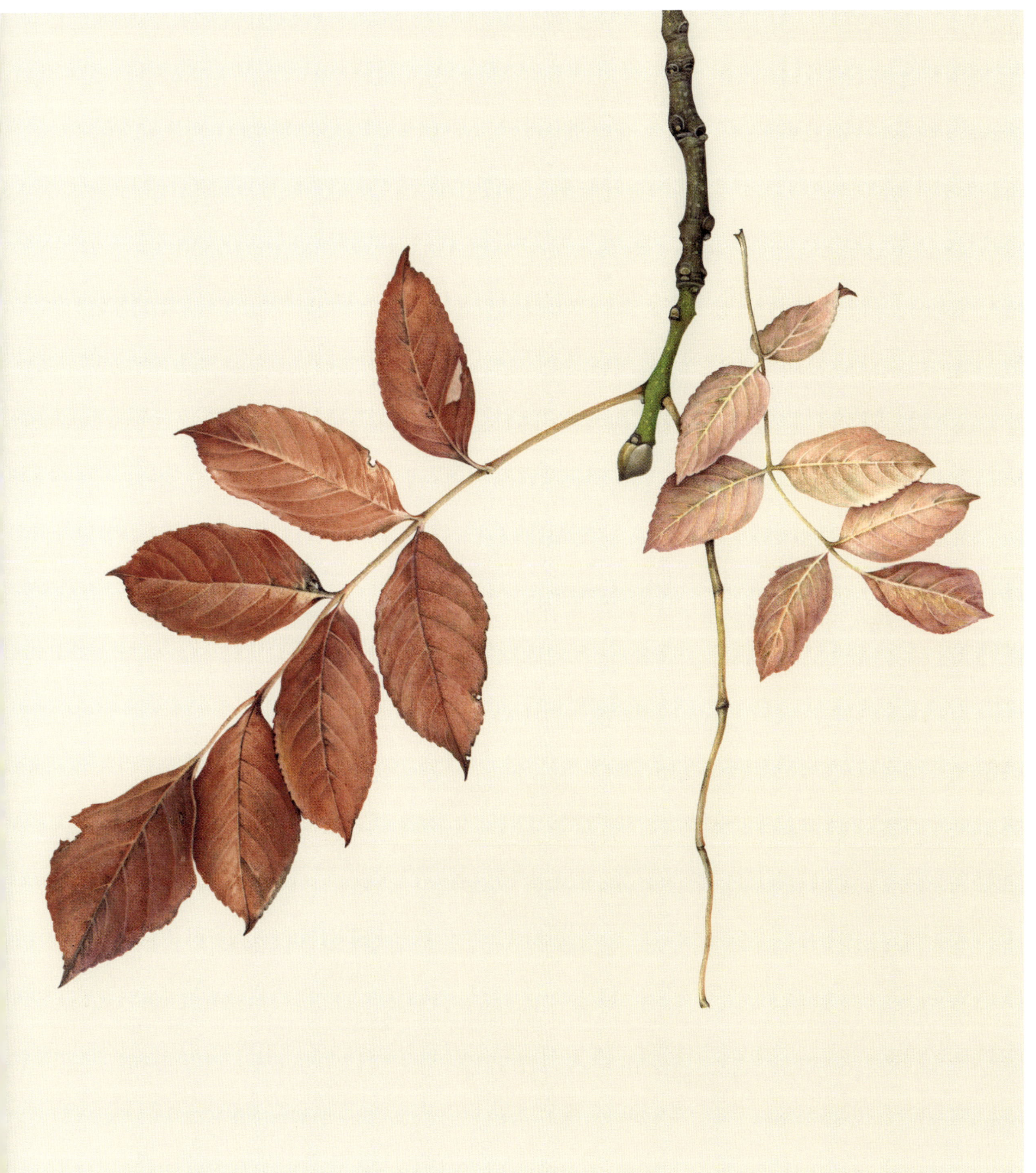

Basic structure

Everything about leaves is focused on their role of producing carbohydrates. Their shape, colour, placement along the stem and venation all play a part. Leaves in higher latitudes tend to be flat and spaced out to catch all the sunlight they can. But look out for forms that help them cope with extreme conditions, or provide a convenient place for some other structure to be placed.

Typically, leaves have an expanded blade (lamina), with a characteristic shape. The upper surface facing the stem is the adaxial and the underside abaxial (think of ad-vance and ab-sent). The margins, tips and bases are what their names suggest. The veins form a pattern, which is usually more prominent on the abaxial side.

A petiole is a stalk connecting the blade to stem. If there is no petiole, the leaf is sessile (meaning that it attaches directly to the stem without a stalk). The angle between stem and petiole is the axil, an important area where new growth arises. A node is the place where the leaf is attached to a stem. At the node and beneath the petiole there may be a pair of leafy, spiny or scaly stipules.

PARTS OF A LEAF

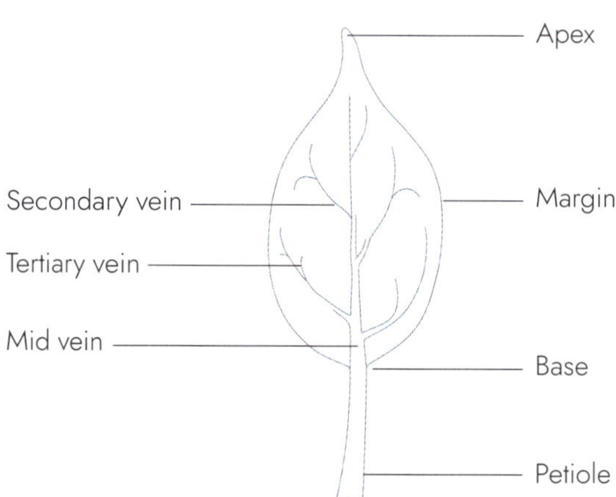

BELOW Spindle (*Euonymus europaeus*) by Sarah Howard has green leaves attached to the stem by a petiole, opposite veins, and a pale coloured abaxial side.

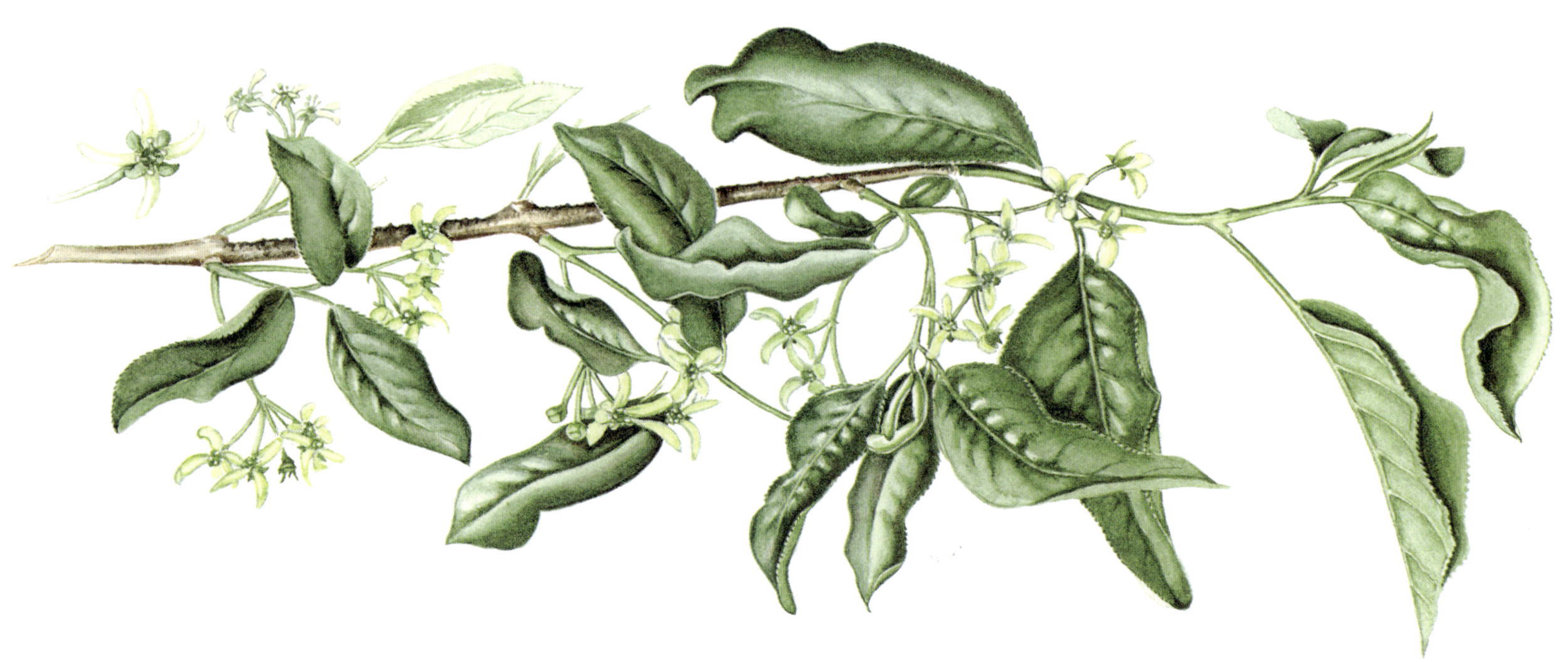

The blade: thickness and colour

If you want to paint a 'typical' version, work out why a leaf might be red, pink, purple or blueish. Sometimes a colour other than green might be natural. The rosette leaves of the beautiful Chilean bromeliad, *Fascicularia bicolor*, turn red at the base to attract pollinators when a flower bud appears. A light blue glaucous colour in hot places provides protection from excess light. A change of colour ranging from pink to purple might be a sign of stress, or a change of colour when a plant decays in autumn.

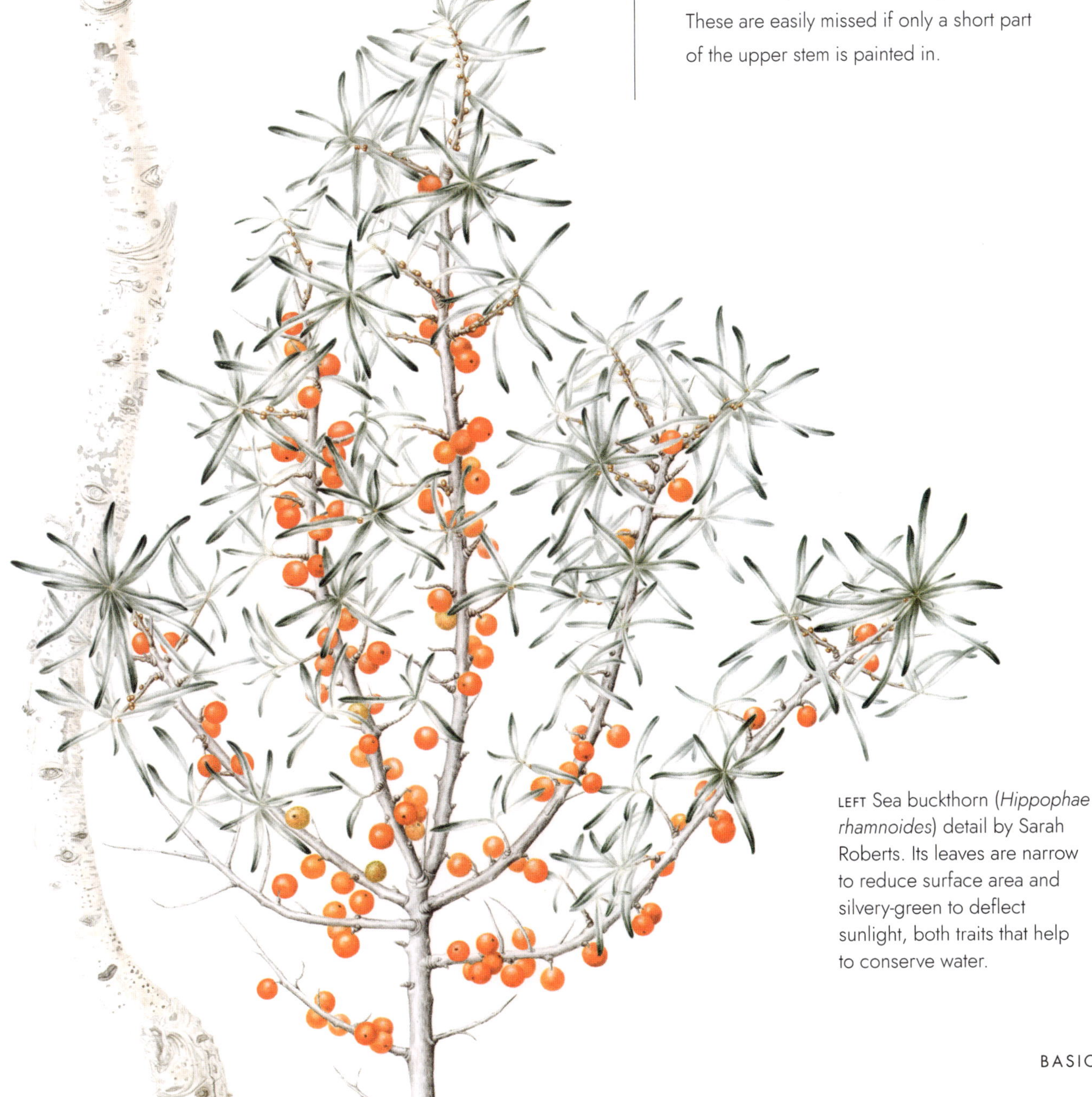

LEFT Sea buckthorn (*Hippophae rhamnoides*) detail by Sarah Roberts. Its leaves are narrow to reduce surface area and silvery-green to deflect sunlight, both traits that help to conserve water.

The blade: folding and margins

In a transverse section, the leaf is not always flat. In drought-prone regions, the shape from margin to margin shows how water is channelled into, or out of, the plant's stem.

If there is a groove running down the middle, or it is concave, it is canaliculate, like a canal. The margins may be sharply folded and have a raised edge like the keel of a boat, as seen in some leaves of species in the genera *Kniphofia* or *Gladiolus*.

LEAF FOLDING

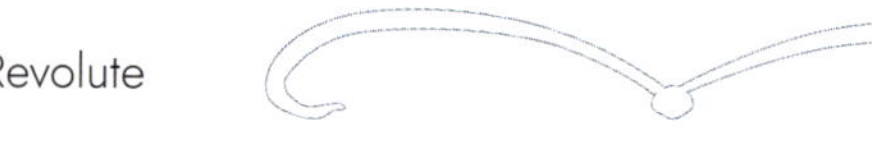

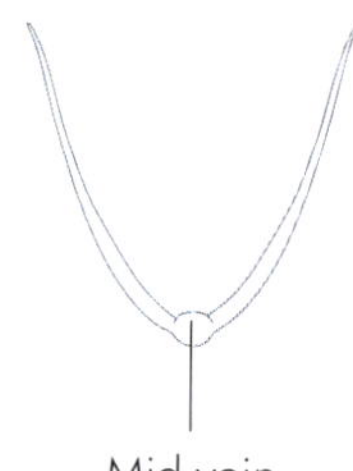

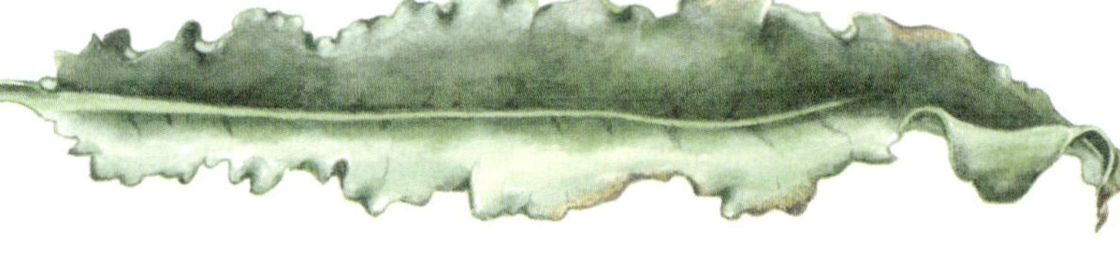

LEFT *Kniphofia schimperi* by Sarah Howard. Many *Kniphofia* have long, slender, keeled leaves, with rough margins caused by minute stiff hairs.

ABOVE Curled dock (*Rumex crispus*) by Sarah Howard. As its name suggests, this has crisped (wavy) leaf margins, but the rest of the blade is smooth.

The margins might be involute or revolute. *Primula* leaf margins face downwards (revolute). The margins of butterwort leaves (*Pinguicula* sp.) turn upwards (involute). Study the leaves of a lemon tree (*Citrus × limon*), which can fold almost in half, especially in bright sunlight when they become conduplicate. If leaves are pleated like fan palms, they are plicate. The margins may have a thickened edge (marginate). They may be undulate (wavy) or crisped (for example, *Rumex crispus*). A leaf is entire if it has no marginal irregularities.

Other leaf features

Petiole: leaf stalks are not necessarily circular in cross-section. They can also be oval in shape, wavy or have a groove along the top (adaxial) side. They may be swollen or have a swollen pulvinus at the base. When they fall off, a leaf scar remains, within which a distinctive arrangement of vascular bundles might be seen.

Axil: this is the angle formed between the stem and the upper leaf or petiole where axillary leaf or flower buds develop. The angle depends on the aspect of the leaf, whether hanging up or down, or close to the stem (see page 78).

Terminal leaf bud: this is the last bud at the top/apex of a stem, and contains all that is needed for further growth.

Stipules: these might not be seen because they fall off quickly, are not present, or not easily distinguished from a leaf, such as the fused stipules at the base of *Alchemilla mollis* leaves. Nevertheless, they can be diagnostic in roses and a common feature in some families. Leafy stipules provide extra surface area for photosynthesis (as in the edible pea, *Pisum sativum*). They can also be modified into spines to protect new growth (*Vachellia*) or tendrils for climbing (*Smilax aspera*).

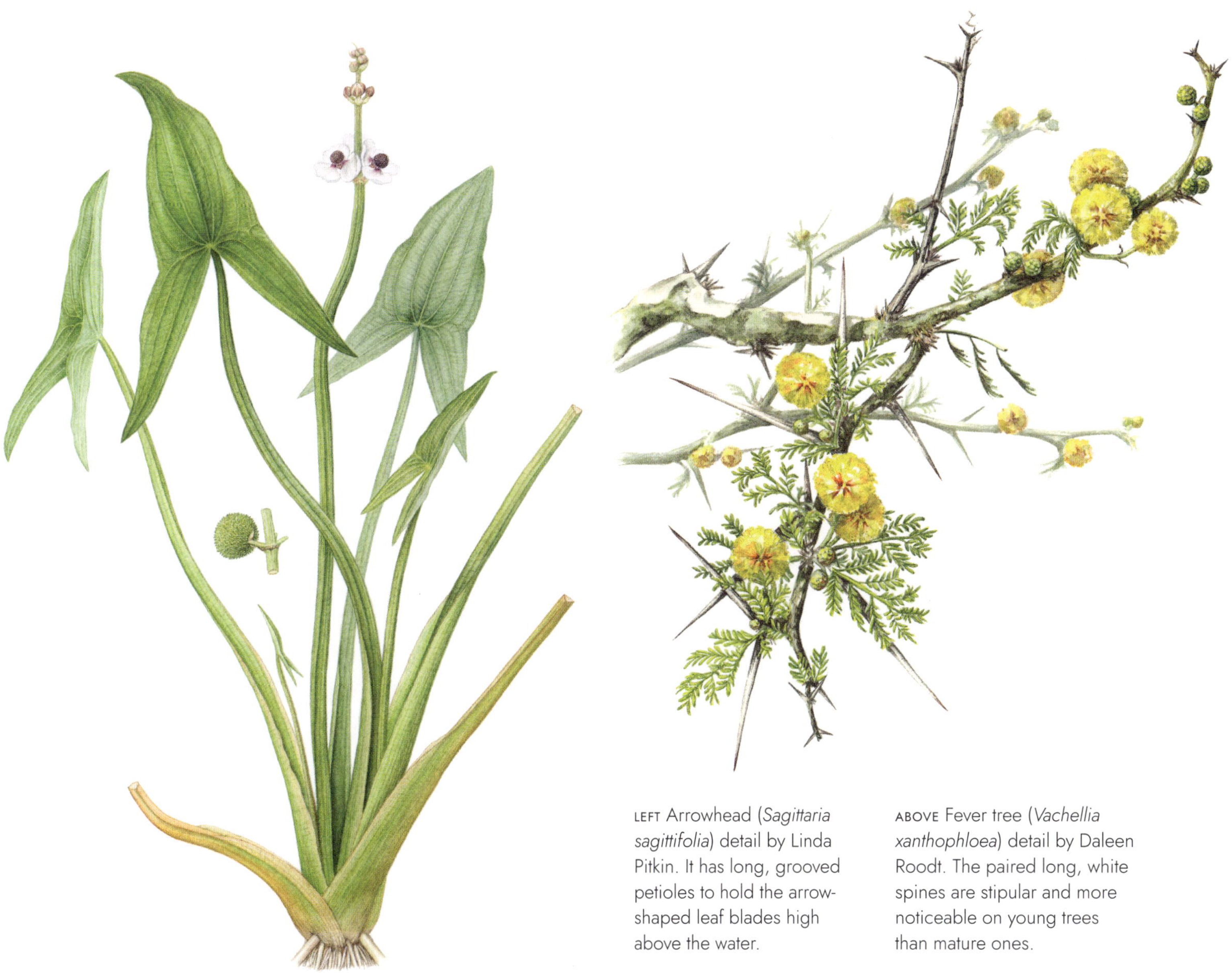

LEFT Arrowhead (*Sagittaria sagittifolia*) detail by Linda Pitkin. It has long, grooved petioles to hold the arrow-shaped leaf blades high above the water.

ABOVE Fever tree (*Vachellia xanthophloea*) detail by Daleen Roodt. The paired long, white spines are stipular and more noticeable on young trees than mature ones.

Leaf arrangements

How leaves are arranged along the stem (phyllotaxis) is an important factor in how they catch the light. Most plants have either a basal rosette of ground-hugging leaves or are cauline, which means they are arranged along a stem. Sometimes plants have both. Leaves arise at nodes, and the space between them is called the internode.

The arrangement of leaves on a stem may be opposite, alternate or spiral. The opposite arrangement is based on a pair of leaves at each node, one on each side of the stem, for example, the *Caryophyllaceae* family (chickweeds, pinks and campions). When successive pairs of opposite leaves are set at right angles to the last, they are called decussate, for example, the *Lamiaceae* family (mints and nettles). A pair (or more) of leaves on opposite sides of the same node, appearing like a ring of leaves around the stem, is a whorl or verticil – for example, the genus *Galium* (bedstraws). Sometimes it's unclear what is leaf and what is stipule.

An alternate arrangement is based on one leaf at each node. The simplest gives the appearance of two rows with alternately set leaves, for example, many *Campanula*.

A spiral pattern has one leaf at nodes that spiral around the stem. Depending on the interval between nodes, the effect may look like leaves alternately set on each side of the stem, or like three rows (as seen in sedges), or they may simply appear as a spiral up the stem. There is usually a pattern which conforms to a Fibonacci sequence, if you can find it. Be sure to record the changing aspect (perspective) of the leaves, whether front on, side on or showing the undersides.

When the arrangement is looked at by the number of rows, the words used are monostichous, distichous, tristichous and so on ('stichos' means row in Greek).

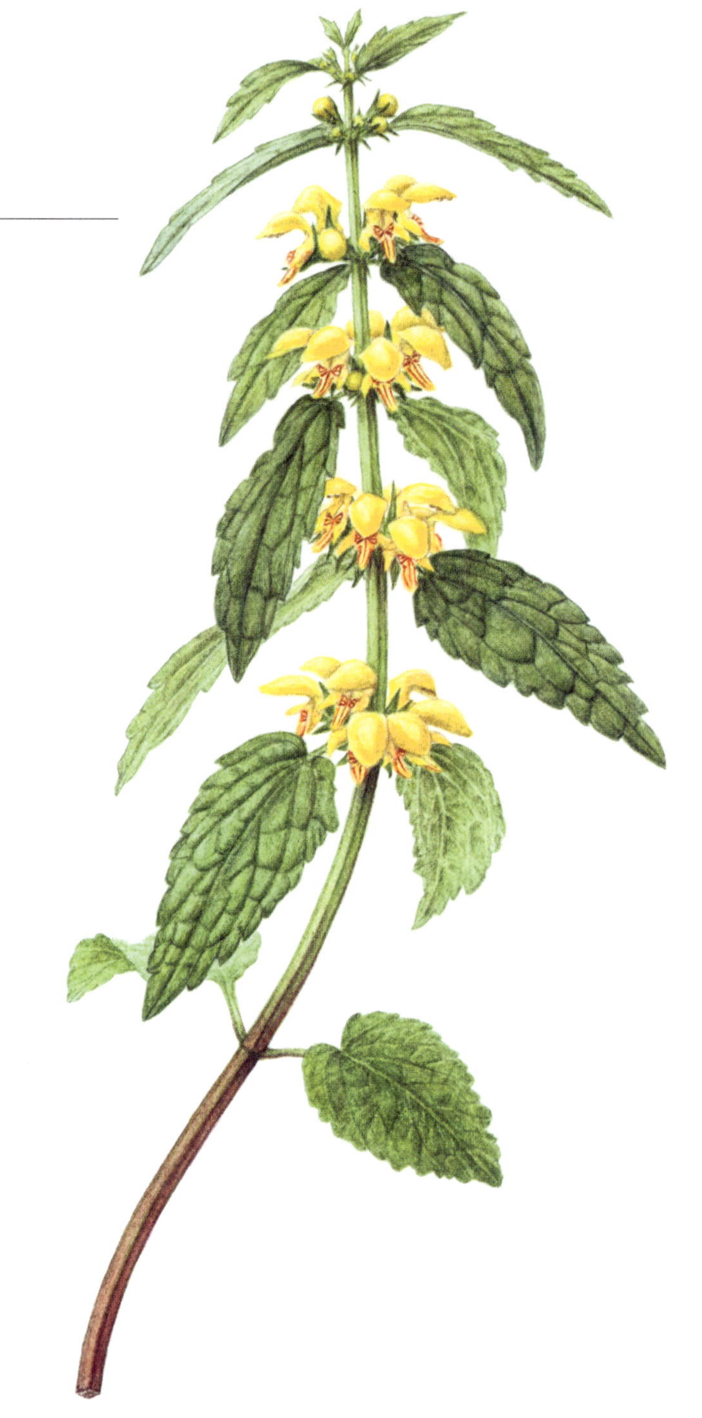

ABOVE RIGHT Yellow archangel (*Lamium galeobdolon*) by Linda Pitkin has a decussate leaf arrangement where each pair of leaves is at right angles to the next.

TYPES OF LEAF ARRANGEMENT

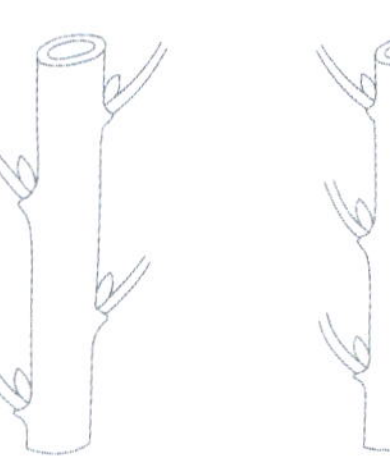

Alternate

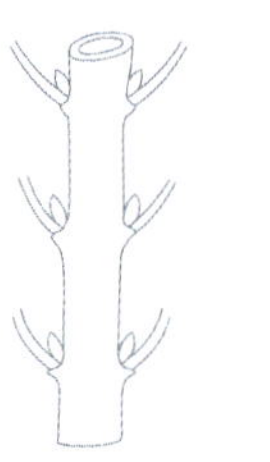

Opposite

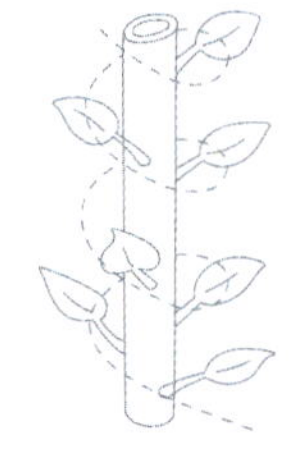

Alternate spirally-arranged

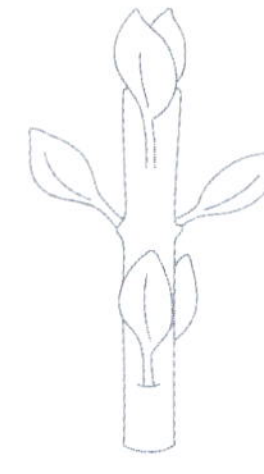

Opposite decussate

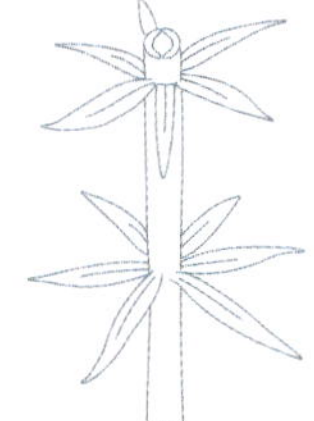

Whorled

Leaf attachment

The stalk attaching leaf to stem is a petiole. If the leaf is attached directly to the stem, it is called sessile. A bundle of leaves at the same point is a fascicle, for example, pine needles.

However, some leaves are attached so that the margins run down the stem (decurrent), for example, thistles. Or they may clasp the stem, like sow thistles. Leaves that clasp the leaf above, forming a plait or braid are equitant, such as moth orchids, or are spathulate, if they clasp the stem, like grasses. Perfoliate either describes a pair of opposite leaves which are joined up and the stem passes through the middle (for example, perfoliate honeysuckle), or one leaf where the stem passes through the base of it.

Artists often imagine leaves attached at a 45-degree angle on the stem, but there are other angles to note. If leaves are set close to the stem, and the axil's angle is acute, they are adpressed. They are ascending if pointing up, reflexed if pointing down, or spreading if pointing sideways.

Some of these angles may only be temporary if the leaves are responding to cold or drought. After a cold spell, the leaves of the *Rhododendron* tend to reflex (point downwards). Leaves, as anyone knows, can droop after picking, which makes sketching in front of a growing plant a necessary initial step.

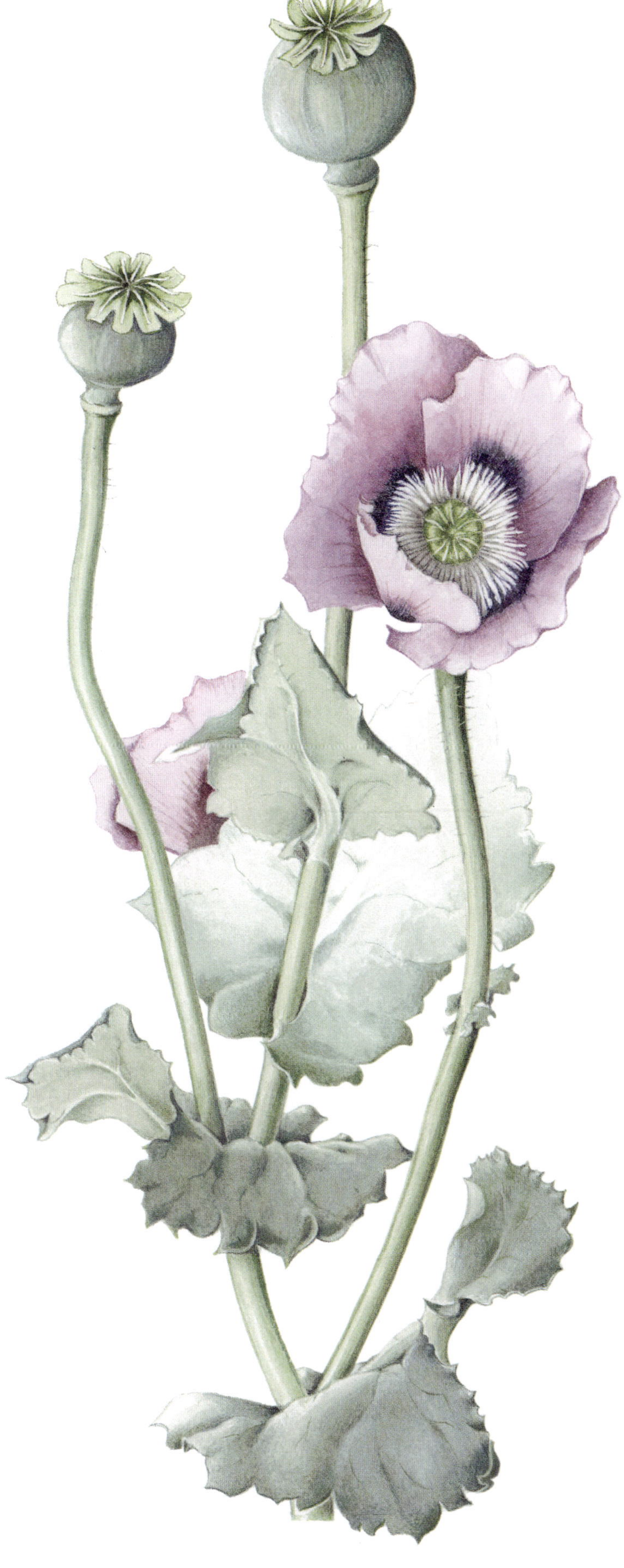

RIGHT Opium poppy (*Papaver somniferum*) by Sarah Howard. The alternately placed leaves have an amplexicaul (clasping) attachment.

Leaf characteristics

Simple leaves

A simple leaf is in one piece, in contrast to compound leaves, which are divided. The shape of leaves is described when they are flattened. Note where the widest part of the leaf is; it could be nearer to the apex, or to the base. The terms listed and illustrated below are sometimes loosely applied by different authors.

- Rounded leaves: circular.

- Oval leaves: elliptic/oval (no points either end), ovate (egg shaped, wider at the base), obovate (wider towards the apex), acuminate (tapers to a long point at the apex), lanceolate (pointed at both ends).

- Spear-shaped leaves: linear (elongated with parallel margins), aciculate (needle-shaped), ensiform (sword-shaped with a sharp point).

- Roughly triangular shaped leaves may be flabellate, deltoid or obdeltoid.

- Terms for shapes describing everyday concepts could include: cordate (heart shaped), spatulate (spoon shaped), palmate/digitate (like a hand or fingers), falcate (sickle shaped, like *Eucalyptus falcata* or *Acacia falcata*), deltate (like a delta), flabellate (fan shaped), reniform (like a kidney), hastate (spear) and sagittate (arrow).

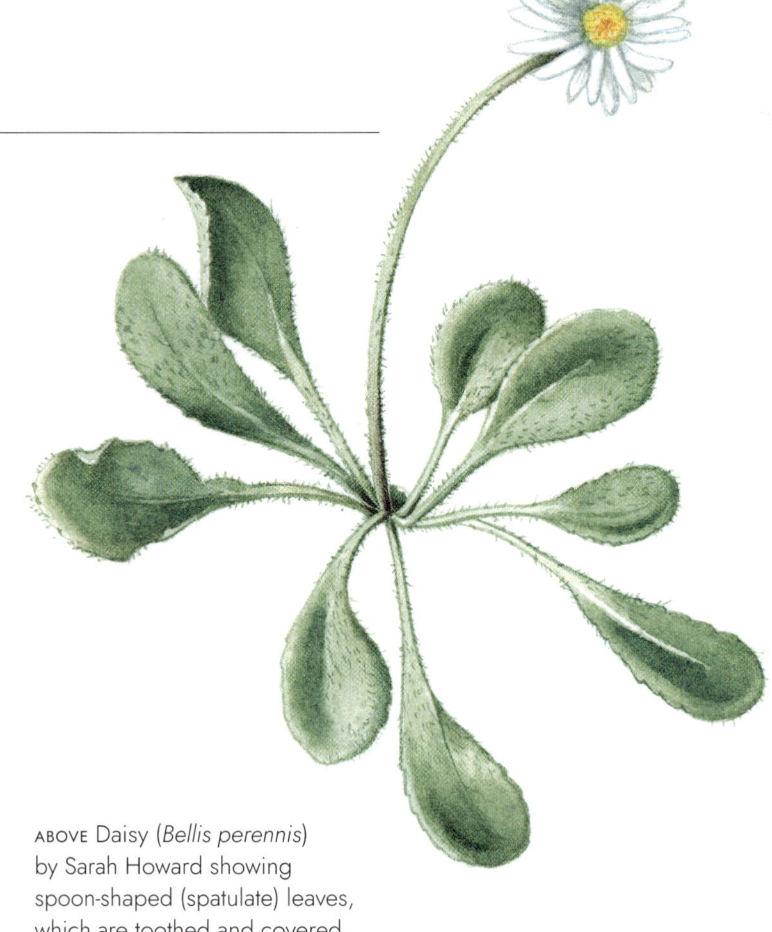

ABOVE Daisy (*Bellis perennis*) by Sarah Howard showing spoon-shaped (spatulate) leaves, which are toothed and covered in soft hairs.

- Monocot leaves are usually long and narrow.

- If the petiole is attached near the centre of the underside of a round leaf such as a nasturtium or water lily, it is known as peltate.

LEAF SHAPES

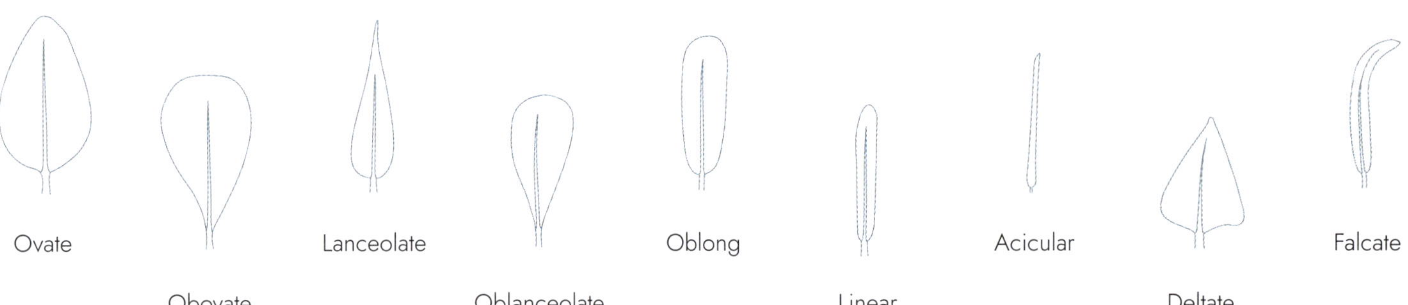

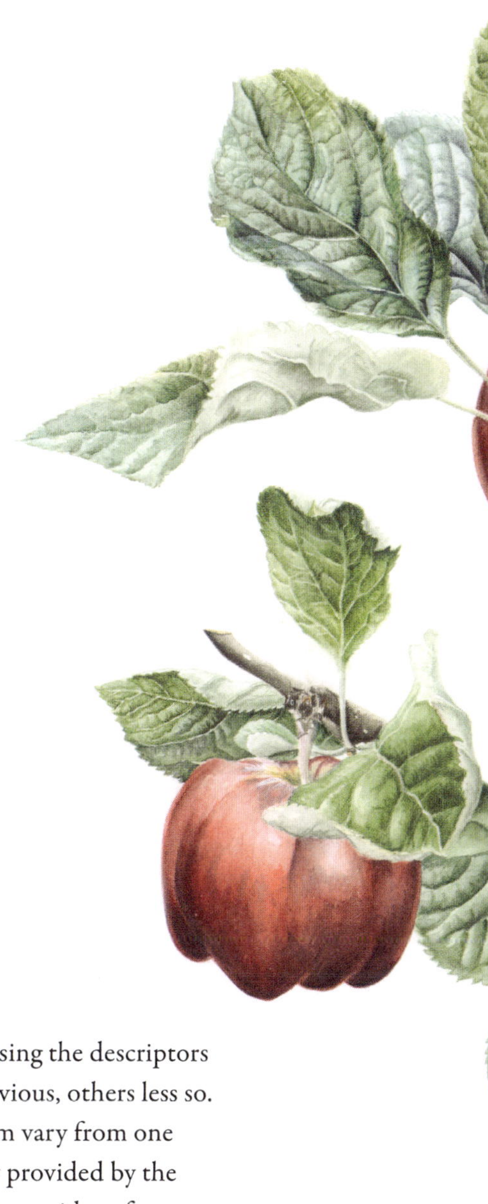

Blade tips

It is best not to be too dogmatic about using the descriptors
for edges, tips and margins. Some are obvious, others less so.
Sometimes the diagrams illustrating them vary from one
book to another. Do rely on the glossary provided by the
author. The examples below, however, give an idea of a
variety of forms.

If there is a small tip extension of the main vein, it is
called mucronate (mucro is Latin for a sharp point, like that
of a sword). If it extends further, it is known as aristate
(from the Latin for awn, found in grasses, see page 157)

LEAF BLADE TIP SHAPES

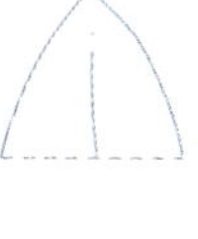

Acute

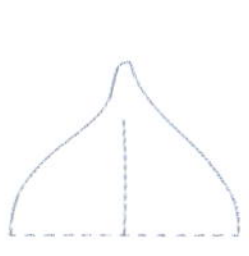

Acuminate

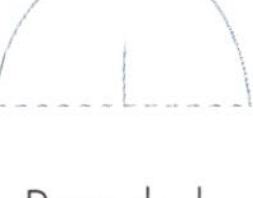

Rounded

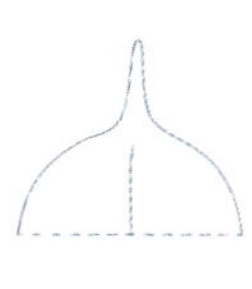

Cuspidate

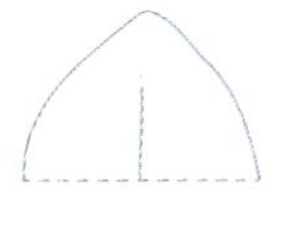

Obtuse

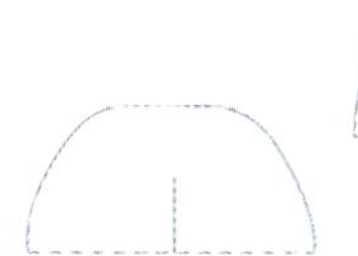

Truncate

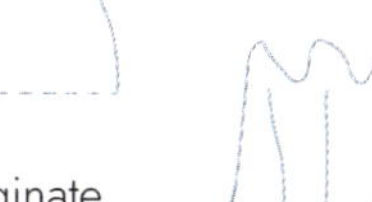

Emarginate

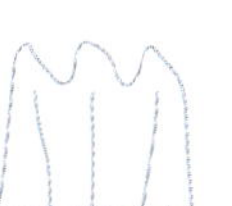

Tridentate

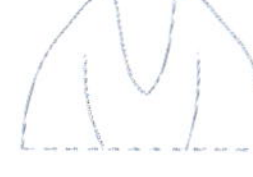

Bifid

Leaf bases

Some base types, such as hastate and sagittate, have already been described in connection with a simple leaf shape. Be careful about the symmetry of the two sides at the base – sometimes they don't match, or one lobe may overlap the other.

Leaf margins

Margins are entire if they are smooth. However, the edges can be anything but smooth. They range from teeth to narrow, hair-like shapes. Along the leaf margins there may be spines (spinose) such as those in *Acanthus*, *Aloe* and *Ilex* (hollies), although mature hollies tend to lack spines. This can make it difficult for an artist to paint a surface that looks smooth on the whole surface of the blade including a sculpted edge without leaving streaks.

Choosing how to work the edges of leaves takes practice and some thought. Ciliate edges are simple enough, because the hairs can be added after the blade is painted, but note whether the hairs are in a shadow or in the light, or against a darker surface. Edges with a more complicated shape might require a puddling technique, unless your normal technique is dry brush.

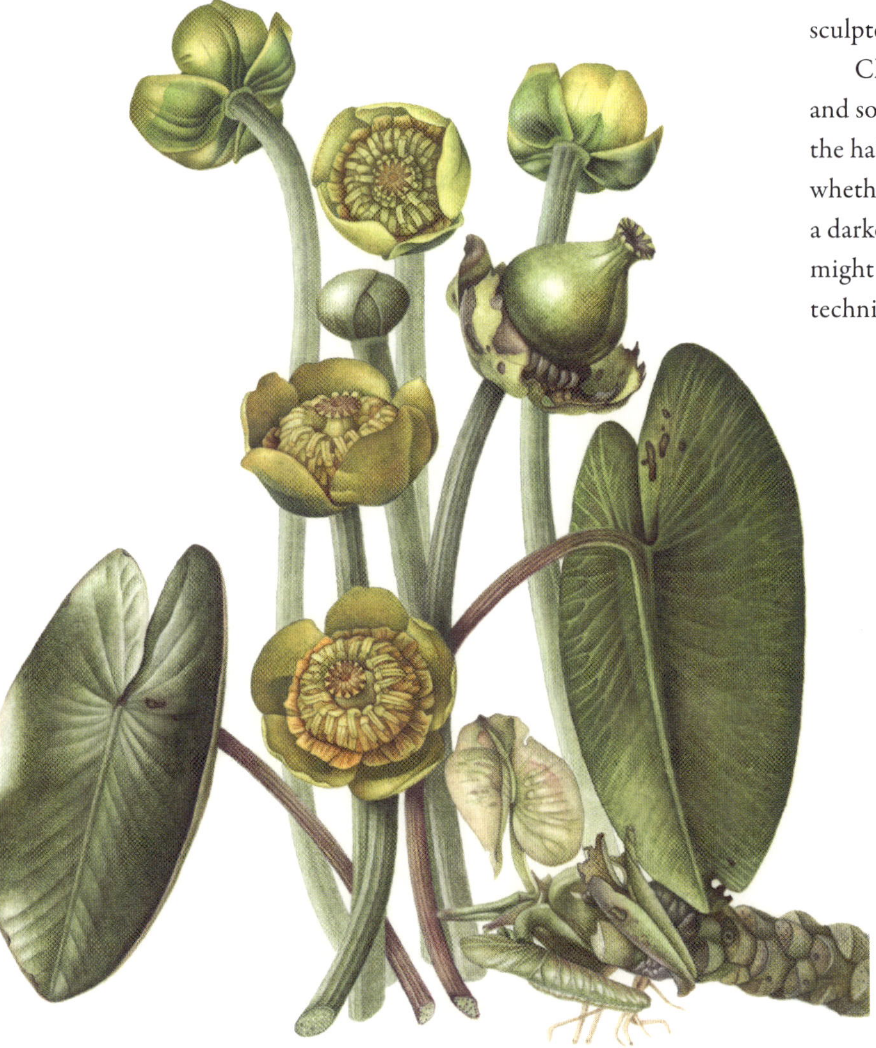

LEAF BASE SHAPES

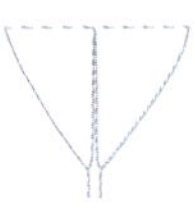
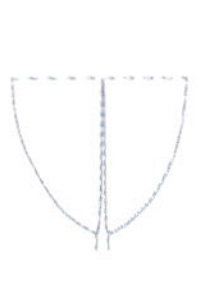
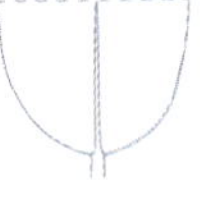
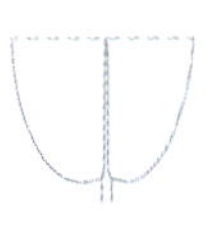
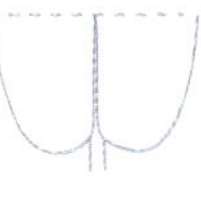
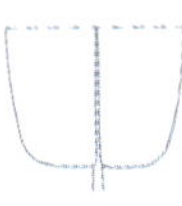
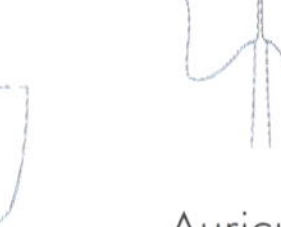
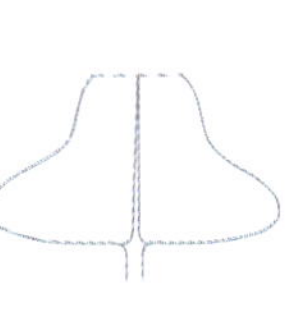
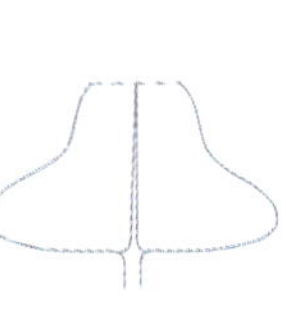

Cuneate   Rounded   Cordate   Auriculate   Sagittate

Obtuse   Subcordate   Truncate   Hastate

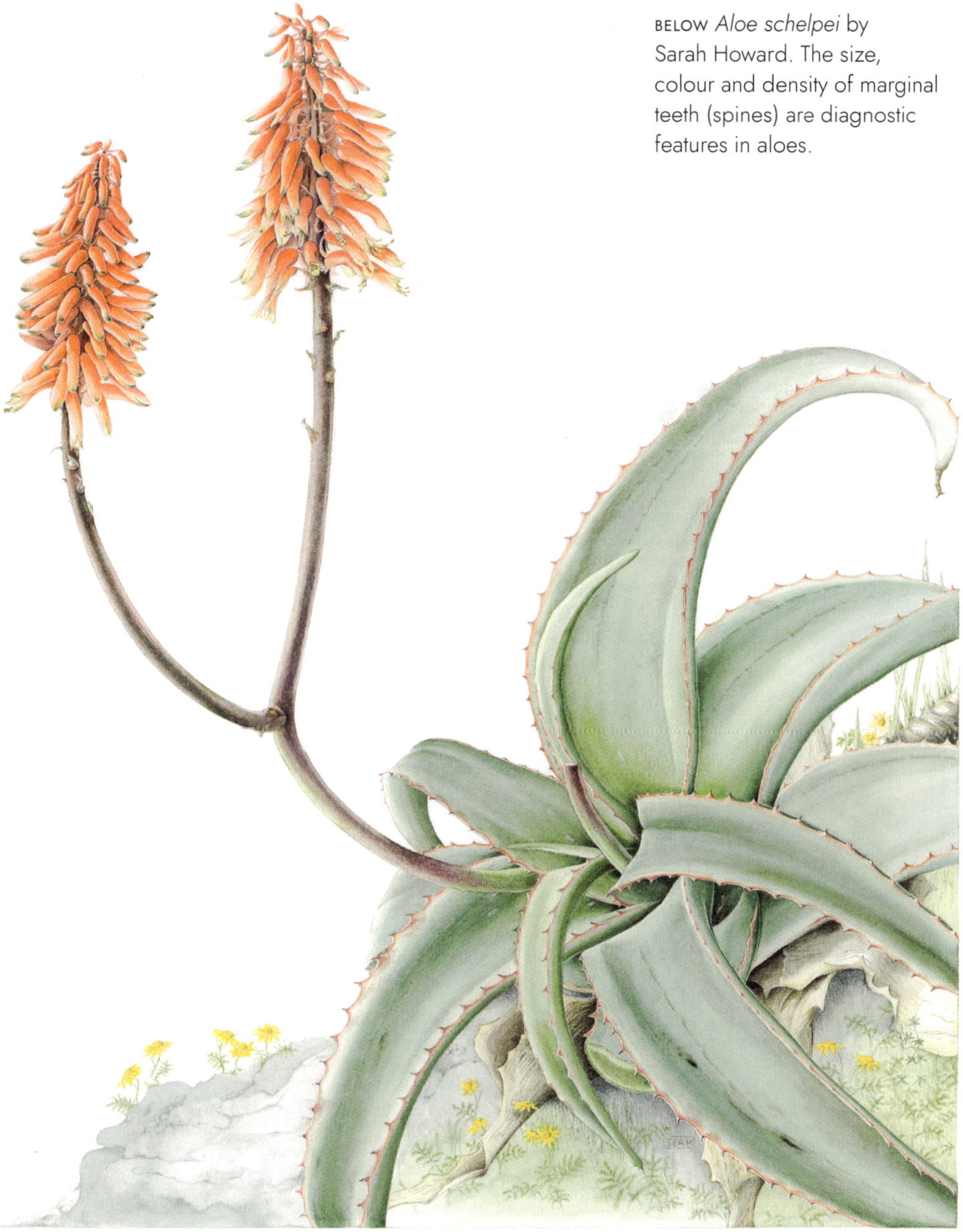

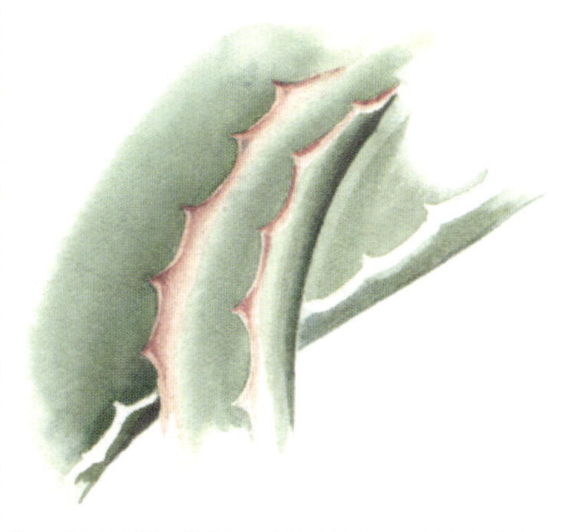

PAINTING ALOE SPINES

This is how to paint *Aloe* spines, which are coloured, and have a margin that catches the light:

- Aloe leaves are long and smooth, so choose a wet technique and large brush. Apply masking fluid to the spines and the edges of the leaves.

- Wet the paper to the edge and swish a first layer of dilute paint close to the edge, leaving a pale area. Allow to dry.

- Swish on another layer of dilute paint to blend the colours, ensuring you leave no streaking. Allow to dry.

- Continue with further layers to intensify the colour on the main part of the leaf to show shadow, allowing to dry between layers.

- Remove the masking fluid. Dampen the spines area and colour the spines with a dry brush.

LEAF MARGINS

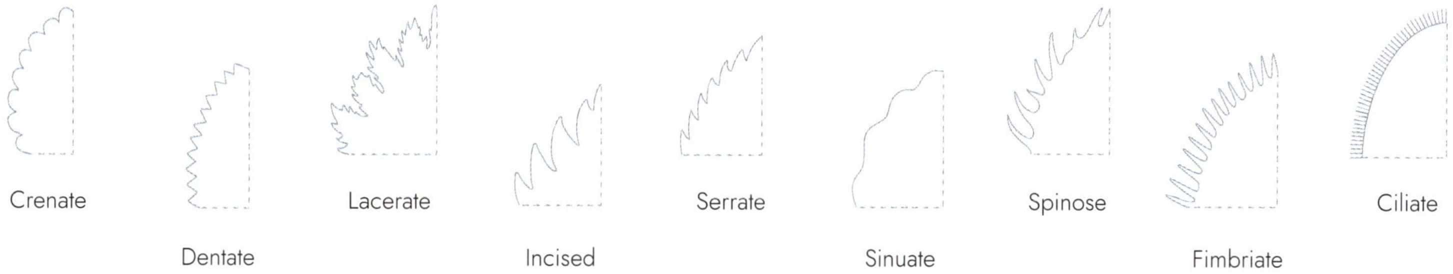

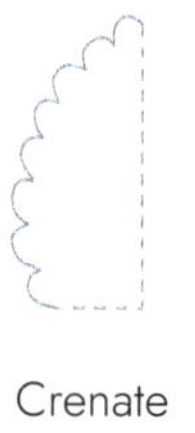

Crenate

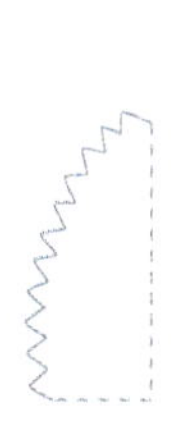

Dentate

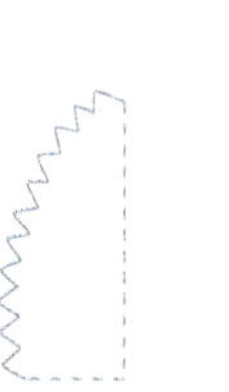

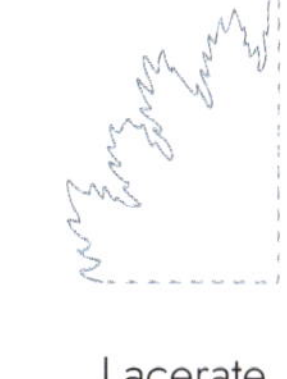

Lacerate

Incised

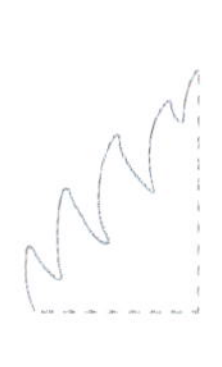

Serrate

Sinuate

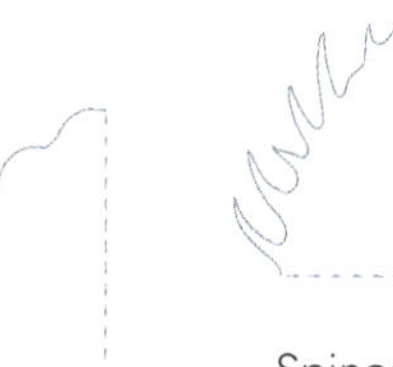

Spinose

Fimbriate

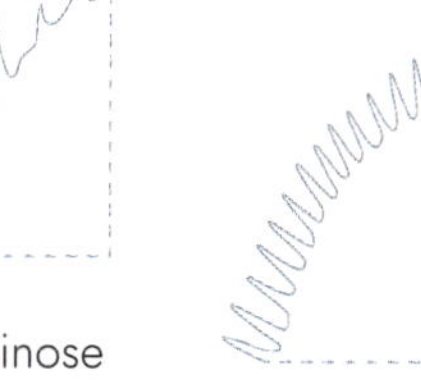

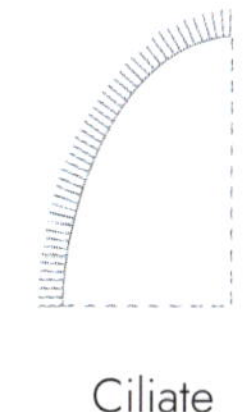

Ciliate

Blade surfaces

Blade surfaces can have hairs, nectaries (organs that secrete nectar) or cushioning. They can be shiny or dull and vary in thickness. A shiny, waxy leaf helps to keep moisture in or deflect water if there is too much. If a leaf is not glabrous (smooth) it may be pulverulescent – covered with a powdery coating made from minute scales of wax formed in the cuticle – or it could be verrucose (bumpy), scabrous (rough to the touch), or pitted with surface depressions. A coriaceous leaf is thick and leathery, like those of a *Magnolia* or *Rhododendron*.

Some leaves in the potato family can have sharp prickles along the midrib or over the blade. If the veins are deeply indented, leaves can look like cushions and the veins themselves may hardly be visible.

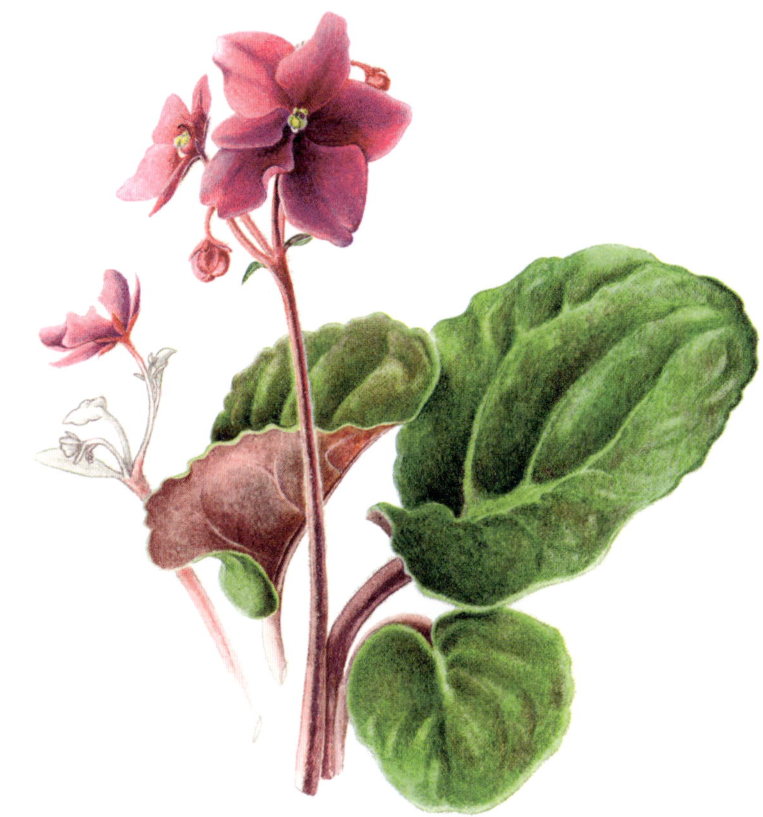

Hairs

The general word for hairiness is indumentu or indument. Hairs provide protection against ultra-violet rays or cold night temperatures. Careful observation shows whether they are long or short, upright or face different directions, glandular or not, forked or star-like, barbed, and so on. Hairs are usually white, but can be other colours such as red or black. Although individual hairs might be difficult to include in a painting, their overall appearance can be noted. Words for different characteristics sometimes appear in the species name, such as 'hirsute', 'puberulous', and 'arachnoid'.

LEFT Borage (*Borago officinalis*) by Linda Pitkin. The look of intense hairiness has been achieved by leaving pale areas around features that overlie others, such as the buds.

White hairs have a shadow side, so painting the hairs grey on the edges of features that lie against the white of the page gives a fuzzy, hairy look.

ABOVE African violet (*Streptocarpus* sect. *Saintpaulia*) by Sarah Howard. It has cushioned leaves sunk into the surface above, whereas they are prominent below.

Leaf venation

From root to leaf, a system of veins feeds and nourishes the whole plant. Leaf veins are most visible on the underside (abaxial). The most prominent veins are directly linked to the petiole and to the rest of the plant.

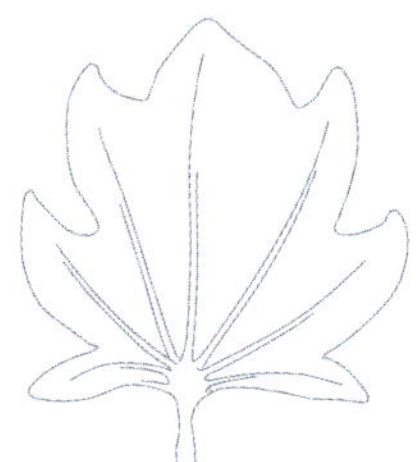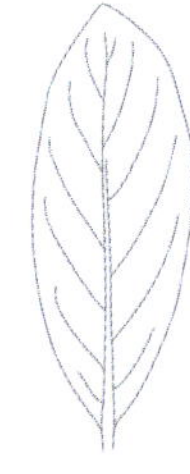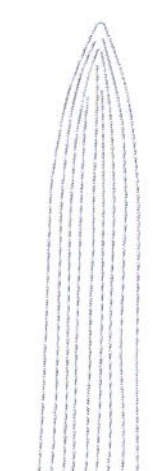

WHAT TO LOOK FOR

- Find the precise 'map' made by the veins.

- Are the veins sunk into the surface, causing a 'cushioning' effect, or are they prominent?

- If you can see them, note the relative widths of primary, secondary and tertiary veins.

- Do the veins skim the margins, or do they head for the edge, especially to a tooth or a lobe?

- When showing a twisted leaf that shows the veins on both sides, ensure they look as though they join up.

- Note the colour of the veins, which are often paler than the blade, or can be red.

- Note whether the veins on either side are a site for hairs, prickles or grooves.

The pattern of veins is typically reflected in the three main patterns of leaf shape: pinnate, palmate or parallel, like most monocots. Sometimes the connections between the veins look less like branches than a net-like reticulation.

In succulent or coriaceous plants, it's often difficult to see any veins other than a midrib and faint secondary veins. If the plant is extremely succulent, there may be no visible midrib at all.

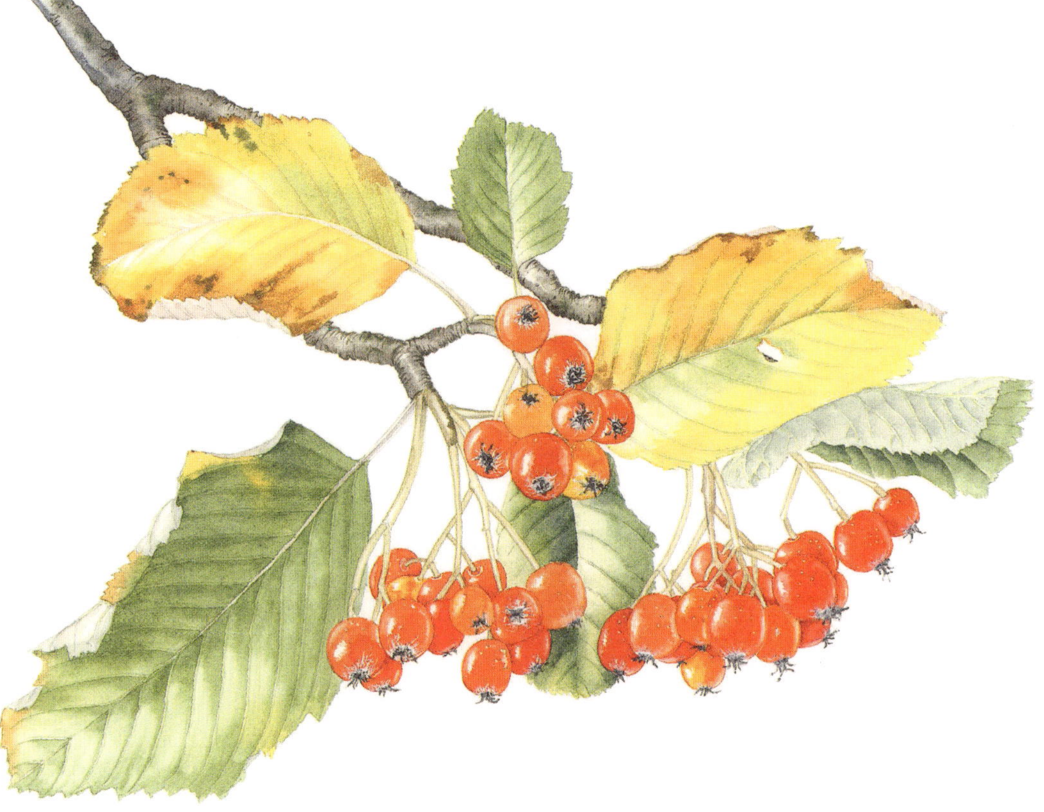

RIGHT Common whitebeam (*Sorbus aria*) by Annie Morris shows its distinctive white colour on the leaf underside, as well as its summer and autumn colours.

Pinnate veins

Pinnate arrangements are usually like a tree with a prominent midvein and secondary branches on either side – or tertiary veins, if they branch again. This arrangement is typical of eudicot plants.

If secondary veins curve before they reach the margin, they are arcuate (for example, *Hosta*). The secondary veins may leave the midrib at an angle, or they may leave at a right angle, like an *Arum* leaf.

It is useful to note the following points when you draw your sketch:

- the number of secondary veins along the midrib.

- the distance between the veins.

Make a leaf rubbing of the more prominent underside to better understand the pattern.

PAINTING VEINS

Before you begin painting, decide how you'll show the narrow and fiddly veins. There are various methods you could use:

- Paint between the more obvious veins, leaving a space to finish off later.

- Use masking fluid so you can wash over the whole leaf first, then remove the masking fluid and wash over the vein areas later.

- Use a paint with good lift-off properties and swish a wet brush along the veins, lifting off paint as you go (keep some paper handy to absorb the wet pigment).

Your decision depends on how big the veins are, the width of the intervals between the veins, how light the veins are compared with the general leaf colour, and whether they 'bleed' into the main colour, or are clearly defined.

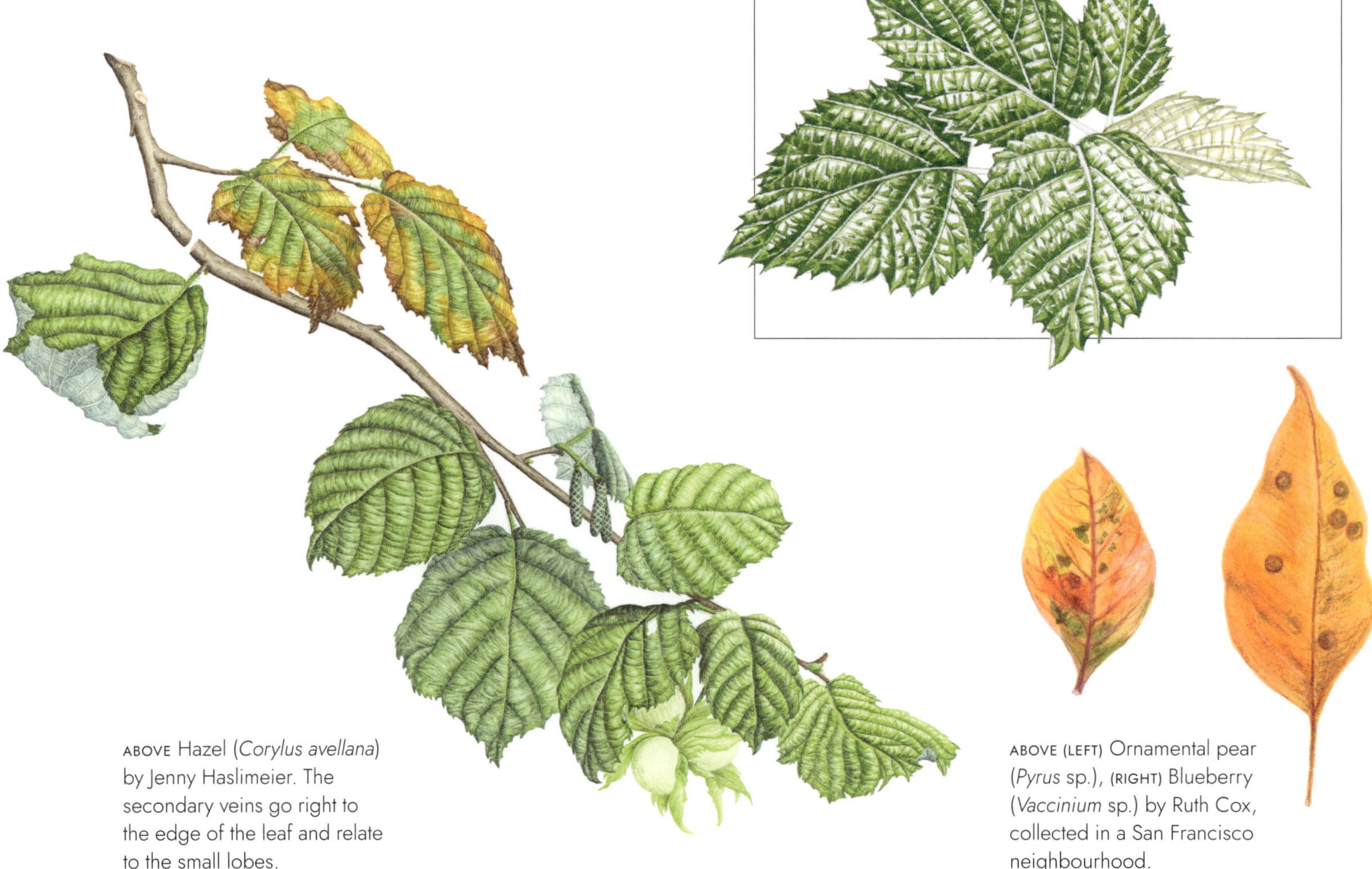

ABOVE Hazel (*Corylus avellana*) by Jenny Haslimeier. The secondary veins go right to the edge of the leaf and relate to the small lobes.

ABOVE (LEFT) Ornamental pear (*Pyrus* sp.), (RIGHT) Blueberry (*Vaccinium* sp.) by Ruth Cox, collected in a San Francisco neighbourhood.

Palmate veins

When three or more main veins fan out from one point at the base of the leaf, it is a palmate arrangement. Palmate venation often relates to palmately lobed leaves, such as maple or fig leaves. The advantage of having three or more main veins is possibly that they are an insurance against insect damage or breakage. If one is broken, there are others to carry the nutrients.

Red maple and sugar maples have three to five lobes and five main veins. A striped maple has three lobes and three main veins. Paperbark maple (*Acer griseum*) has three lobes and three veins.

However, note that the horse chestnut (*Aesculus hippocastanum*) and the red horse chestnut (*Aesculus pavia*) have palmate leaflets, but each leaflet is pinnately veined. The commonly seen white clover (*Trifolium repens*) has a palmate arrangement, though each individual leaflet has pinnate venation.

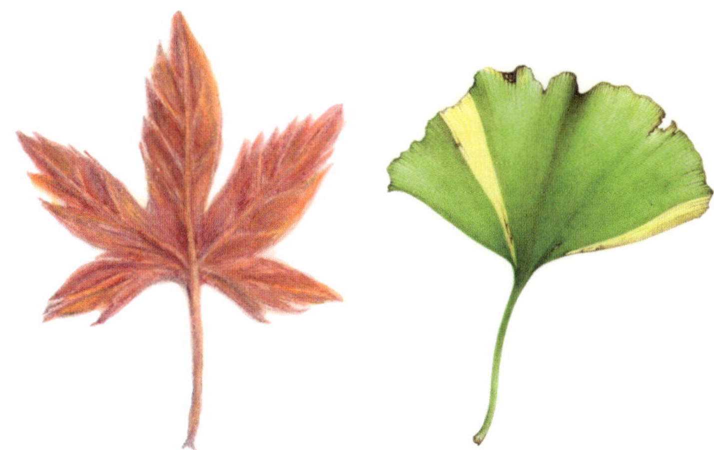

Another variant is a fanned venation found in the ancient gymnosperm *Ginkgo biloba*. Two parallel veins enter each blade from the petiole, and immediately fork repeatedly towards the apex of its deltate shape, making the leaf look like a fan – flabellate. The slight dip in the middle of some leaves indicates two lobes, as the name suggests.

UNDERSTANDING VEINS

Count the number of lobes. This can be tricky if they are both shallow and toothed. Establish where the secondary veins really go; some head for the sinus, splitting before they get there, whereas others go into the lobes. Persevere with these conundrums by selecting a perfect, typical leaf and making an outline. Measure between the main lobes as well as between base to lobe. Map the veins within, ensuring symmetry on both sides. Lastly, draw in the subsidiary lobes and toothed margins.

ABOVE LEFT Palmate veins of red Japanese maple (*Acer palmatum*) by Ruth Cox.

ABOVE RIGHT Fan venation of ginkgo leaf (*Ginkgo biloba*) detail by Lucilla Carcano.

RIGHT Common stork's bill (*Erodium cicutarium*) by Isobel Bartholomew. The leaves are divided into leaflets, each pinnately veined, having a central vein and secondary veins from which branches lead to the lobes.

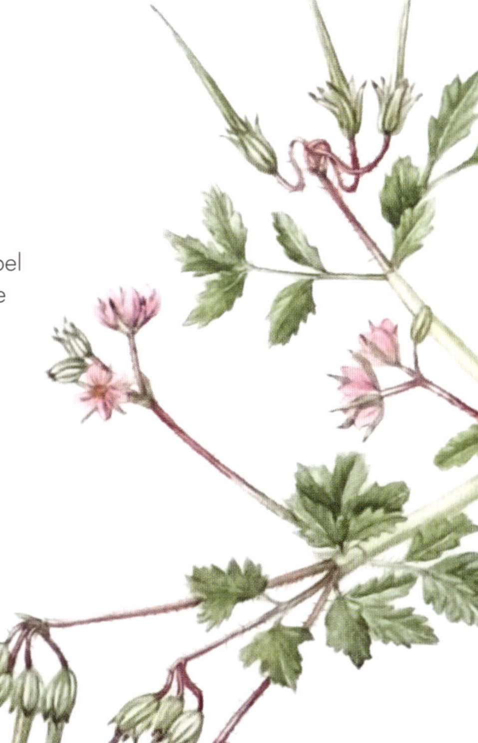

PALMATE LEAF

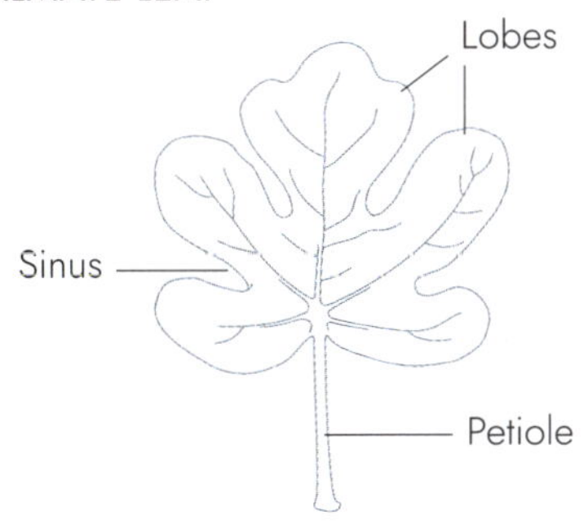

Parallel venation

This is where the veins travel in parallel from one point at the base, before gently meeting at the apex. This form is typical of the monocot group of flowering plants, which usually have elongated, slender leaves.

The main veins do not intersect, but are connected to each other by small secondary veins, usually hardly noticeable, though sometimes they are tessellated, such as *Clivia* leaves.

Marginal lobes

Many leaves are lobed in a consistent pattern, usually relating to the pattern of veins, whether palmate or pinnate. The lobes can be shallow or deeply incised. Maple tree species can have shallow lobes or deeply incised palmate leaves, and oaks may be pinnately lobed either shallowly or deeply, depending on the species.

There are also lobes that create interesting shapes. Lyrate lobes makes the leaf look like a lyre, common in the cabbage family. A hastate leaf looks like the head of a spear. A sagittate leaf looks like the head of an arrow.

Pedate refers to palmate leaves and derives from the Latin word for feet, where a lower lobe may be subdivided and turned back.

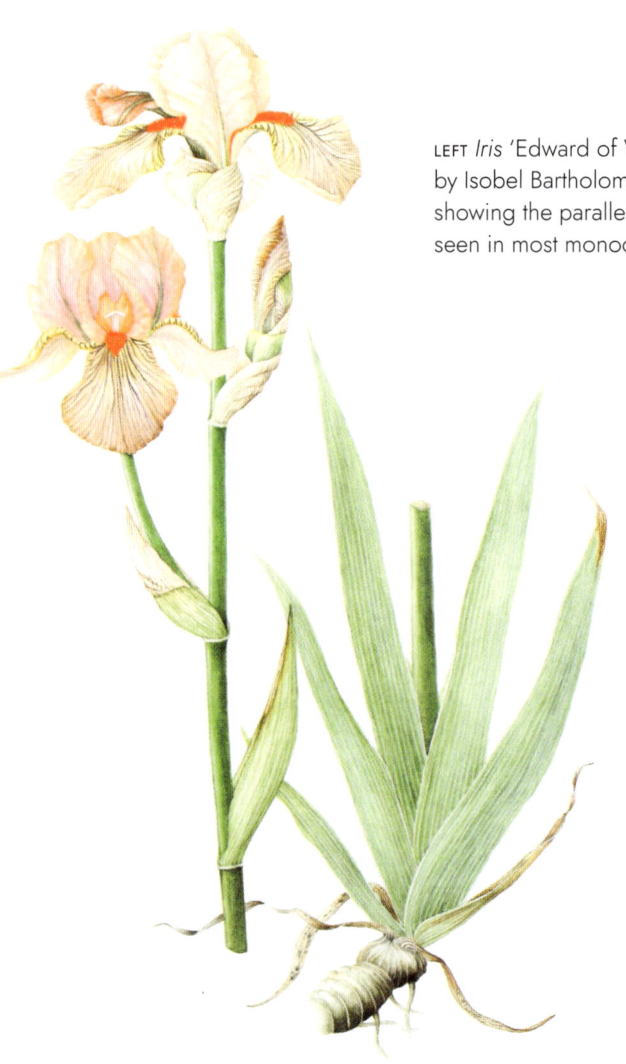

LEFT *Iris* 'Edward of Windsor' by Isobel Bartholomew showing the parallel veins seen in most monocot leaves.

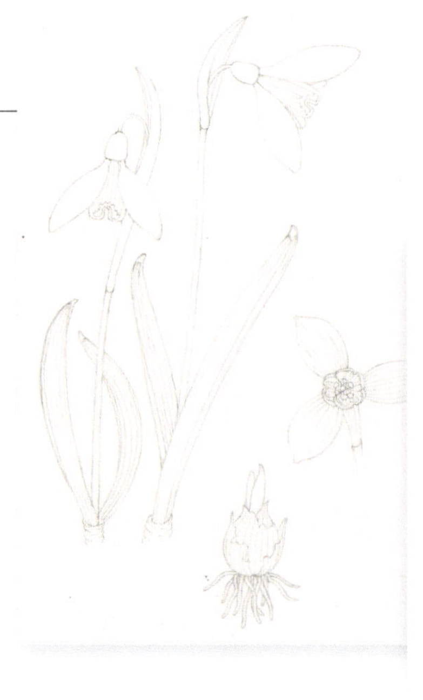

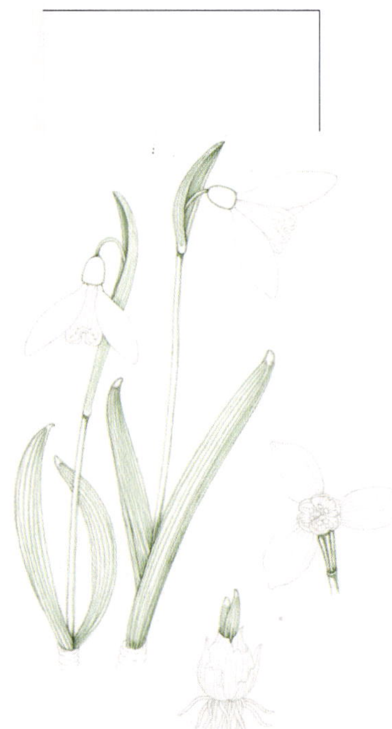

HOW TO GET THE LINES PARALLEL

- Count the number of parallel veins first. There will normally be a midvein and an even number either side of it. Even if there are too many veins to count, still draw in a middle vein first.

- Practise drawing your brush from bottom to top to ensure you have enough wet pigment to carry you through. Doing it this way round means your pigment will peter out towards the top, creating a lighter touch.

- Paint in the midvein, following your pencil line.

- Insert veins in the middle to the left and right of the midvein, and so on. This way, you will get an even spacing when there are lots of parallel lines.

Compound leaves

When the lobes of a leaf become so deeply incised as to reach the central vein, they form separate leaflets or folioles. The smaller leaflets are visibly separate from each other and together form the leaf. The two basic types are pinnate and palmate.

Look for the bud at the base of the petiole. It will tell you that all growth above, whether divided or not, is the leaf. In a compound leaf there will be no buds at the base of the leaflets. This will ensure that you are looking at a compound leaf, rather than a length of separate leaves.

The diagram below provides terms for the parts of a compound leaf in both pinnate and palmate forms. A petiolule is a subsidiary petiole. The rachis is the axis on which pinnate leaflets sit. Strictly, a palmate compound leaf does not have a rachis, as all the veins arise from the base of the leaf.

Two similar sounding terms which can be used incorrectly are trifoliate, which refers to three leaves, and trifoliolate, which refers to three leaflets. The bogbean (see page 135) has trifoliolate leaves, as does the *Trillium* (see page 53).

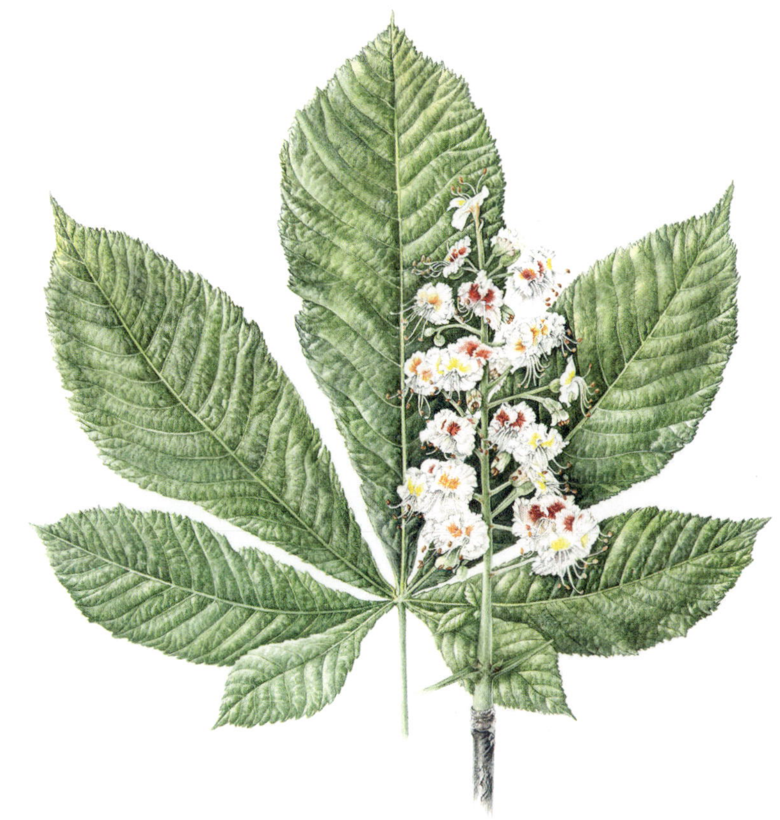

BELOW Rowan (*Sorbus aucuparia*) by Nicola Macartney has several opposite pairs of pinnate leaflets originating along the length of the rachis.

ABOVE Horse chestnut (*Aesculus hippocastanum*) by Maggy Fitzpatrick has seven palmate leaflets derived from the same point on the stem.

PINNATE COMPOUND LEAF PARTS

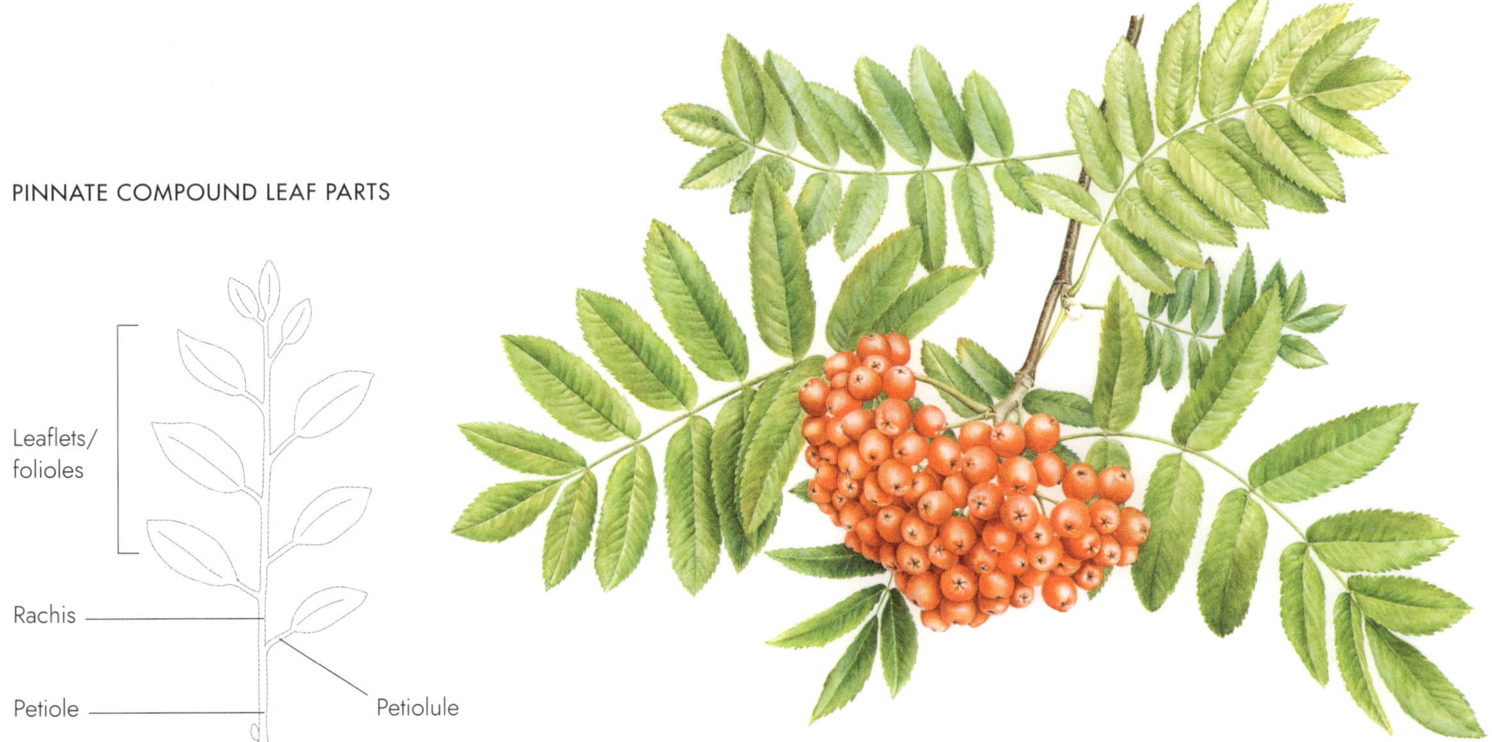

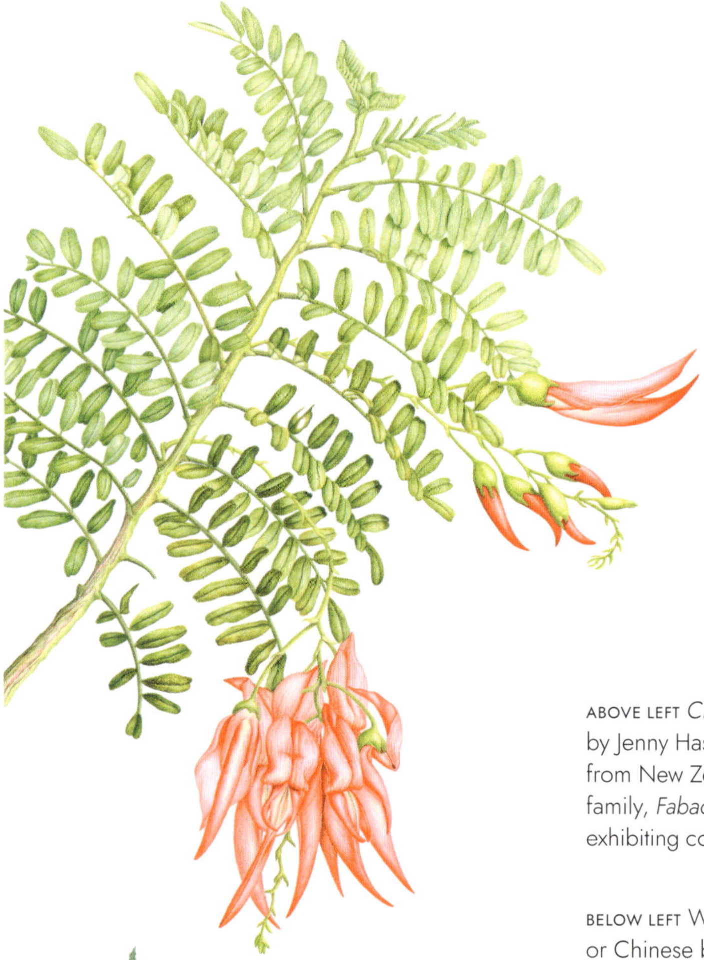

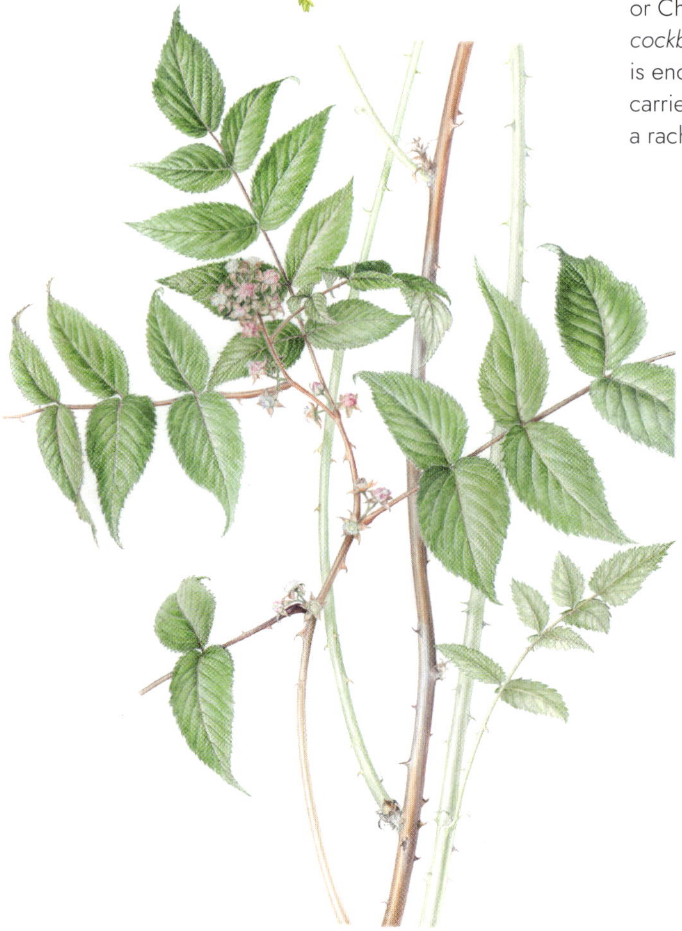

Compound pinnate leaves

The leaflets are arranged along a rachis, which in turn is attached to the stem by the petiole, as described above (or is sessile). The arrangement can be of a pair of opposite or alternate leaves. If the leaflets have stalks, these are called petiolules.

Be vigilant; don't mistake a row of leaves along a thin shoot for a pinnate compound leaf. Check that there is a bud at the base of the petiole for a single leaf. If not, it is a leaflet on a rachis.

ABOVE LEFT *Clianthus puniceus* by Jenny Haslimeier, a plant from New Zealand in a family, *Fabaceae*, usually exhibiting compound leaves.

BELOW LEFT White stemmed or Chinese bramble (*Rubus cockburnianus*) by Janet Dyer is endemic to China and carries several leaflets on a rachis.

WHAT TO LOOK FOR

- Lateral leaves on a rachis may be divided again to form bipinnate leaves, like the *Clianthus*, or tripinnate if divided again.

- Count the number of leaves. Some may have an even number; others may have an odd number with the last set at the pinnacle of the rachis.

- Determine whether, at every level, there are stalks and how they are attached. Sometimes there can be slight twisting at the attachment point.

- Are there wings (lateral extensions of the stem) on the rachis?

COMPOUND PINNATE LEAVES

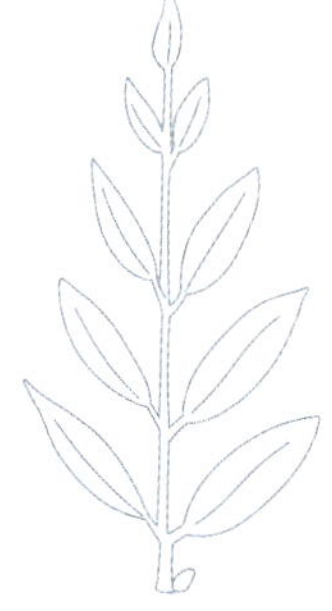

Pinnate odd-numbered

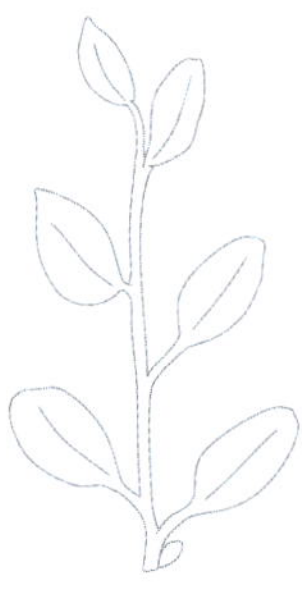

Pinnate even-numbered

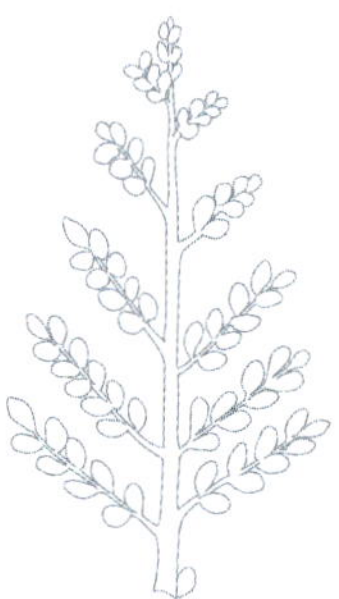

Bi-pinnate

Compound palmate leaves

Strictly, there is no rachis in a compound palmate leaf, for leaflets are grouped at the end of a petiole. There can be any number from two or more leaflets. The basic structure, like pinnate leaves, could in turn be divided into two or three, each linked to the petiole by a petiolule or stalk. If using the term palmate, it is best to qualify it with the word 'compound' to distinguish between its venation or its shape – features usually related anyway.

The word digitate, meaning fingered, is used interchangeably with palmate. The leaves of the shrub schefflera look more like slender fingers than an *Aesculus hippocastanum* (horse chestnut), where the word palmate might be more appropriate.

Again, be vigilant: palmate compound leaves consist of separate leaflets and can look like deeply lobed simple leaves (see page 87). It is not compound if there is no gap between the leaflets.

Trifoliolate, or ternate, means a leaf in three parts, each a leaflet. Trifoliolate leaves are seen in wood sorrels (*Oxalis* sp.), broom (*Cytisus* sp.) and clovers (confusingly called the *Trifolium* genus). Remember that trifoliolate is not the same as trifoliate. The Latin for 'leaf' is folium, and 'little leaf' is foliolum.

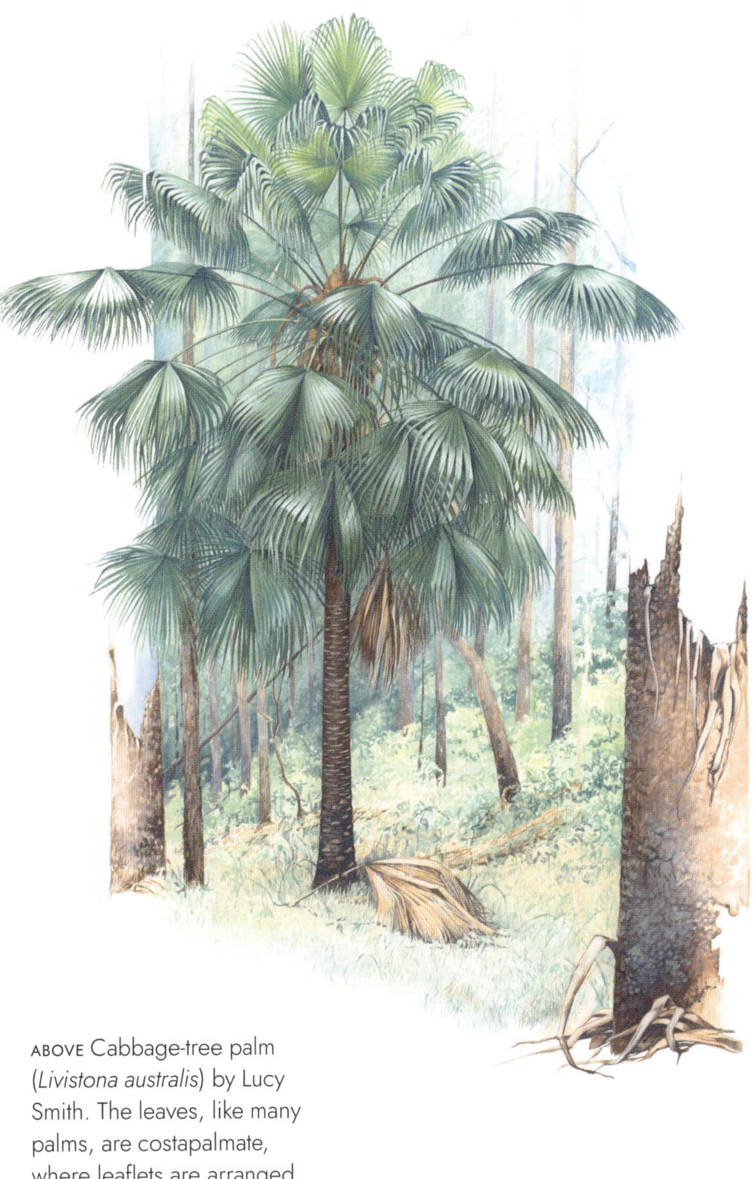

ABOVE Cabbage-tree palm (*Livistona australis*) by Lucy Smith. The leaves, like many palms, are costapalmate, where leaflets are arranged palmately but inserted into an extension of the petiole.

COMPOUND PALMATE LEAVES

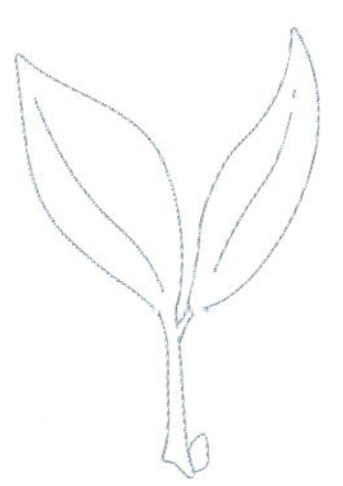

Bifoliolate

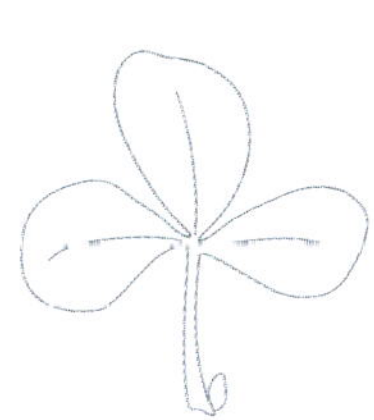

Trifoliolate/ ternate

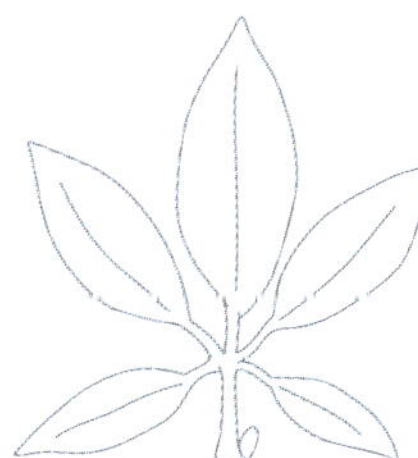

Palmate

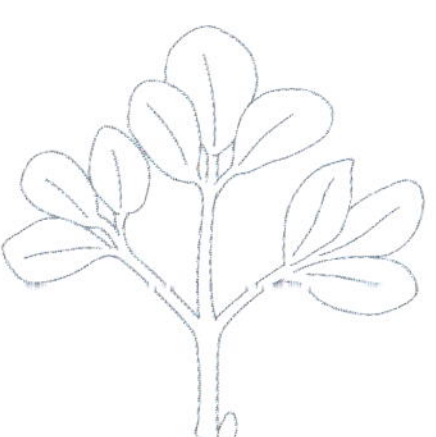

Biternate

Painting a horse chestnut leaf

The horse chestnut leaf shape is iconic. This one
was painted in September when the leaves were
just starting to die, and so there are some
interesting crunchy brown sections at the ends
of the leaves, and patches throughout. Although
the leaf shape is relatively simple, there are many
washes, darkened areas and patches left light to
depict its bumpy structure. Areas of shadow are
also very important in order to show that the
leaves roll back on themselves.

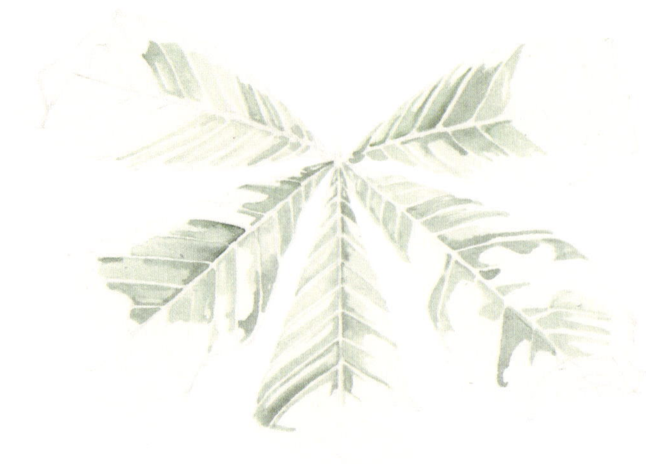

1 | Draw the outline of your leaf in pencil.
Mix a light wash of sap green and olive
green. Apply clean water to wet the paper,
then use a wet-on-wet technique to add the
colour to the green areas. The colour will
spread throughout the wet areas, but be
careful to leave the veins white between
the green sections.

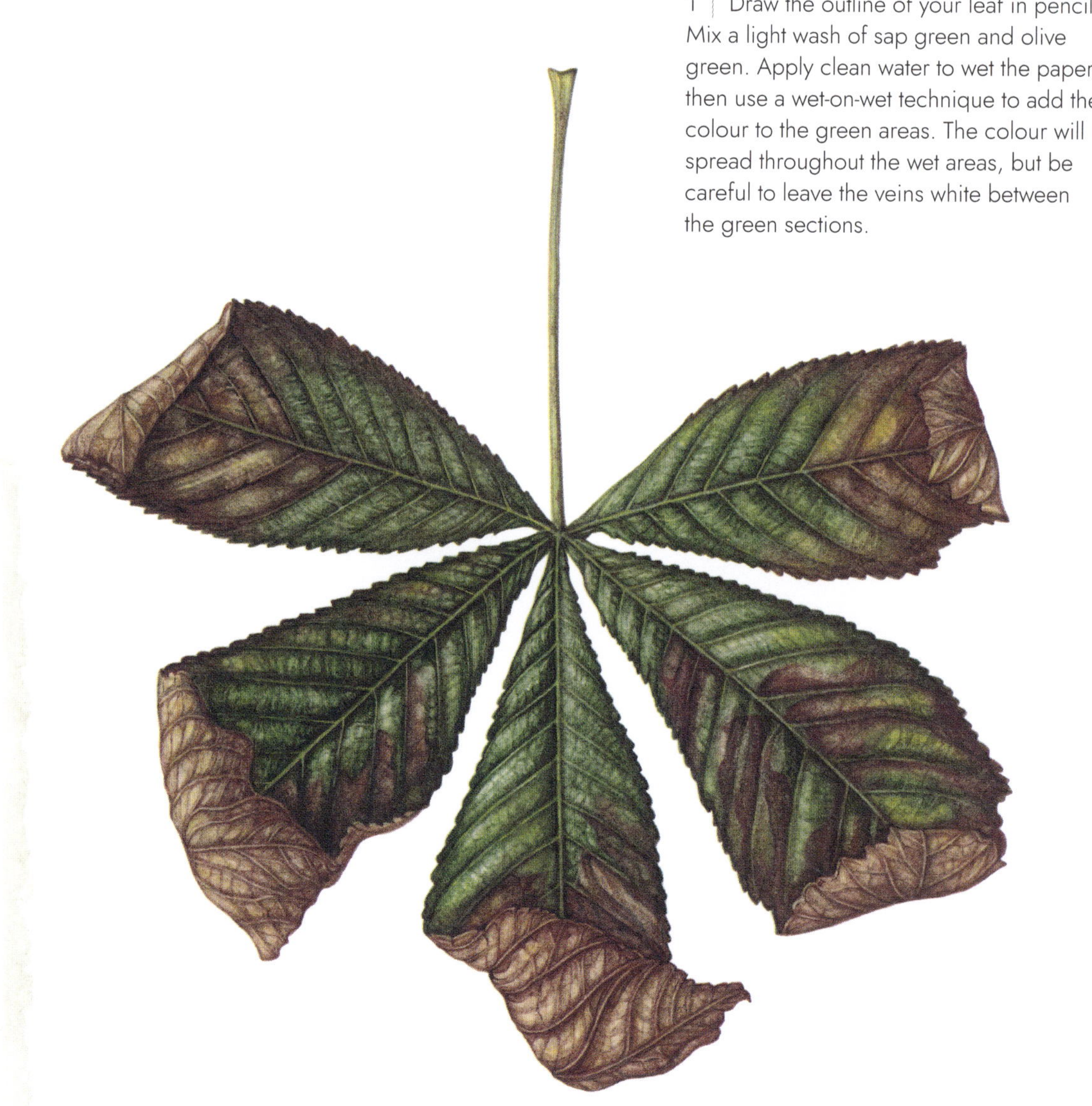

PAINTS USED:

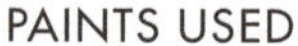 Sap green

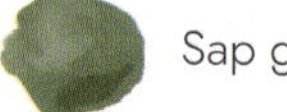 Olive green

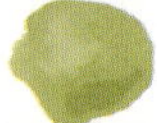 Indian red

 Burnt umber

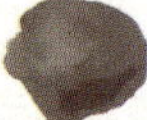 Lemon yellow

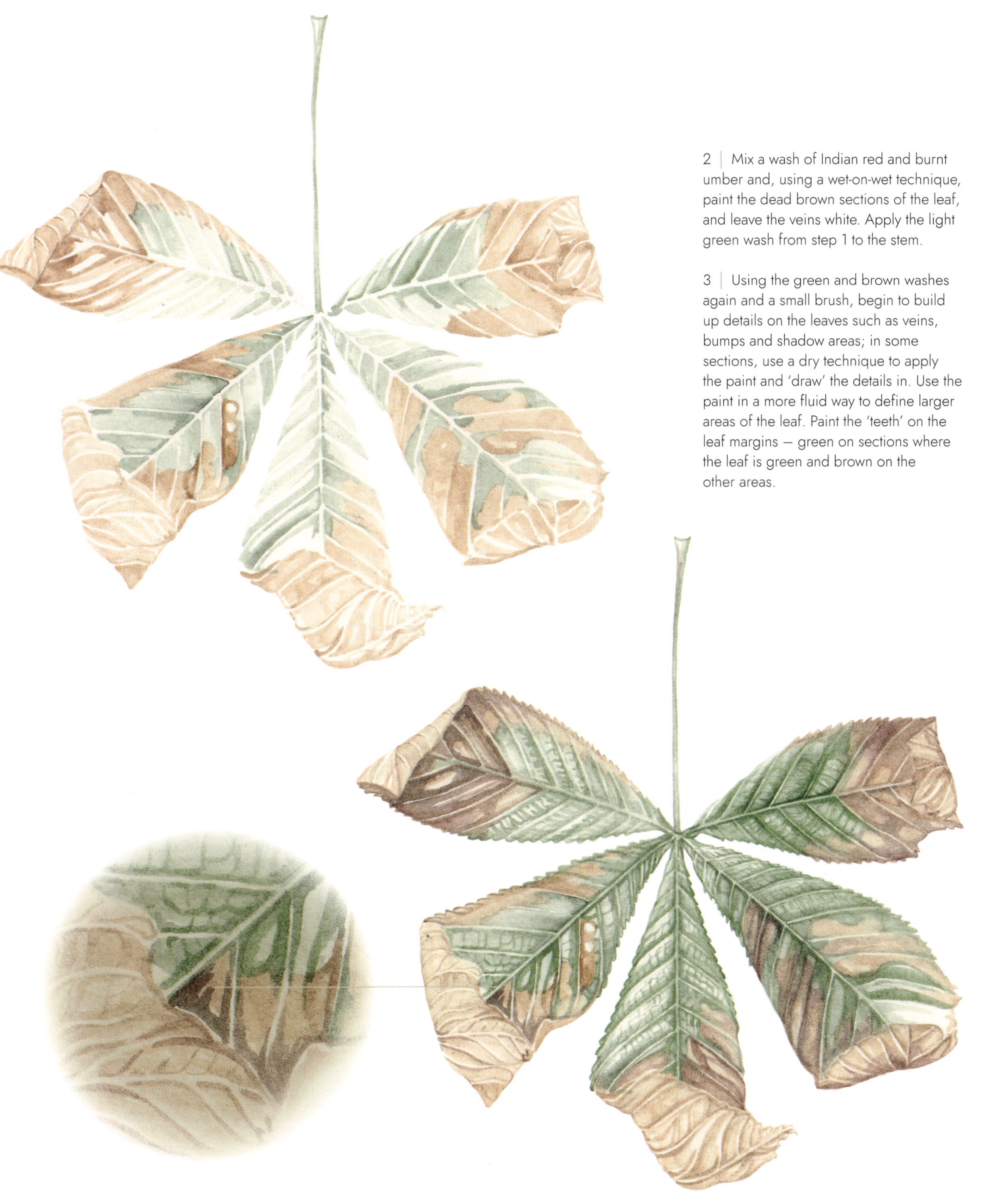

2 | Mix a wash of Indian red and burnt umber and, using a wet-on-wet technique, paint the dead brown sections of the leaf, and leave the veins white. Apply the light green wash from step 1 to the stem.

3 | Using the green and brown washes again and a small brush, begin to build up details on the leaves such as veins, bumps and shadow areas; in some sections, use a dry technique to apply the paint and 'draw' the details in. Use the paint in a more fluid way to define larger areas of the leaf. Paint the 'teeth' on the leaf margins — green on sections where the leaf is green and brown on the other areas.

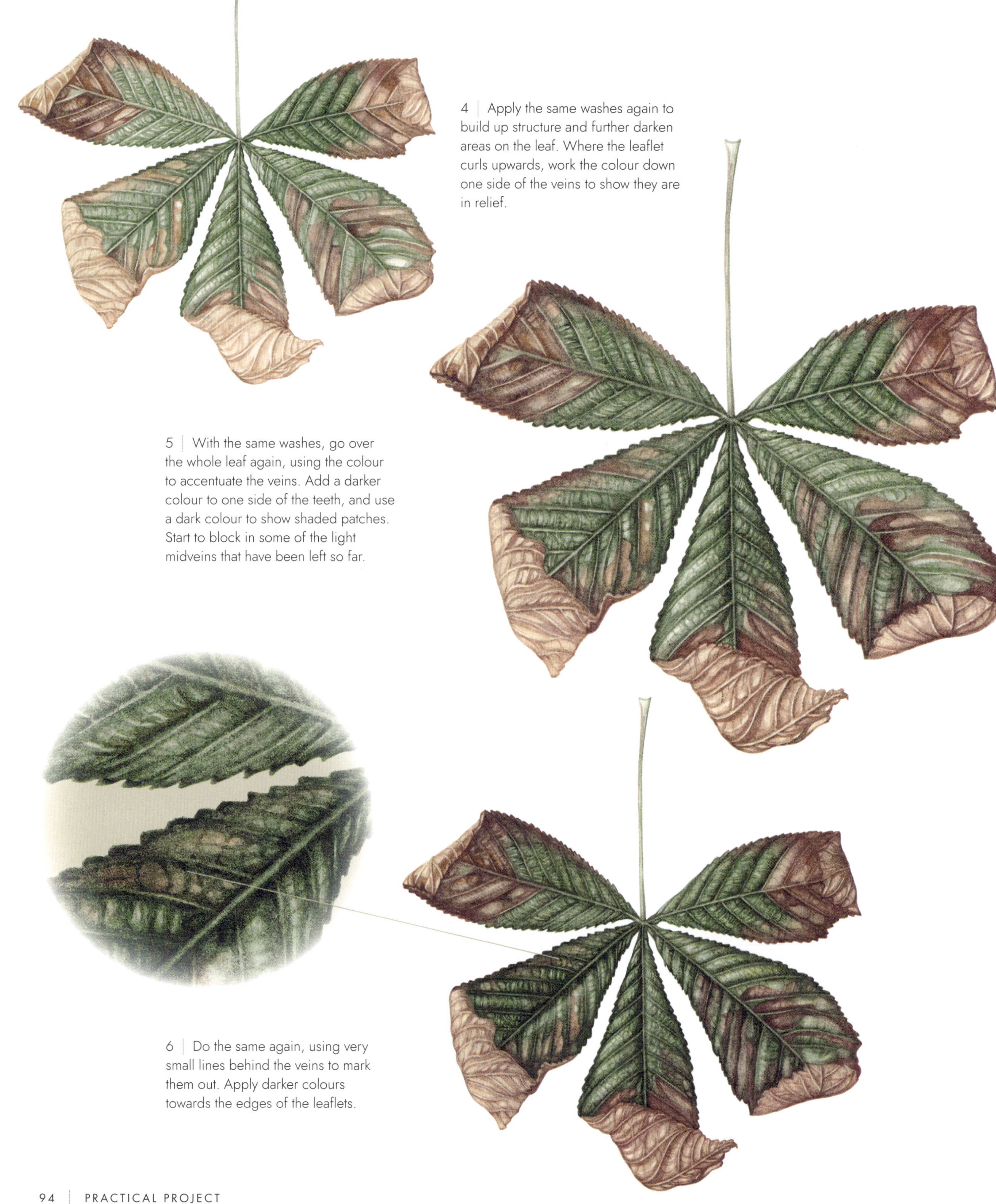

4 | Apply the same washes again to build up structure and further darken areas on the leaf. Where the leaflet curls upwards, work the colour down one side of the veins to show they are in relief.

5 | With the same washes, go over the whole leaf again, using the colour to accentuate the veins. Add a darker colour to one side of the teeth, and use a dark colour to show shaded patches. Start to block in some of the light midveins that have been left so far.

6 | Do the same again, using very small lines behind the veins to mark them out. Apply darker colours towards the edges of the leaflets.

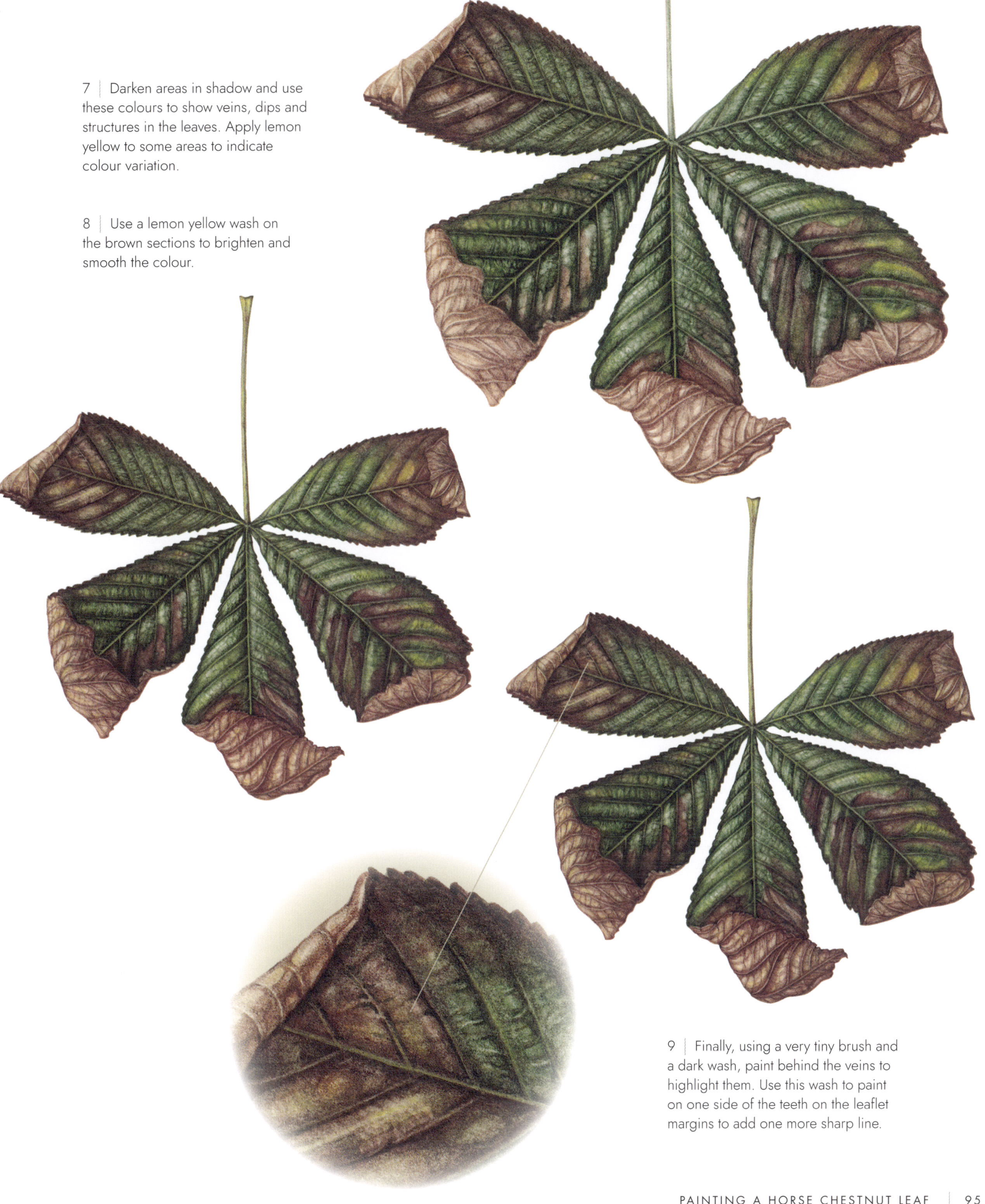

7 | Darken areas in shadow and use these colours to show veins, dips and structures in the leaves. Apply lemon yellow to some areas to indicate colour variation.

8 | Use a lemon yellow wash on the brown sections to brighten and smooth the colour.

9 | Finally, using a very tiny brush and a dark wash, paint behind the veins to highlight them. Use this wash to paint on one side of the teeth on the leaflet margins to add one more sharp line.

Arrowhead plant (*Syngonium podophyllum*)
by Maria Alice de Rezende

Alice de Rezende's *Syngonium podophyllum* is a dramatic portrayal of a plant often seen in houses around the world because of its strongly sagittate and often variegated leaves, adding to its ornamental value.

This portrait, however, avoids the clichés and focuses on its upper leaves and fruit, not normally found in houseplants. *Syngonium podophyllum* can be a vine several metres tall. Only its lower leaves are simple and arrow-like, and it never fruits in cultivation — nor are the leaves much variegated in the wild.

Wild beauty

Alice introduces us to the wild plant, by using ripening fruits to draw the eye to a comfortable angle where adventitious roots secure the plant to an unseen tree. The reds offset the green compound leaves, and grooved petioles lash out dramatically in three directions towards the right of the page in search of light.

This version of *Syngonium podophyllum* was painted from a specimen in the Atlantic Forest on the eastern seaboard of Brazil near Rio de Janeiro. The wild plant introduces Alice's audience to the plant's beauty, rather than its utilitarian value. She shows its true integrity as a climber or a creeper, reaching towards the light and adapting its leaves to suit the conditions. Only the upper leaves are divided, their petioles wrapped around the stem in an alternate phyllotaxis.

Alice's painting is a visual story where we imagine a distant and unseen juvenile below. Within sight is the adult above, seemingly poised above our heads, with fruit about to burst open.

RIGHT A common Brazilian houseplant reveals its wild aspect. Size: 69 × 50cm (27$\frac{1}{4}$ x 19$\frac{3}{4}$in), watercolour on Fabriano Artistico watercolour paper.

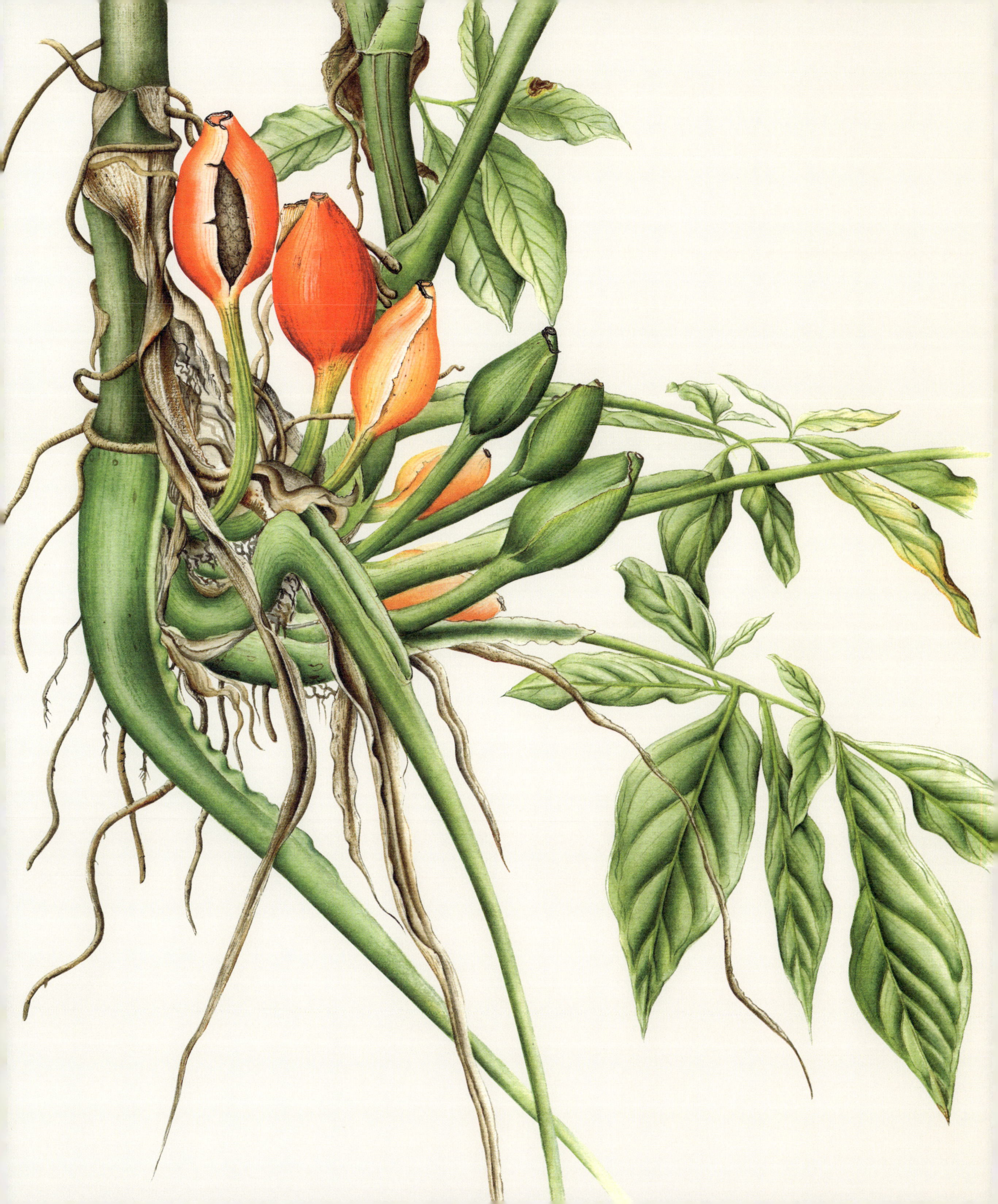

Storage leaves

Storage leaves hold water and nutrients to help plants survive dry periods. They can be below ground or above ground, where they are called succulent leaves.

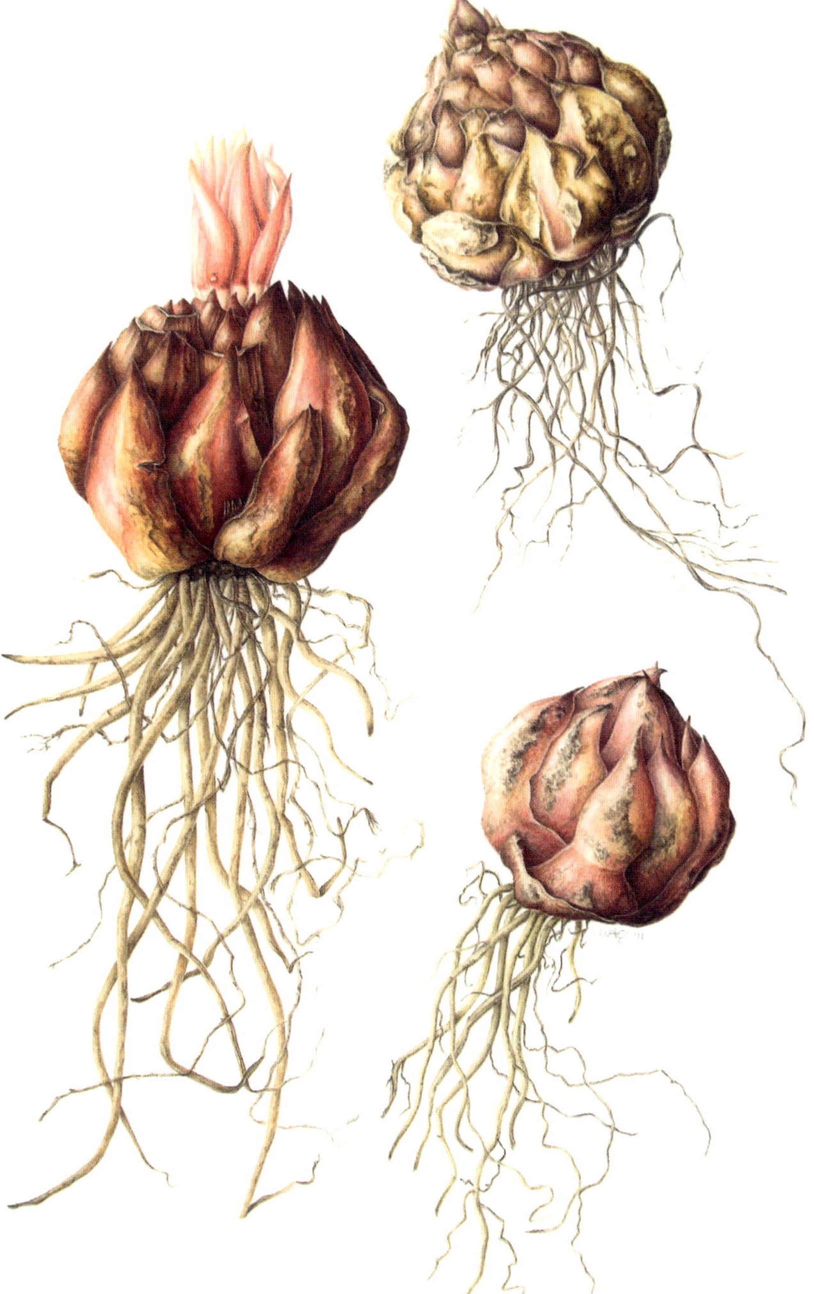

Bulbs

Bulbs are formed of swollen underground leaf scales, which represent leaves. These sit on a disc called a basal plate, which represents the stem. Adventitious roots grow from the stem plate to anchor the bulb. Bulbs are common in monocots, but only one eudicot genus, *Oxalis*, has true bulbs.

Note the texture of the bulb's exterior and the way the scales sit. Tunicate bulbs have a papery membrane that protects tightly arranged scales within, for example, allium, tulips, hyacinth, narcissi or reticulated irises. Non-tunicate bulbs, such as lilies and some fritillaries have looser scales without the protective 'tunic'.

Baby bulbs growing from the base of the bulb are incorrectly called bulbils. Strictly speaking, bulbils are little bulbs that grow in the axils of above-ground leaves, or in the place of flowers. The invasive few-flowered garlic, *Allium paradoxum*, is a menace because the many bulbils that develop at the base of its 'few' flowers drop off and easily reproduce in the ground.

The rich transparent colours of a tunicate bulb, such as an onion, repay a painting method in layers of different transparent paints. The sheen of the colours underneath will shine through the surface colour.

BULB PARTS

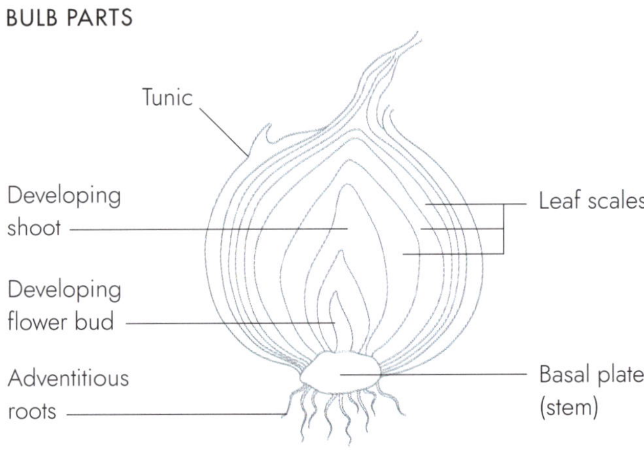

Above ground

The phenomenon of swollen leaves above ground is widespread among plant families in both eudicots and monocots, especially if they grow in dry conditions.

Leaves may look cylindrical, as in the houseplant *Sansevieria cylindrica* in the asparagus family. They may look like the scales that seem to climb the stems in stonecrop species or be peltate, like *Umbilicus rupestris*, a navelwort, both of which grow in dry walls or rock.

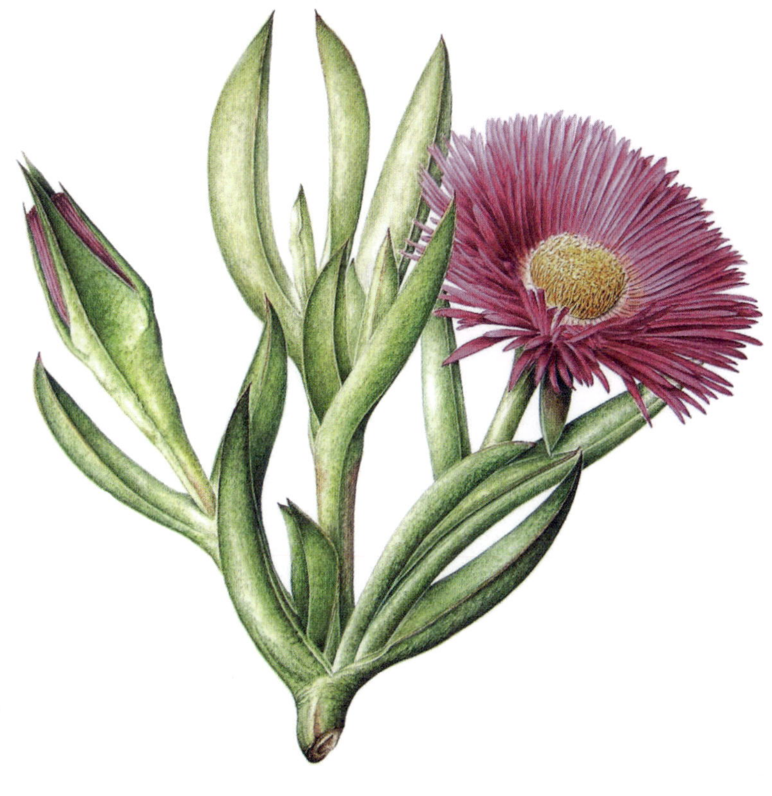

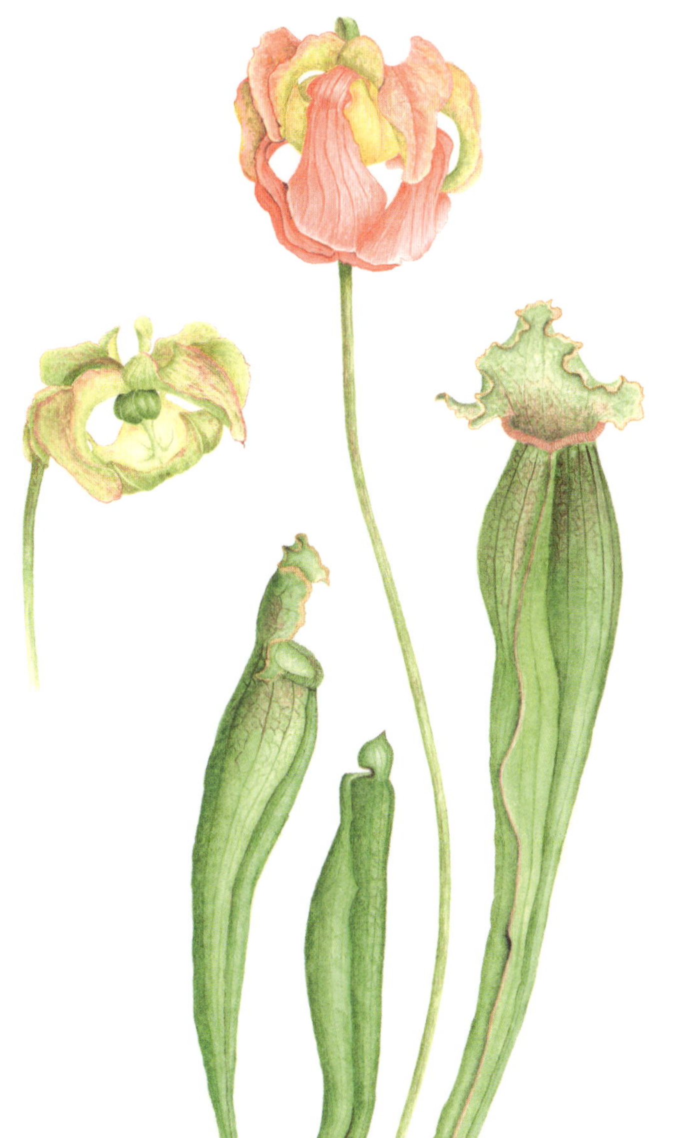

BELOW Mitchell's pitcher plant (*Sarracenia* x *mitchelliana*) by Vivienne Taylor has leaves formed into a hollow cylinder in which to trap nutritious insects.

RIGHT Pigface (*Carpobrotus* sp.) by Leigh Ann Gale originates in dry sandy South African habitats and stores water in its succulent leaves.

The swollen petiole base of the Florence fennel variety of *Foeniculum vulgare* is a part of a leaf that can be succulent.

Houseplants such as *Aloe*, *Kalanchoe* and *Echeveria* give plenty of scope for painting the various succulent leaf forms in the comfort of a studio. Cut the leaf in cross-section to check its shape. Explore the thickness of the cuticle and have a look at what is inside. Often, the leaves are so swollen that you will only see a gelatinous mass. If it is an *Aloe*, be careful not to lick your fingers – you will soon realize why its vernacular name in many languages means 'bitter'.

Other features of leaves

Check the following leaf modifications:

- Tendrils: these could replace a leaflet at the apex of an odd-numbered pinnate leaf, for example, common vetch.

- Pitchers and traps: modified leaves in which to trap insects.

- Bulbils: small bulbs that grow in the axils of leaves.

Painting a hosta leaf

The hosta leaf is broad and somewhat flat with the veins running vertically up it. This is a great chance to apply the paint in a loose and vibrant way, using broad brushstrokes, without losing the structure of the leaf. The colour is applied wet-on-wet to start with, and then the shape of the leaf is built up in between the veins using darker tones on one side.

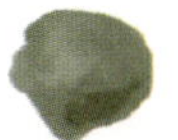

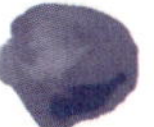

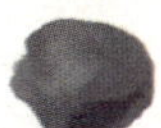

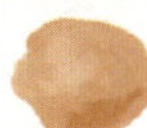

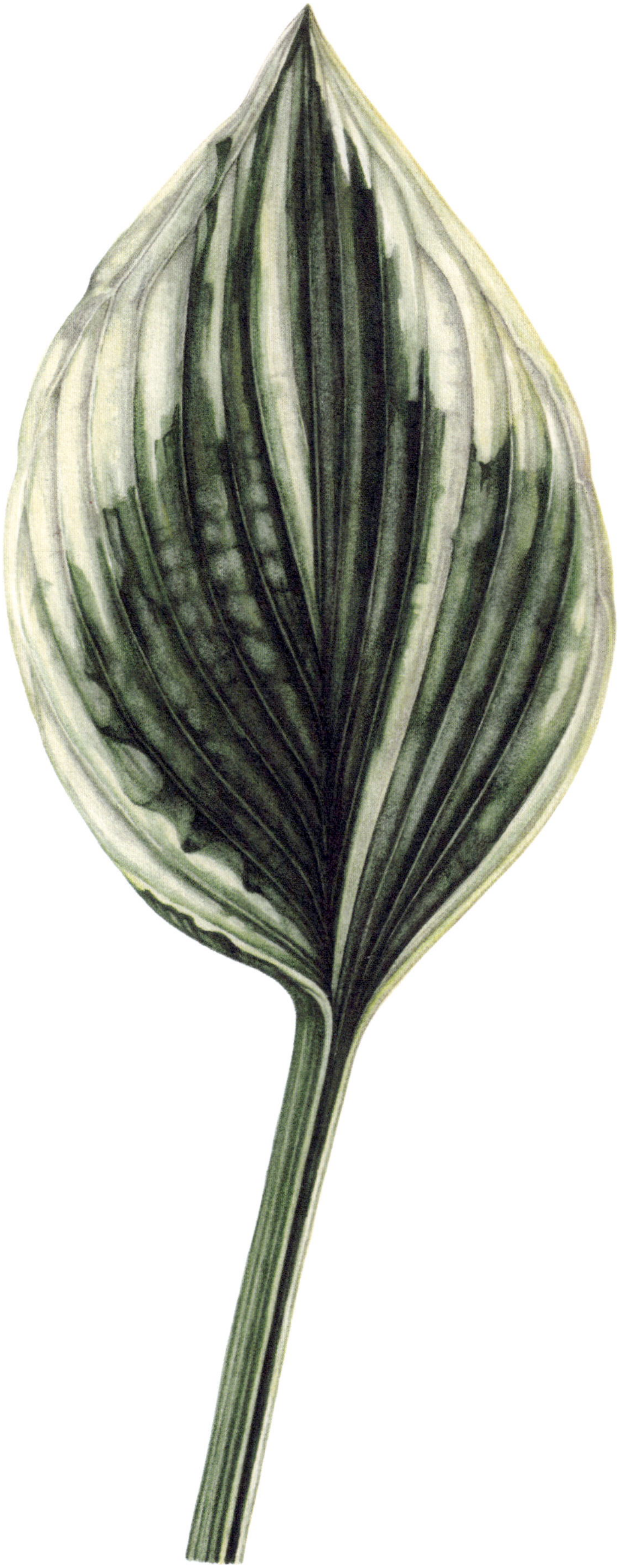

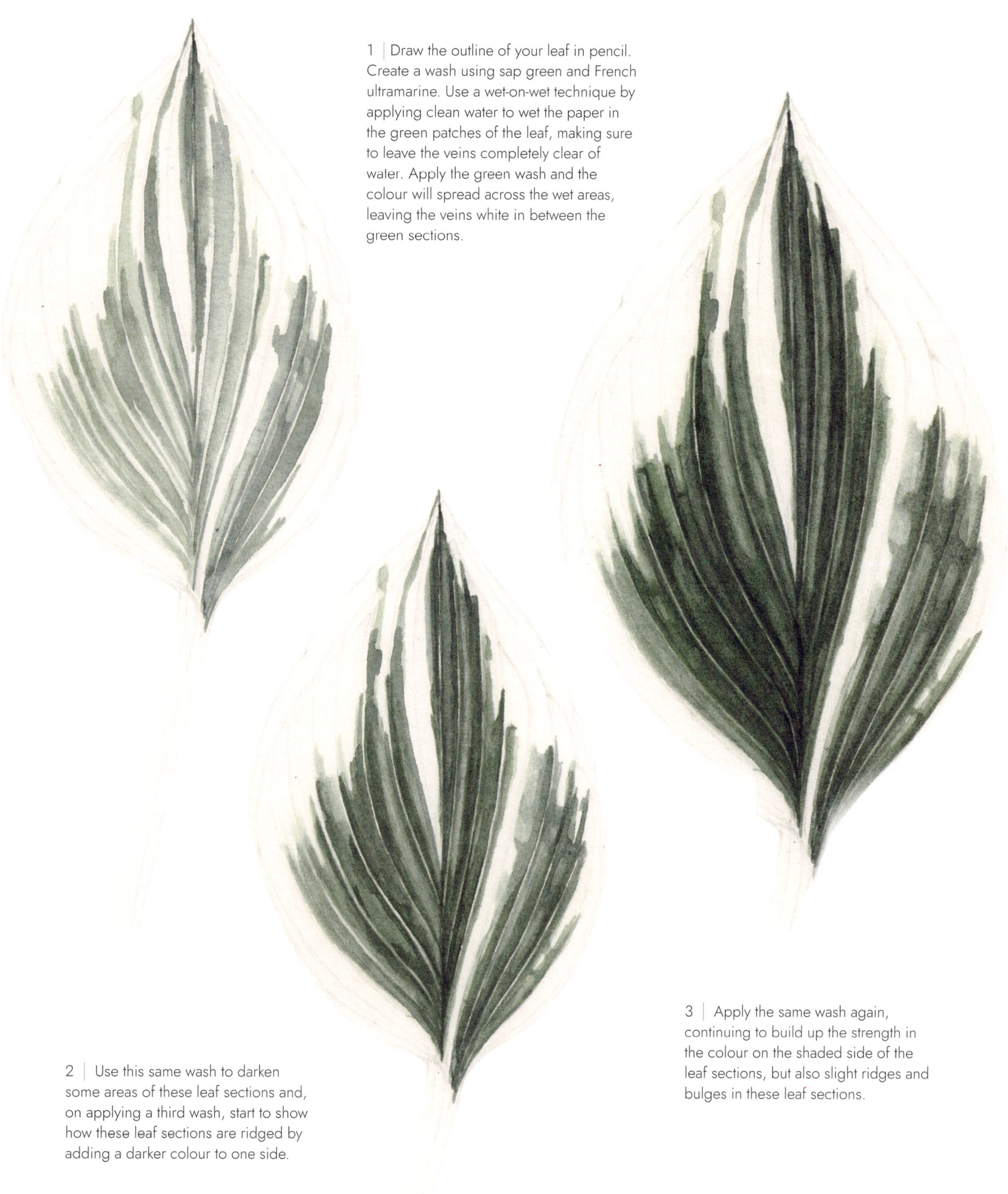

1 | Draw the outline of your leaf in pencil. Create a wash using sap green and French ultramarine. Use a wet-on-wet technique by applying clean water to wet the paper in the green patches of the leaf, making sure to leave the veins completely clear of water. Apply the green wash and the colour will spread across the wet areas, leaving the veins white in between the green sections.

2 | Use this same wash to darken some areas of these leaf sections and, on applying a third wash, start to show how these leaf sections are ridged by adding a darker colour to one side.

3 | Apply the same wash again, continuing to build up the strength in the colour on the shaded side of the leaf sections, but also slight ridges and bulges in these leaf sections.

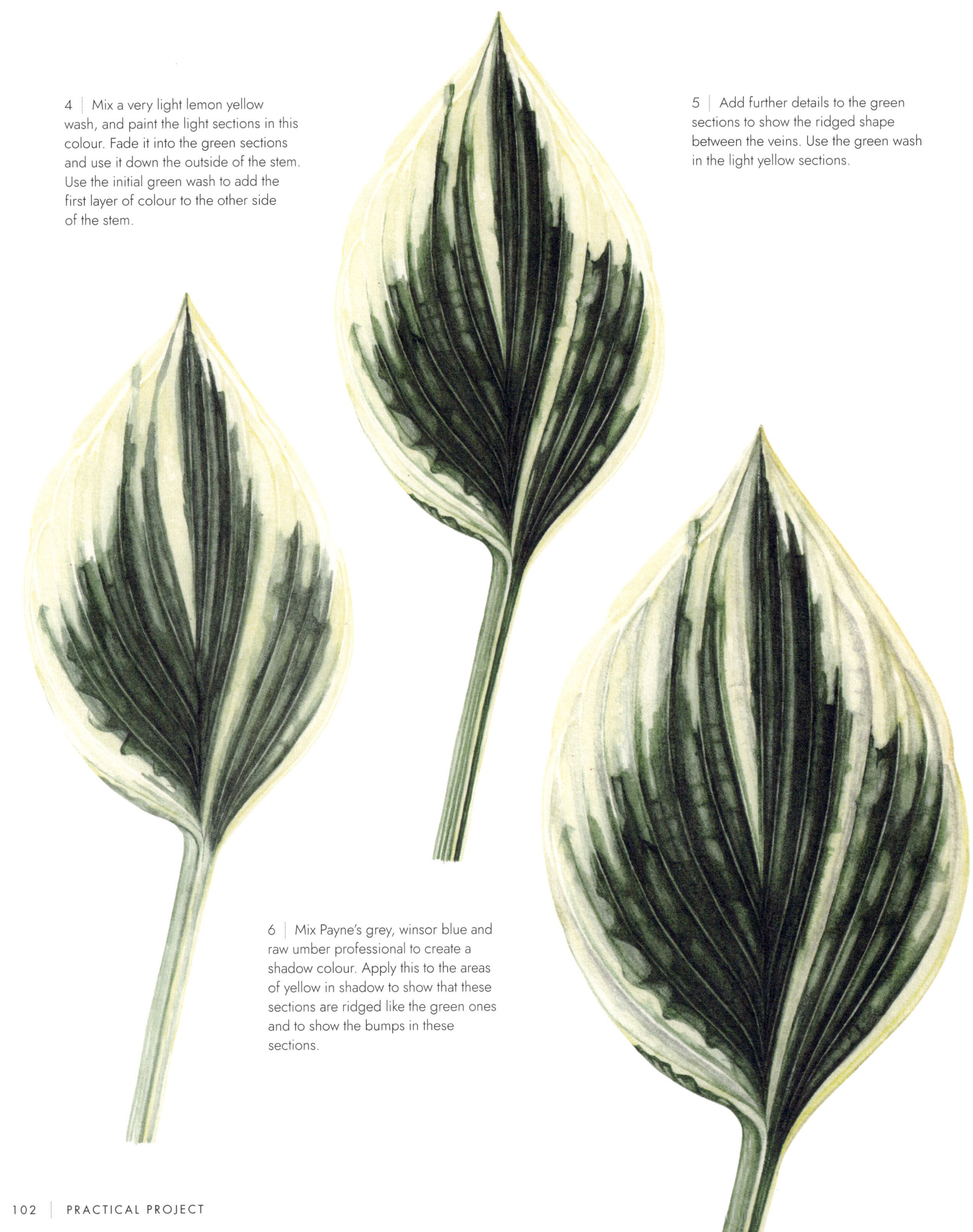

4 | Mix a very light lemon yellow wash, and paint the light sections in this colour. Fade it into the green sections and use it down the outside of the stem. Use the initial green wash to add the first layer of colour to the other side of the stem.

5 | Add further details to the green sections to show the ridged shape between the veins. Use the green wash in the light yellow sections.

6 | Mix Payne's grey, winsor blue and raw umber professional to create a shadow colour. Apply this to the areas of yellow in shadow to show that these sections are ridged like the green ones and to show the bumps in these sections.

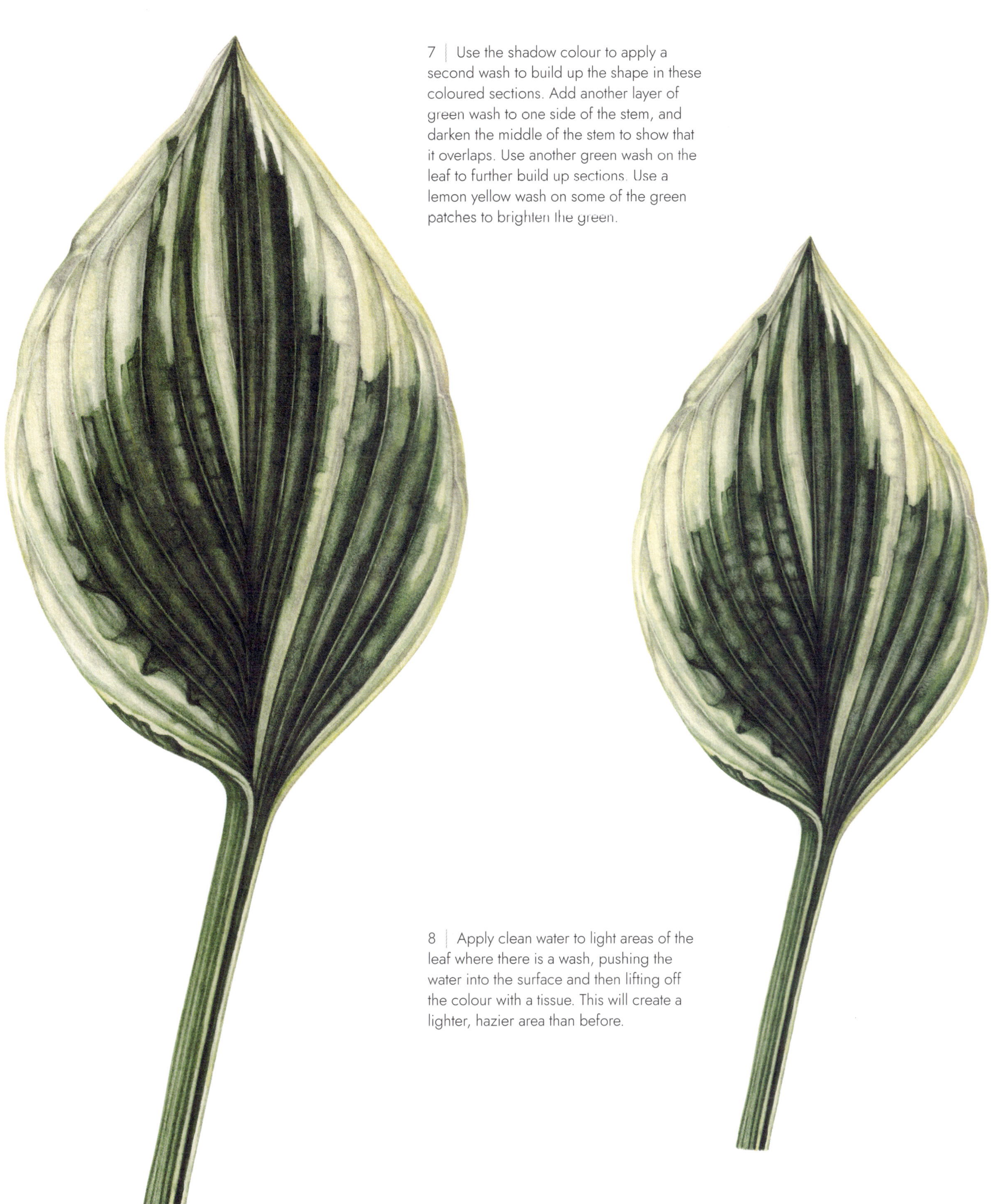

7 | Use the shadow colour to apply a second wash to build up the shape in these coloured sections. Add another layer of green wash to one side of the stem, and darken the middle of the stem to show that it overlaps. Use another green wash on the leaf to further build up sections. Use a lemon yellow wash on some of the green patches to brighten the green.

8 | Apply clean water to light areas of the leaf where there is a wash, pushing the water into the surface and then lifting off the colour with a tissue. This will create a lighter, hazier area than before.

Cyprian plane (*Platanus orientalis* var. *insularis*)
by Sandra Doyle

This is a good tree to sit under during the heat of the day. A breeze through the delicate translucent leaves whispers a welcome, and a cluster of tiny achenes parachutes to the ground to make new trees.

A tree with history

Sandra Doyle has created a masterpiece with her painting, *In the Shade of the Hippocrates Tree*. It rolls history, botany and aesthetics into her one painting of *Platanus orientalis* var. *insularis* at the Royal College of Physicians in London. This tall tree, with its wide shady canopy, is descended from one on the Greek island of Kos where the physician, Hippocrates, used to teach his students in classical Greece. It is remarkable to think the tiny achenes floating to earth are several generations' descendants of the original 4th century BCE tree.

Light and shade

Sandra's composition emphasizes the tree's welcome shade by implying that there is more tree outside the top and side margins. The zigzag pattern down the centre calms yet stimulates. The eye switches from side to side, following the line of the dominant branch alongside the falling achenes. Perhaps the effect reflects the stimulating oratory of a master physician? Or perhaps the pattern makes us peer more closely to see what the fruit and seeds look like?

Her painting suggests thick clusters of leaves by using aerial perspective to make the back leaves lighter, and by showing casting shade on the foreground ones.

The botany is scrupulously followed. Not only are the deeply incised, palmately lobed leaves beautifully clear but, on close observation, the tiny parachutes taking the achenes to another life are also visible. The clustered fruit from which they fall are small, yet the minute bumps of each achene are well defined.

Most of the illustration is painted in watercolour. Sandra used layers of transparent colours to build up the required intensity. Only where she needs opacity, where the leaves cross to prevent light shining through, does she resort to Polychromos colour pencil layered with the paint. She finds this method is less likely to dull the vibrancy of the transparent colours.

RIGHT The welcome shade of a plane tree. Size: 63 × 44cm (24³⁄₄ × 17¹⁄₄in), watercolour and coloured pencil on paper.

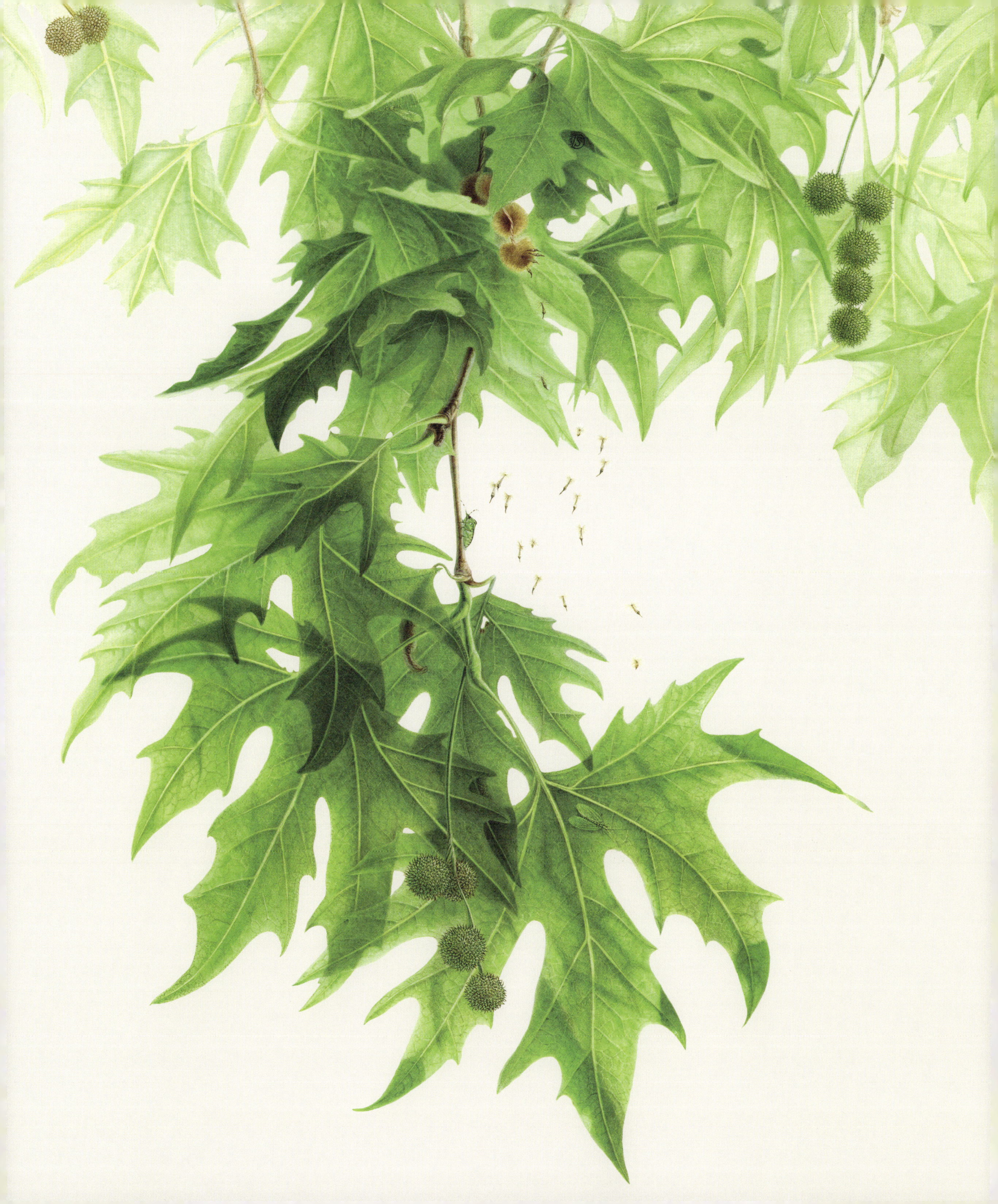

Chapter five

Flowers

Flowers contain the elements needed for sexual reproduction. Their anatomy and colour are closely matched with the demands of pollinators, or suited to self-pollination if necessary.

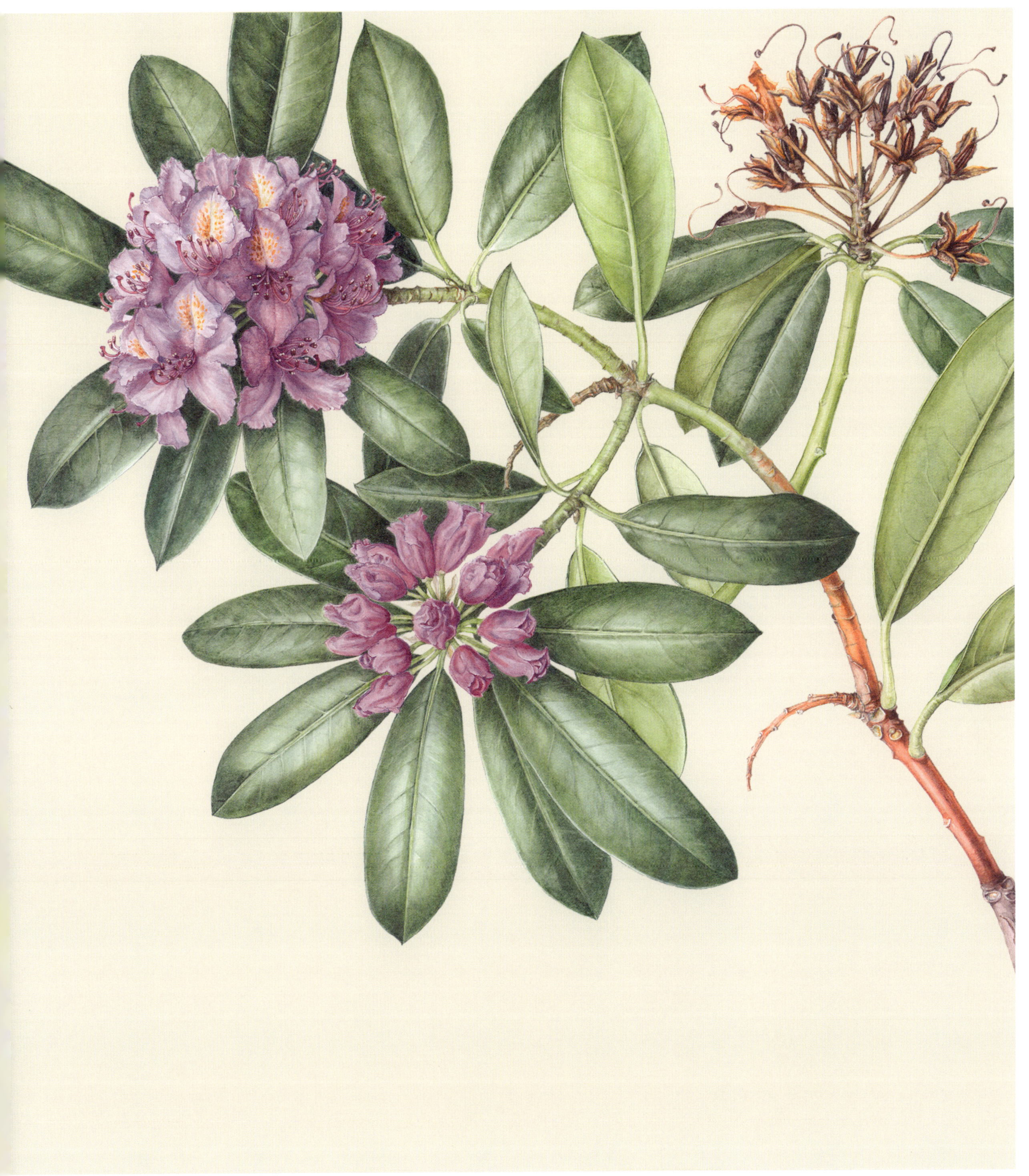

Basic flower parts

Flowers have evolved so that plants can reproduce themselves. They must be attractive to pollinators, and provide for genetic exchange, thereby allowing efficient reproduction. There is a bewildering variety of flower forms, which can lead you astray if you try too hard to shoehorn a theoretical (and hierarchical) set of terms into what is seen.

Flowers are set on stems called pedicels either singly (solitary flowers) or in clusters (inflorescence). There may be a leafy bract, or bracteole, to protect the bud. The terms and descriptions in this chapter apply to both solitary flowers and clusters, which might help when trying to unravel something unusual.

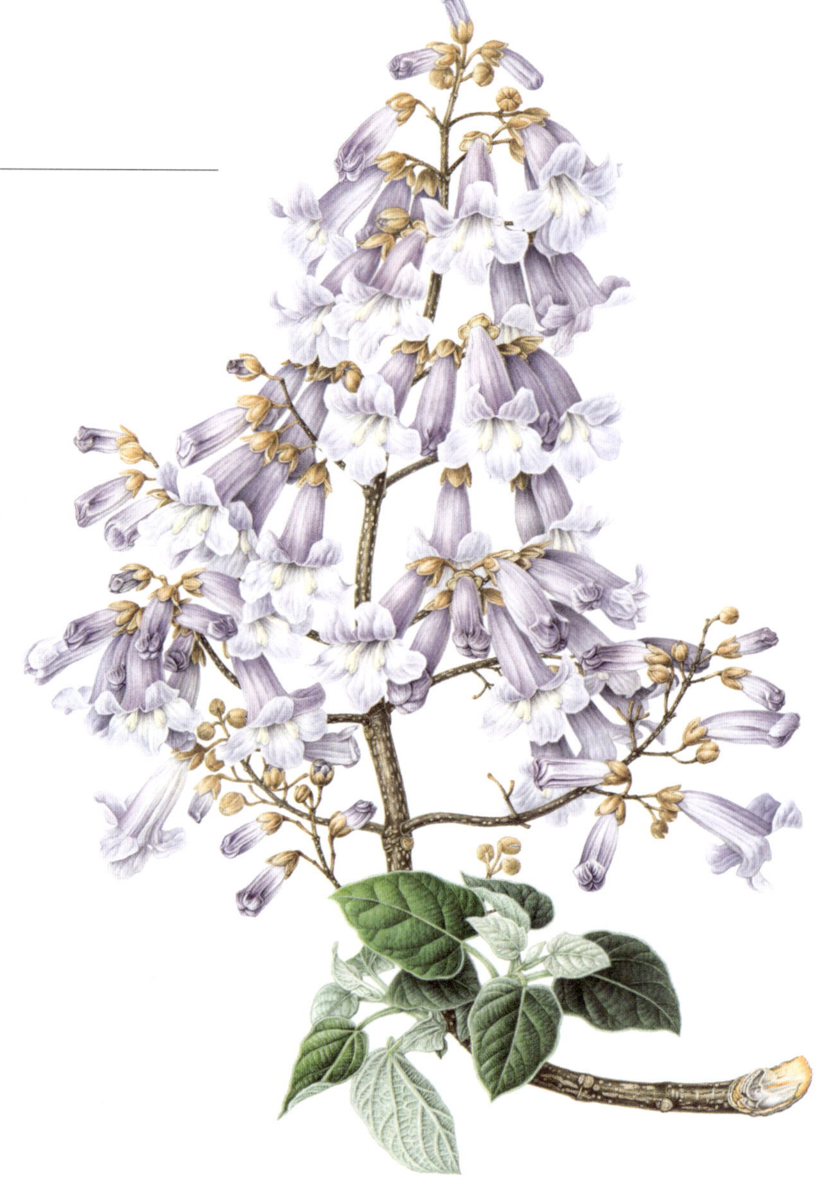

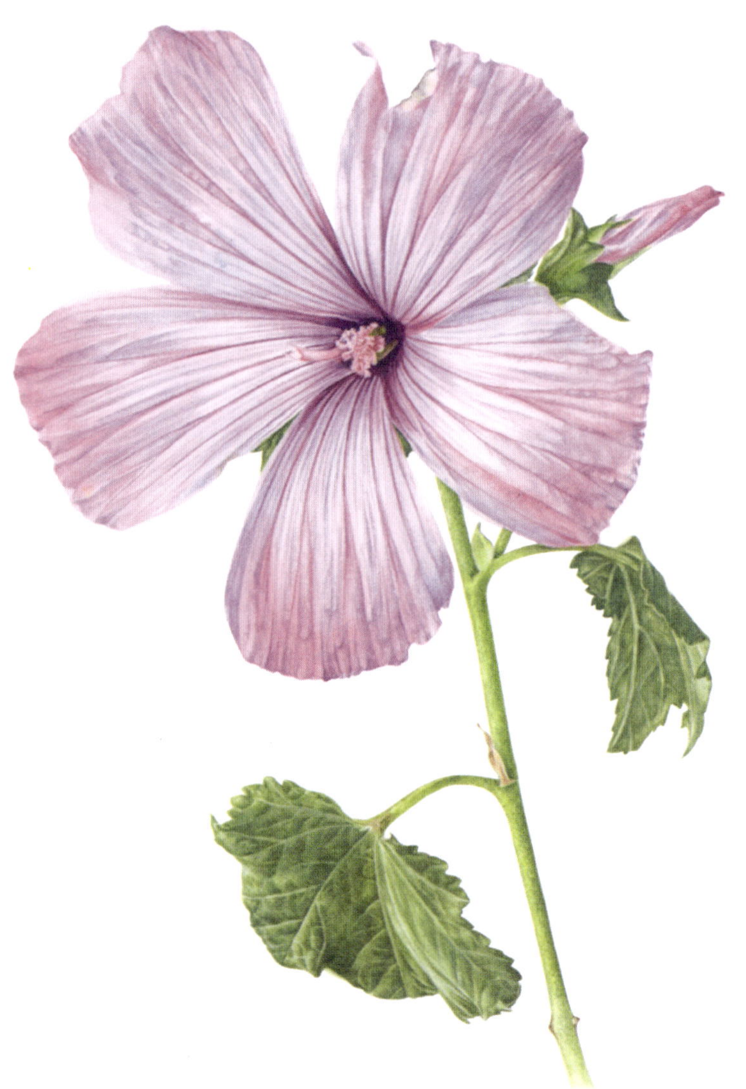

The floral parts are essentially in four whorls: the sepals and petals (collectively perianth) on the outside; the male stamens (collectively androecium), consisting of filament and anthers; and the female gynoecium consisting of the ovary, style and stigma on the inside. The corolla is a collective name for the second whorl – the petals.

LEFT Spanish mallow (*Malva hispanica*) by Toni Dade. The five petals of its solitary flowers are set in an even radial shape.

ABOVE Foxglove tree (*Paulownia tomentosa*) by Mieko Ishikawa. The flowers are set in a cluster, or inflorescence, and its individual flowers are uneven or zygomorphic.

Appearance

It is important for the artist to decide when to 'freeze' an image of a flower in time. From the bud stage, through anthesis (the fertile period) to decay, parts can change at different stages in their development or show male and female forms. It is best to open several flowers at varying stages to show how they might change inside.

In a typical flower, most parts are well developed for fertilization, but there may be stunted or 'missing' parts.

Sometimes petals change colour to signal to pollinators they have been fertilized. Evening primroses (*Oenothera* sp.) open their white or yellow flowers in the evening, but these fade to pink or orange by morning once moths have visited them.

Barley flower anthers fall off soon after they emerge, which makes them difficult to count, as well as a challenge for artists deciding how to include them.

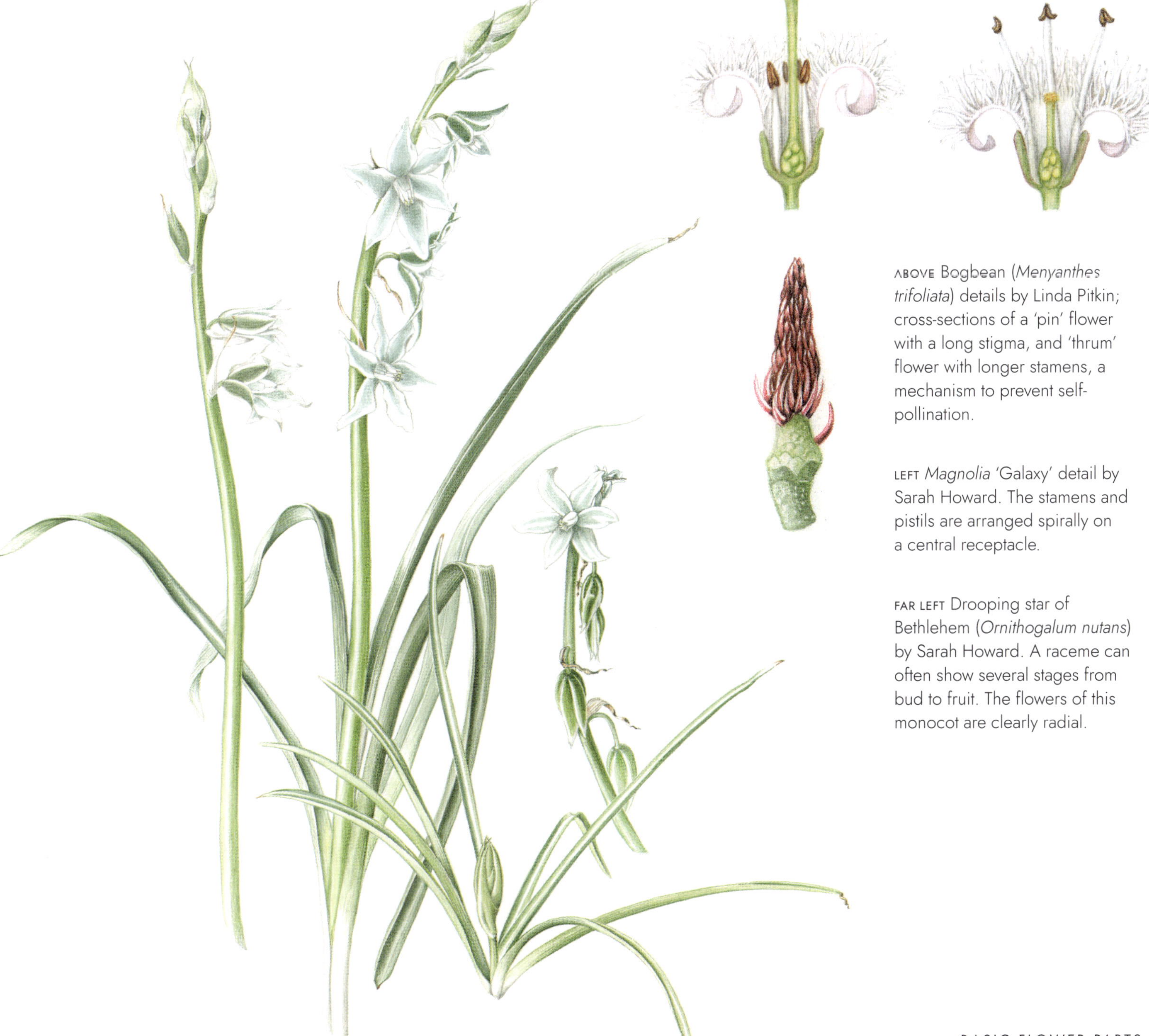

ᴬᴮᴼᵛᴱ Bogbean (*Menyanthes trifoliata*) details by Linda Pitkin; cross-sections of a 'pin' flower with a long stigma, and 'thrum' flower with longer stamens, a mechanism to prevent self-pollination.

ʟᴇꜰᴛ *Magnolia* 'Galaxy' detail by Sarah Howard. The stamens and pistils are arranged spirally on a central receptacle.

ꜰᴀʀ ʟᴇꜰᴛ Drooping star of Bethlehem (*Ornithogalum nutans*) by Sarah Howard. A raceme can often show several stages from bud to fruit. The flowers of this monocot are clearly radial.

Symmetry

Symmetry is a guiding principle in flowers. Most flowers are actinomorphic (evenly radial and symmetrical from all sides). Some are bilaterally symmetrical or zygomorphic (each of two sides a mirror image of the other, like a human body), for example, *Fabaceae* or *Lamiaceae*. Zygomorphic flowers tend to have specialized pollination structures, having co-evolved more closely with insects, birds and bats.

Radial symmetry

Seen from above, radially symmetrical flowers look circular if the petals are reflexed. They can be cut across the diagonal in any direction to make two symmetrical halves. Their radial shape, with anthers and styles sticking out of the centre, is ideal for a variety of crawling pollinators. Many different features can be observed:

- Note the position of the calyx lobes or sepals in relation to the petals. If they are set between the petals, some of them will be seen. If they are set behind each petal they might be hidden. If they are reflexed, they might not show at all. If they need to be shown, it is fine to remove a petal (sometimes they fall off anyway).

- Establish which petal, if any, overlaps another and in what order, as it is easy to miss if a petal is mistakenly pushed in another direction: it is valvate if petals meet without overlapping; it is imbricate if each petal overlaps the next (imbricate may involve one petal being on the outside of others that overlap); or two valvate petals on the outside, two on the inside, with one overlapping another.

Four-petalled flowers, like those in the mustard family (*Brassicaceae*), are radial but can be cut in only two directions, so they are biradial (or bisymmetrical). Trimerous monocots such as trillium and alliums can be cut in only three directions.

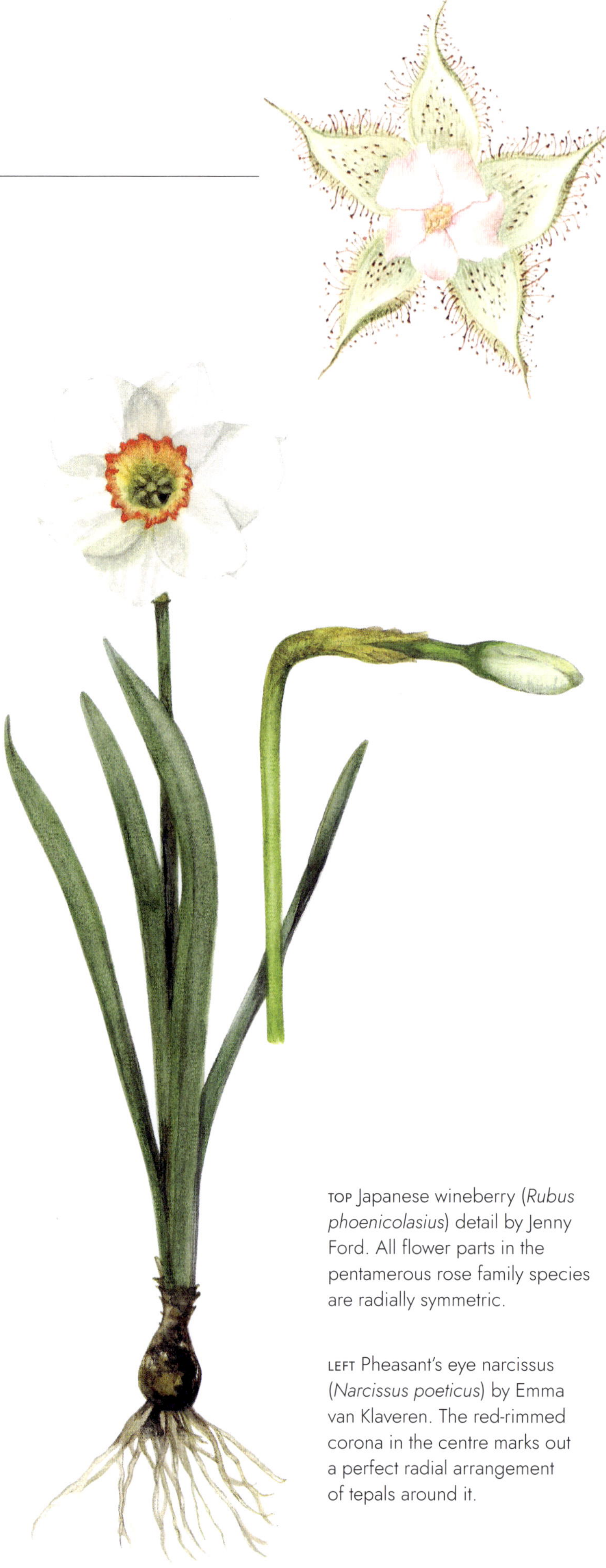

TOP Japanese wineberry (*Rubus phoenicolasius*) detail by Jenny Ford. All flower parts in the pentamerous rose family species are radially symmetric.

LEFT Pheasant's eye narcissus (*Narcissus poeticus*) by Emma van Klaveren. The red-rimmed corona in the centre marks out a perfect radial arrangement of tepals around it.

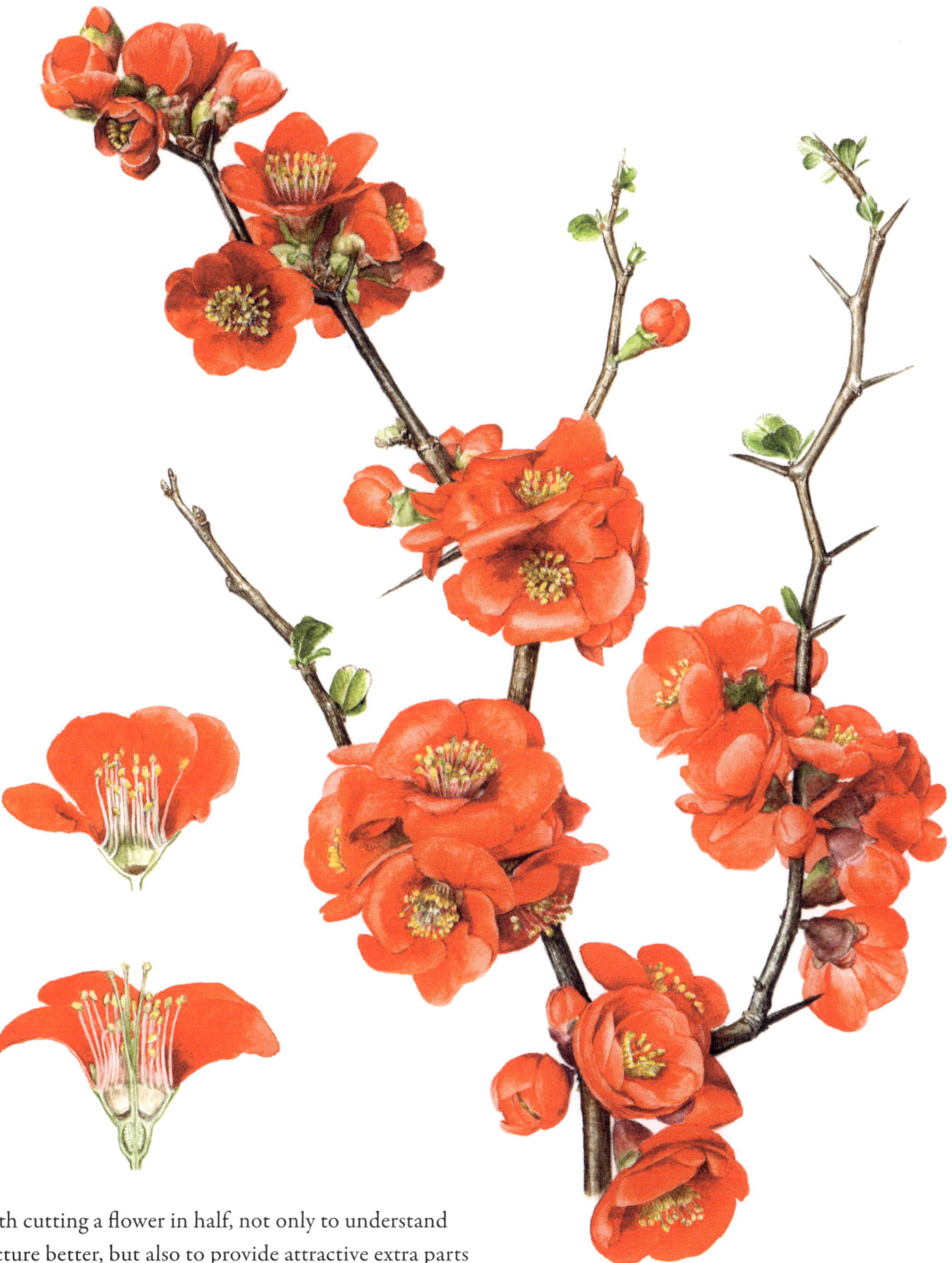

It is worth cutting a flower in half, not only to understand the structure better, but also to provide attractive extra parts that could balance a composition. Occasionally, flowers may have floral parts arranged radially but not equally, in which case a half flower cannot be cut in any direction to produce symmetry. The rock rose (*Helianthemum*) has five radial petals, but it has three large and two small sepals, so a cut can only show half of either petals or sepals.

Examples of radial symmetry are found in daffodils, hollyhocks and dahlias (see practical projects on pages 112, 136 and 146).

ABOVE Japanese quince (*Chaenomeles japonica*) by Roger Reynolds. It has five-petalled, radially symmetrical flowers, typical of *Rosaceae,* with petals that overlap like tiles (imbricate).

Painting a daffodil

Daffodils are a bright, happy reminder that spring is on the way. The flowers have a sculptural shape and are a good example of radial symmetry. At certain angles they are very difficult to capture, but this tutorial shows them in easier positions. It is important to keep the yellows as fresh as possible to make them light and vibrant. A couple of yellows are used to capture the different colours, while the buds have a tint of green before they open.

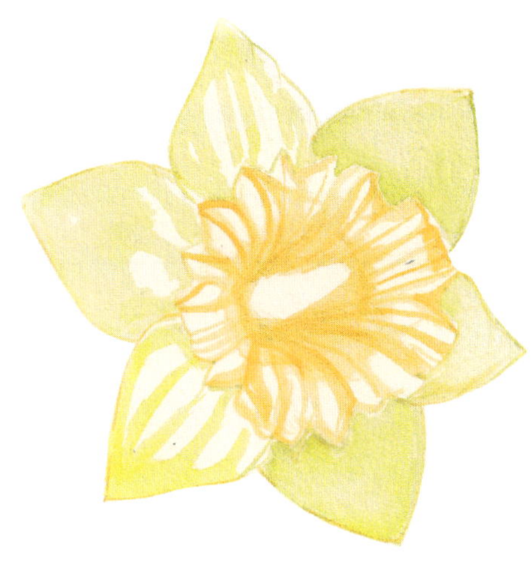

1 | Draw the outline of your flower in pencil. Mix pale cadmium yellow, lemon yellow and a tiny touch of sap green for the first wash of the petals, leaving the lightest areas white. Use a pale cadmium yellow wash on the trumpet, following its form, and blending from the centre outwards. Leave the lightest areas completely white.

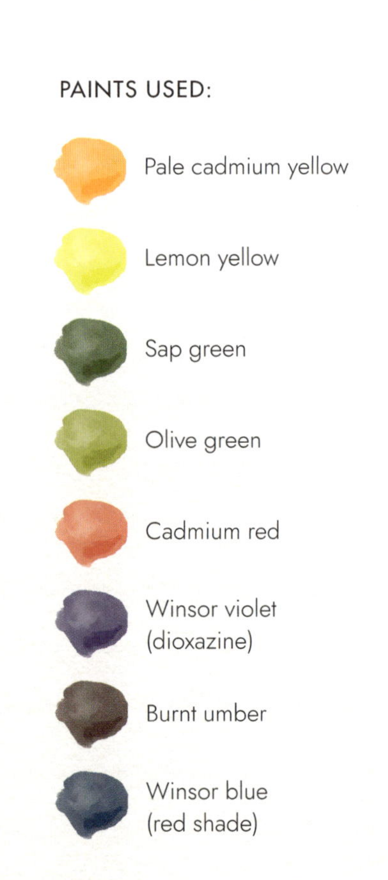

PAINTS USED:

- Pale cadmium yellow
- Lemon yellow
- Sap green
- Olive green
- Cadmium red
- Winsor violet (dioxazine)
- Burnt umber
- Winsor blue (red shade)

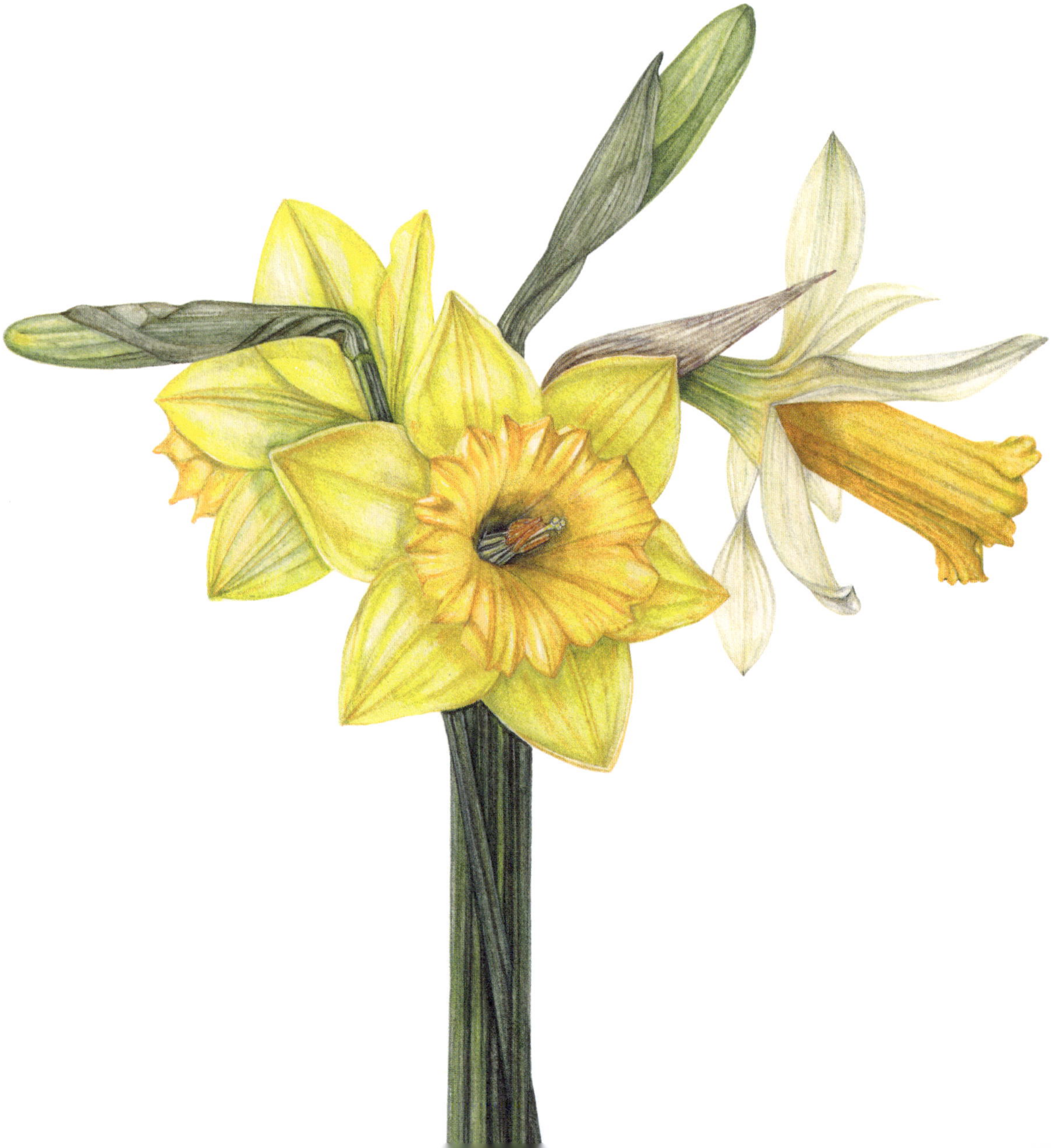

 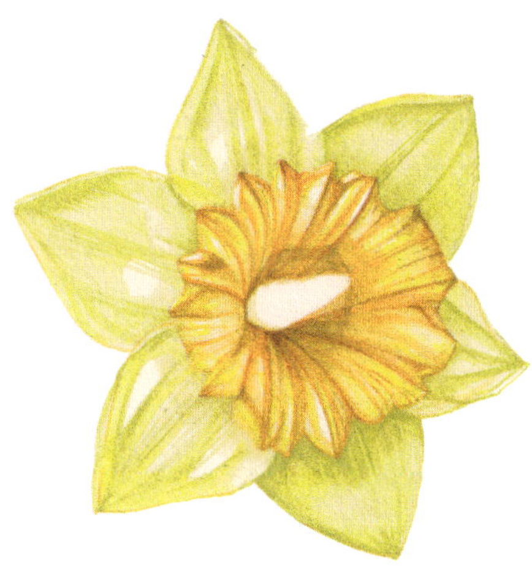 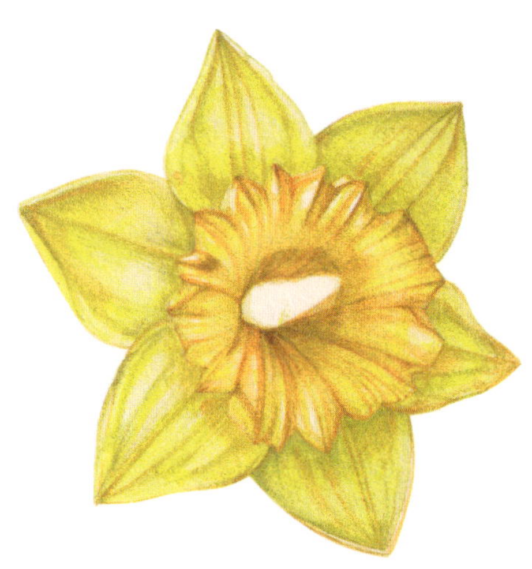

2 | Use a lemon yellow wash for the bright areas of the trumpet, washing the colour over the whole trumpet. Use this wash over the petals, leaving the white areas clear of colour. Apply another wash of pale cadmium yellow to the darker areas of the trumpet. Using a mix of pale cadmium yellow and cadmium red on the darker areas of the trumpet, use a small dry brush for this and 'draw' with the paint.

3 | Use a mix of pale cadmium yellow, cadmium red and a tiny spot of winsor violet to add shade to the trumpet, both on the fluted part and inside, to show depth. This is just the first wash of shadow inside the trumpet, so be delicate. Use the wash from step 1 to draw into the petals, to show the ridges, the shadows behind the trumpet, and where the petals fold up at the edges.

4 | Apply a light lemon yellow wash over the whole surface of the petals to brighten them. Use pale cadmium yellow on top of this to accentuate the details of the petals further.

5 | Mix sap green and olive green together and use a light wash inside the trumpet to darken the shadows. Use a wash of lemon yellow on the stigma and the filaments of the stamen. Use pale cadmium yellow for the anthers. All these parts are small so require dryish paint and

a drawing technique. Use winsor violet in between the stamens for shadows and to differentiate between them. For the first bud, use the wash from step 1 and apply it to the top and bottom of the bud. Mix more sap green into this wash and apply to the centre of the bud.

6 | Make a sap green, olive green and burnt umber wash, and paint this over the spathe (the papery sleeve covering the bud). Wait for this to dry, then use this same wash with a dry brush technique and a small brush to 'draw' in the lines that follow the form of the spathe. In the same manner, use the dark green wash to show dark areas, particularly where the bud petals overlap each other. Paint the trumpet on the next open flower all over with a wash of pale cadmium yellow.

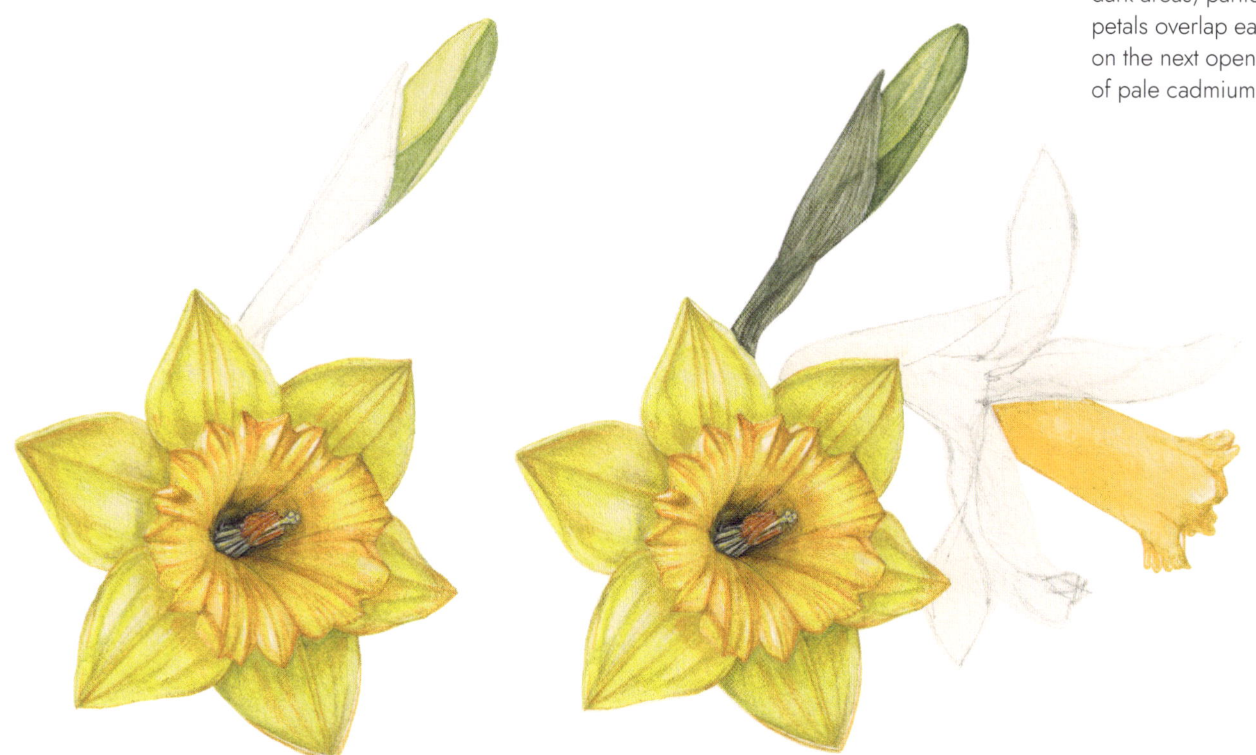

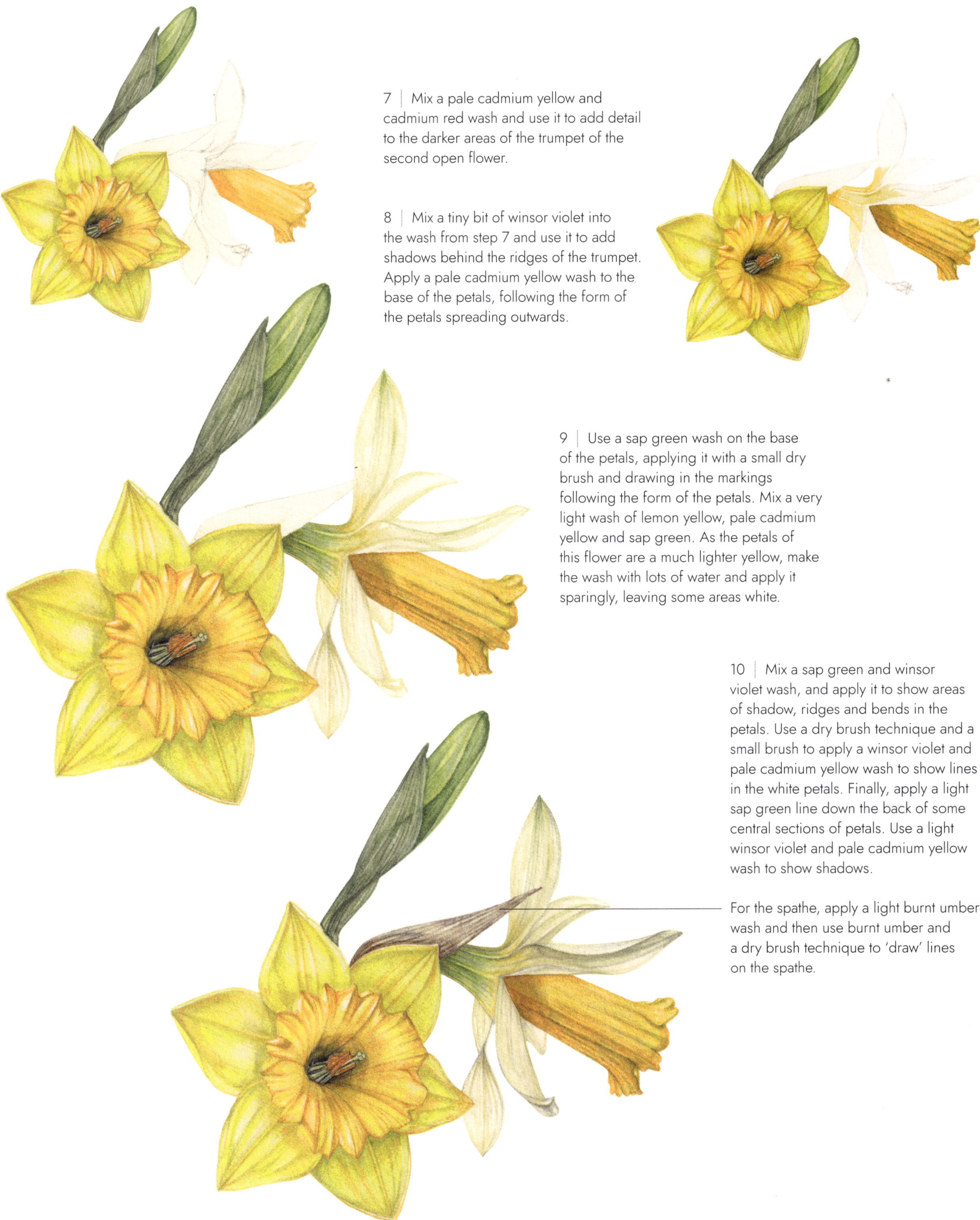

7 | Mix a pale cadmium yellow and cadmium red wash and use it to add detail to the darker areas of the trumpet of the second open flower.

8 | Mix a tiny bit of winsor violet into the wash from step 7 and use it to add shadows behind the ridges of the trumpet. Apply a pale cadmium yellow wash to the base of the petals, following the form of the petals spreading outwards.

9 | Use a sap green wash on the base of the petals, applying it with a small dry brush and drawing in the markings following the form of the petals. Mix a very light wash of lemon yellow, pale cadmium yellow and sap green. As the petals of this flower are a much lighter yellow, make the wash with lots of water and apply it sparingly, leaving some areas white.

10 | Mix a sap green and winsor violet wash, and apply it to show areas of shadow, ridges and bends in the petals. Use a dry brush technique and a small brush to apply a winsor violet and pale cadmium yellow wash to show lines in the white petals. Finally, apply a light sap green line down the back of some central sections of petals. Use a light winsor violet and pale cadmium yellow wash to show shadows.

For the spathe, apply a light burnt umber wash and then use burnt umber and a dry brush technique to 'draw' lines on the spathe.

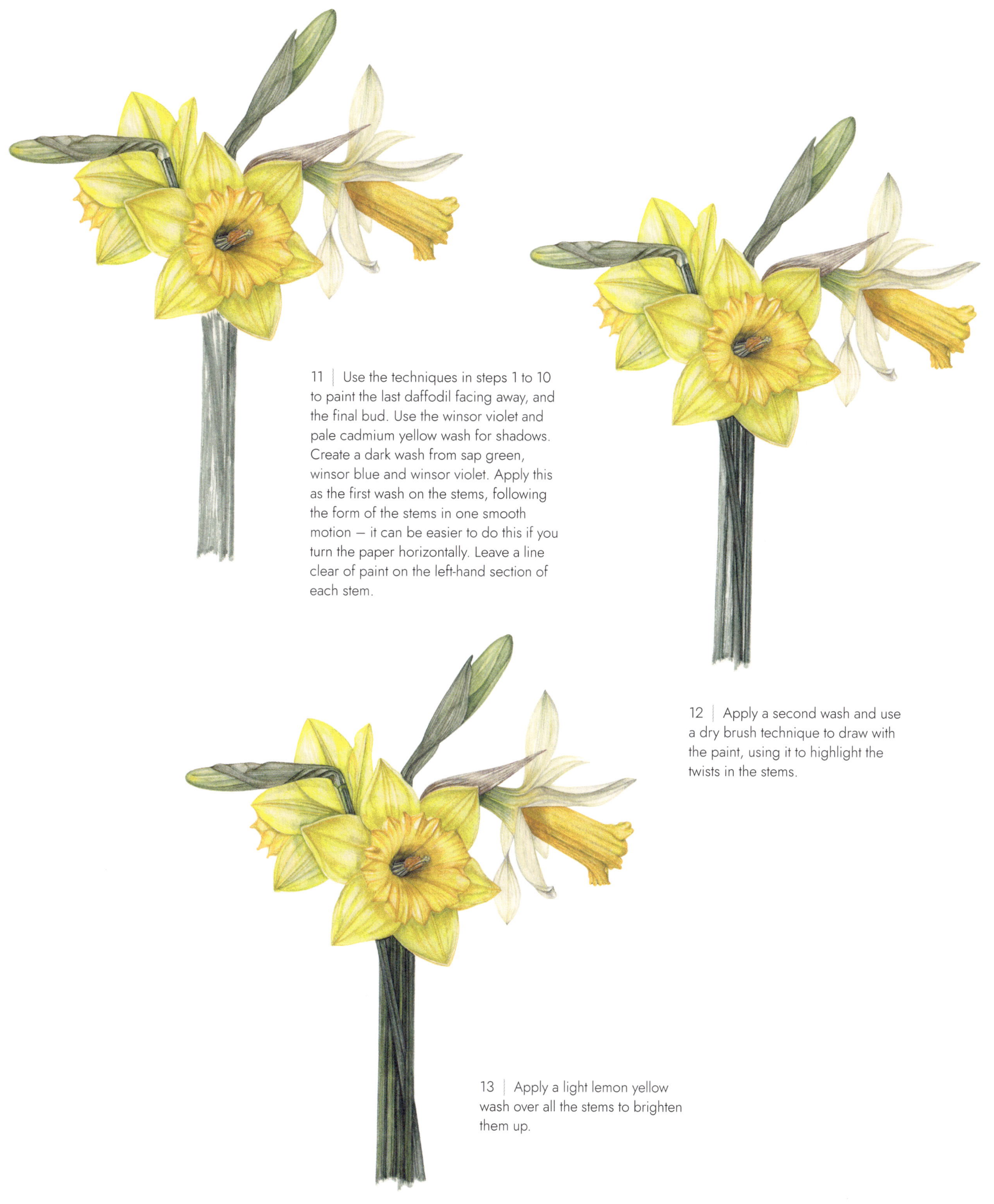

11 | Use the techniques in steps 1 to 10 to paint the last daffodil facing away, and the final bud. Use the winsor violet and pale cadmium yellow wash for shadows. Create a dark wash from sap green, winsor blue and winsor violet. Apply this as the first wash on the stems, following the form of the stems in one smooth motion — it can be easier to do this if you turn the paper horizontally. Leave a line clear of paint on the left-hand section of each stem.

12 | Apply a second wash and use a dry brush technique to draw with the paint, using it to highlight the twists in the stems.

13 | Apply a light lemon yellow wash over all the stems to brighten them up.

Ulmo (*Eucryphia cordifolia*)
by Gülnur Ekşi Bona

Eucryphia cordifolia is part of a collection of outstanding paintings commissioned for the book *Plants from the Woods and Forests of Chile* (RBG Edinburgh, 2022) by Martin Gardner et al. Three Turkish artists, trained by Kew Gardens' Christabel King, including Işik Güner and Hülya Korkmaz, and British artist, Mary Benstead, worked on the paintings. Above all, they demonstrate how art can draw attention to the shocking decline of a habitat — beautiful paintings may be all that is left behind to see.

Painting challenges
Gülnur had two problems: how to depict the tree and how to portray its white flowers. The *Eucryphia* is a large tree native to the Valdivian temperate rain forests of Chile and Argentina, where its relatively small, pure white flowers stand out in February and March. It is also present in milder gardens in temperate climates on the west coasts of the British Isles and North America. Culturally important in Chile, where it is known as 'ulmo', it is famed for its wood and honey production.

The process
First, Gülnur worked outside, sketching a tree in Logan Botanic Garden in southwest Scotland. She drew it from different angles, letting natural light highlight the features. Since natural light changes, she then went indoors, carefully lighting her branch against a simple background. Adding lichens gives the impression that it is outdoors, with the other seasonal parts telling a story.

The cluster of flowers placed against a background of dark leaves emphasizes the white and draws the eye to that area. Meanwhile, a branch effortlessly cuts through the focal point, drawing the eye to another cluster diagonally opposite. This time, the flowers are a footnote to the leathery leaves with their mildly toothed margins. There is little doubt that the picture portrays a tree, with its hanging woody branches providing a glimpse of what more might be seen.

The leaves' smoothness comes from Gülnur's technique of initially using fluid washes. While building up the layers, however, she becomes more controlled, finally using her brush like a pencil to add features and textural details like the many stamens, style arms and the gnarled twig. The outcome reflects the extraordinary connection Gülnur found in *Eucryphia cordifolia*'s visually striking appearance.

RIGHT The pretty white flowers of *Eucryphia* stand out against a background of dark leaves. Size: 49.5 x 35cm (19^1/$_2$ x 13^3/$_4$in), watercolour on paper.

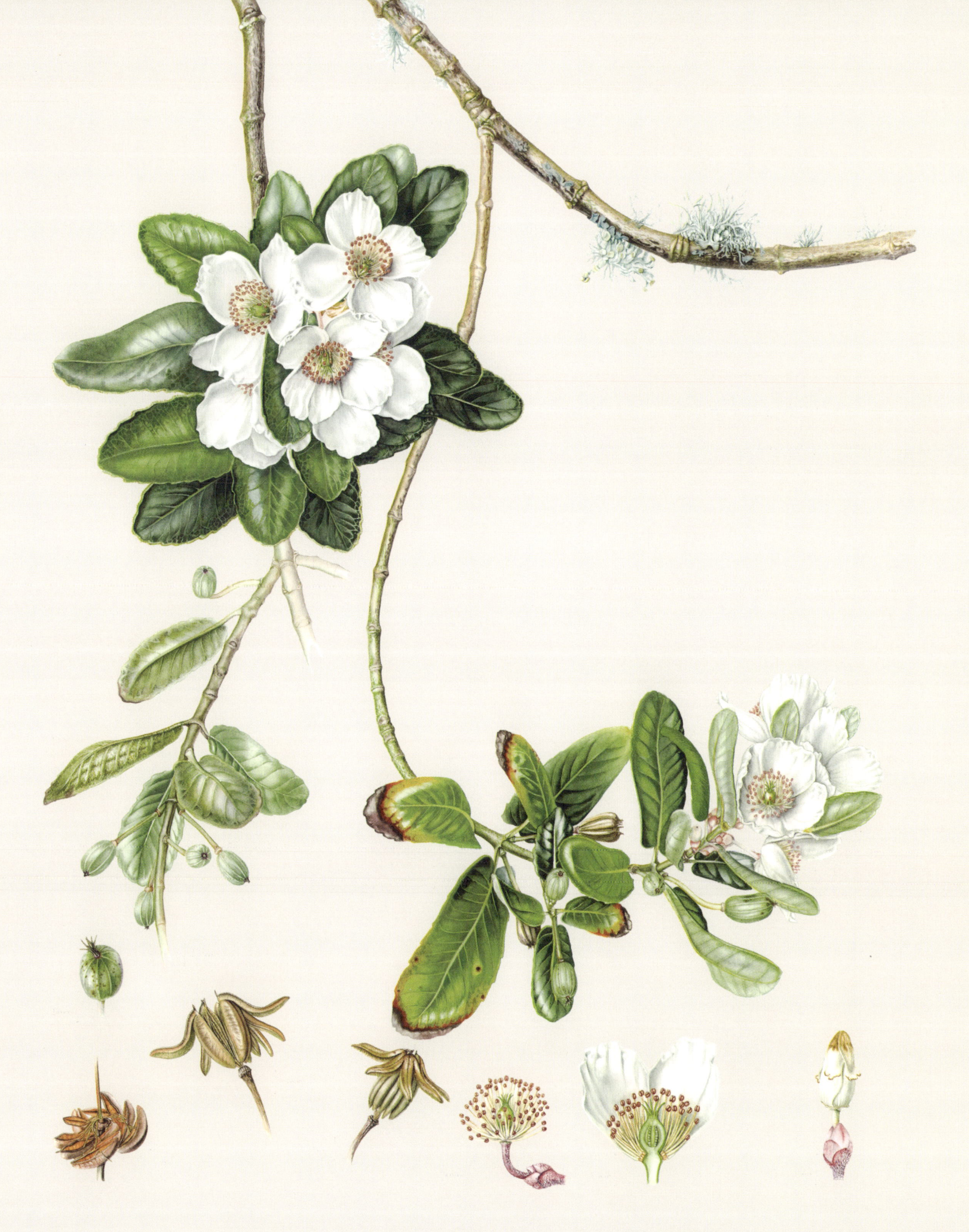

Bilateral symmetry

Flowers that are bilaterally symmetrical (zygomorphic) have unequal-sized petals that can only be cut longitudinally in one direction to produce mirror images. Sometimes it is not obvious, but a close look at pansies and speedwells shows this trait.

Zygomorphism developed in conjunction with a proliferation of insect life and the needs of specialist pollinators. The parts that evolved often have specialist terms (see pages 154–157). They occur widely in plant families but especially in peas, beans and mimosa (*Fabaceae*), mints (*Lamiaceae*) and orchids (*Orchidaceae*). But many other families also display zygomorphic symmetry, such as violas (*Violaceae*), irises (*Iridaceae*), nasturtiums (*Tropaeolaceae*) and speedwells (*Plantaginaceae*).

Zygomorphic flowers typically open for longer than actinomorphic flowers and are visited less often by pollinators. Are the two connected? Do flowers open for longer to allow for the fewer insects visiting the flower? When they do, does pollination happen more efficiently? Artists might look carefully at the fertilizing mechanisms in the flower and the pollinator, and decide for themselves.

WHAT TO LOOK FOR

- Are the petals of uneven sizes?

- Do petals appear to be in two distinct parts, like a lip (labium), or fused petals with lobes?

- Are keels present? These are formed by two lower petals and are sometimes hidden by lateral petals until they have been 'tripped' by a pollinator.

- Are there any spurs, formed from any part of the flower (see also Nectaries p142)?

Flowers in the pea family (*Fabaceae*) are characterized by their keels. They have petals called standards that open first, two lateral petals or wings, and two in between which cleave together to form a keel. The stamens and pistils are kept inside the keel, often in a horizontal position.

BELOW Pansies by Alison Harley. Their petals are often not of an equal size and the uppermost two petals overlap.

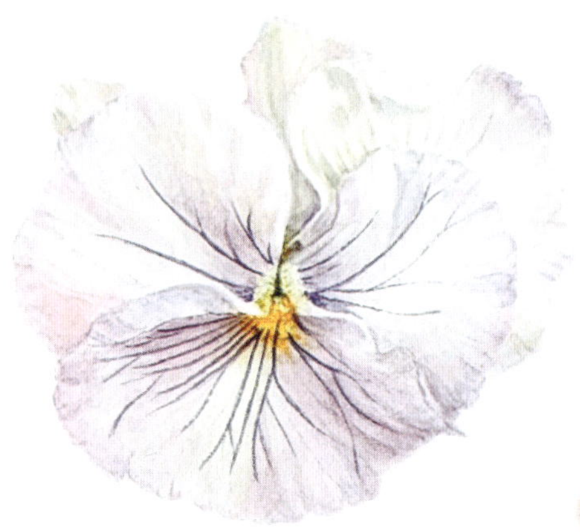
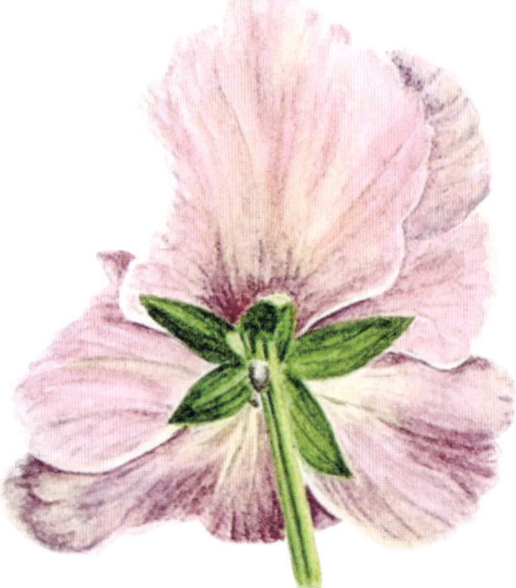
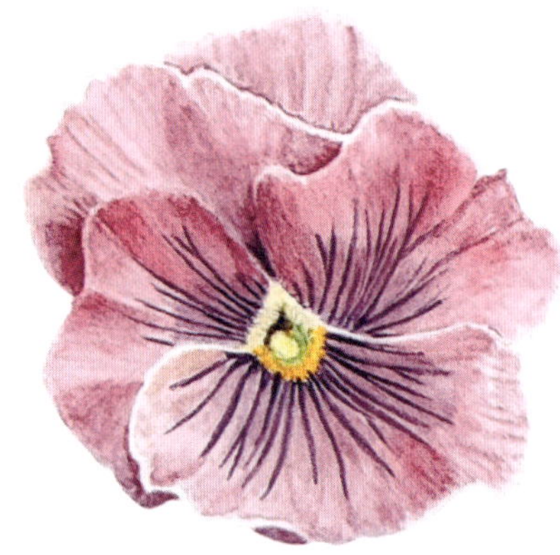

Salvias and mints (a term that gave its old family name of *Labiatae* – now *Lamiaceae*) typically have a lower and an upper lip, usually with two lobes on the upper lip, and three on the lower. The flowers can be described as labelliform, a term also used for the labellum of an orchid. In these flowers, the stamens and pistil are often curved to suit the shape.

Asymmetrical flowers, such as magnolia, cannot be cut in two identical halves as their tepals are arranged in whorls.

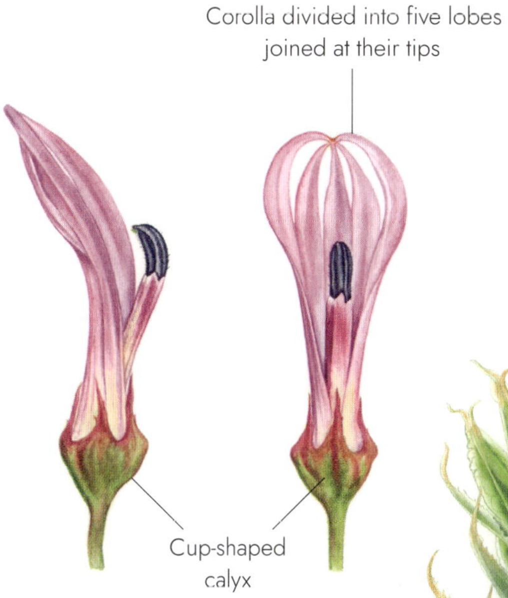

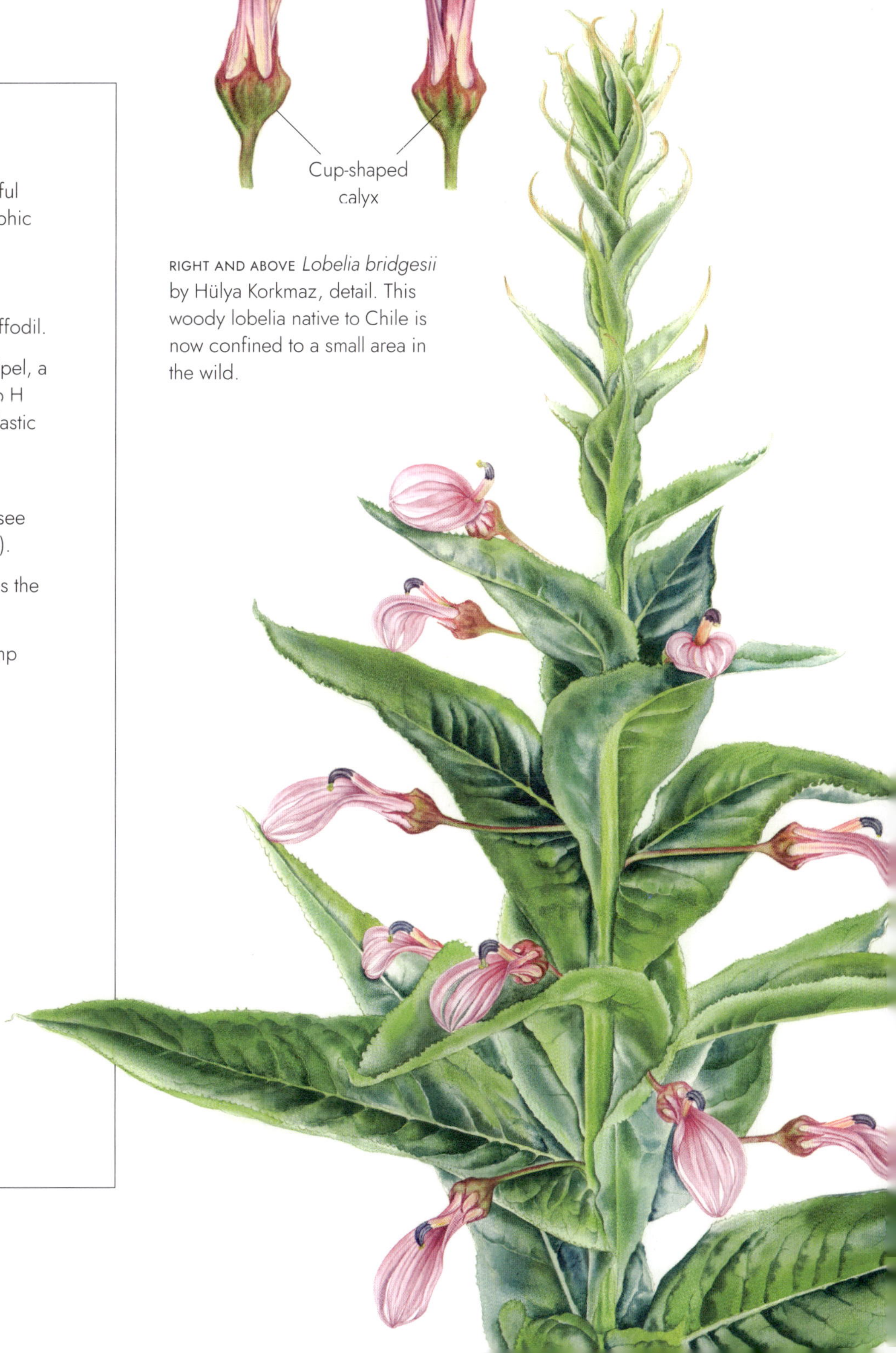

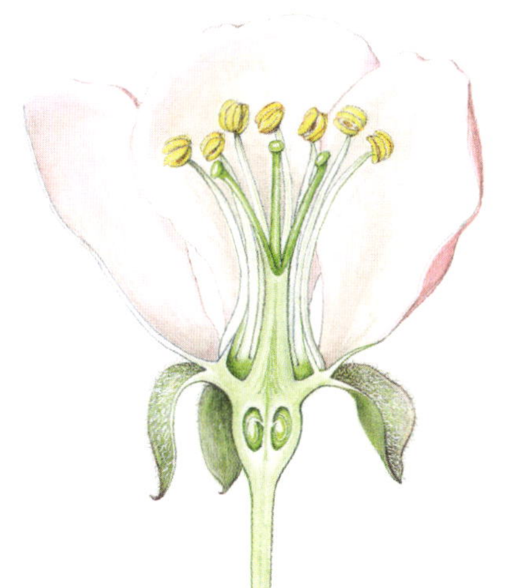

DRAWING PREPARATIONS

Here are some tips on drawing half-flowers, useful to understand either actinomorphic or zygomorphic flowers, or to add to your painting:

- Ensure you have plenty of material.

- Practise on a large fleshy flower, such as a daffodil.

- You will need a very sharp razor blade or scalpel, a magnifying glass or simple microscope, sharp H pencils, tweezers for lifting the parts, and a plastic cutting mat.

- Decide where you want to cut, depending on whether you are dealing with radial, biradial (see page 110) or zygomorphic planes (see above).

- Make your cut through the centre, so it catches the ovary through the middle.

- Put both halves into a dish and cover with damp tissue, as the material will dry out quickly.

RIGHT AND ABOVE *Lobelia bridgesii* by Hülya Korkmaz, detail. This woody lobelia native to Chile is now confined to a small area in the wild.

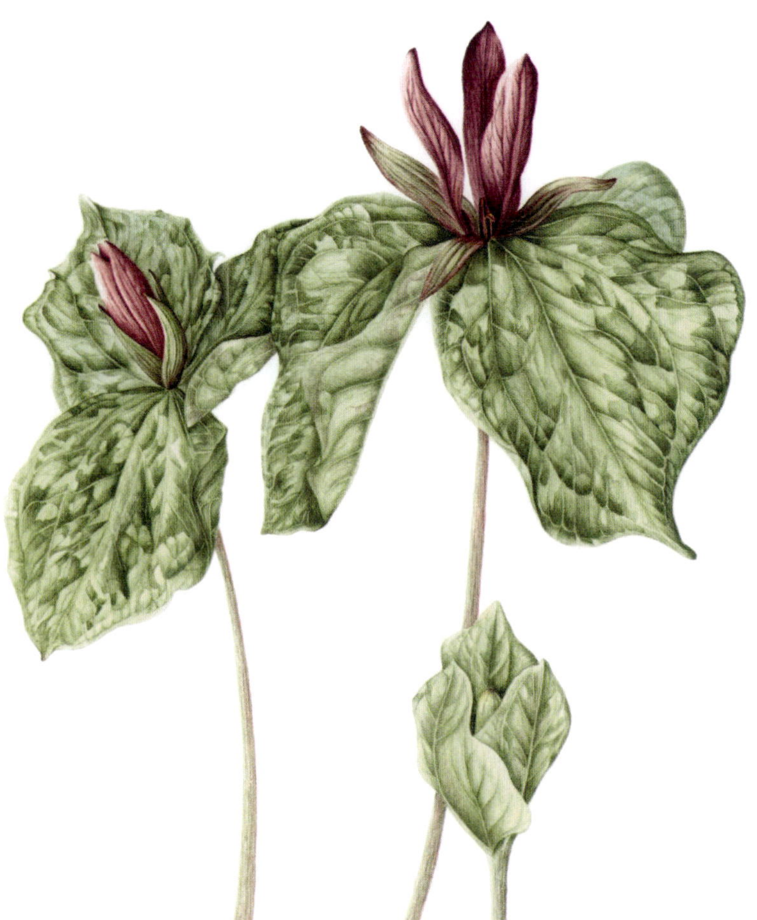

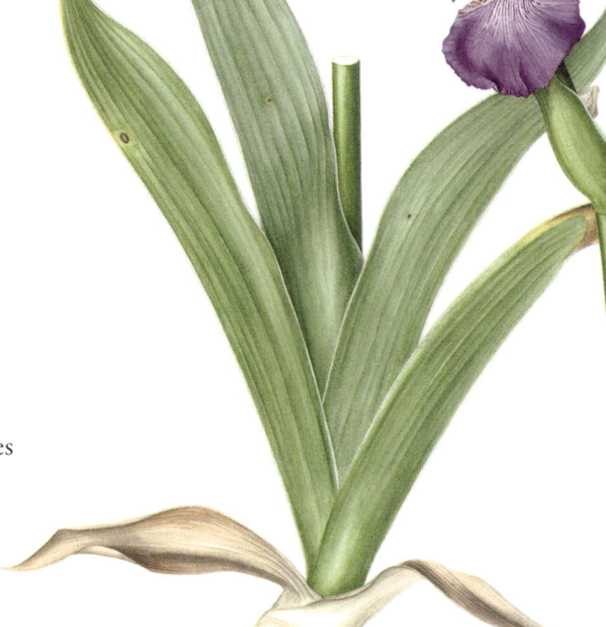

Monocot flowers are unmistakable with their parts in threes (trimerous), and long, slender parallel-veined leaves. But their flowers can be both radially and bilaterally symmetrical. Tulips and trilliums are actinomorphic, whereas irises and gingers are zygomorphic.

Often their sepals look like petals, so the term tepal refers to both. Tulips have six tepals: effectively three sepals and three petals. Stamens and style arms of monocot flowers also follow the trimerous pattern.

The parts of an iris flower are divided into three 'falls' (the sepals) that hang down and provide a colourful and often patterned landing pad for insects. Partly inside them are three standards — upright petals — joined at the base and set on top of the ovary. The three stamens are hidden within, whilst the three petalous style arms (attached to the ovary) appear out of the tube in the centre of the flower.

Members of *Araceae*, such as Jack-in-the-pulpit and arums, provide dramatic architectural shapes for painting. They have in common a spadix, which is a spike of flowers on a fleshy axis enclosed in a spathe, and their stamens are joined in a synandrium (see pages 152 and 170).

The calyx

The calyx (combined term for the sepals) is the outermost whorl of the flower. The term perianth is used for both the calyx and the corolla combined.

Sepals provide protection for the flower bud. They are usually green and provide extra photosynthesis for the plant. If they are coloured, they can be attractive to pollinators, while prickly ones can deter unwelcome visitors. The calyx is attached close to the base of the petals and may consist of two rings; the outer one is called an epicalyx and occurs in some mallows.

Some flowers have no sepals or they quickly fall off, while in other flowers the sepals are the main feature for an artist.

A fused calyx can have many shapes, each with a descriptive name: for example, cupular (cup-shaped), campanulate (bell-shaped), bi-labiate (two lipped), infundibular (funnel-shaped), tubular (tube-shaped) and so on.

WHAT TO LOOK FOR

- Sepals can be fused or separate.

- The number of sepals is often, but not always, the same as the number of petals.

- Sepals can lie flat against the petals, be reflexed or recurved.

- The shape, length and width of sepals are variable.

- Sepals can be hairy, whereas other floral parts are not.

- Lobes at the top of a fused calyx can represent individual sepals.

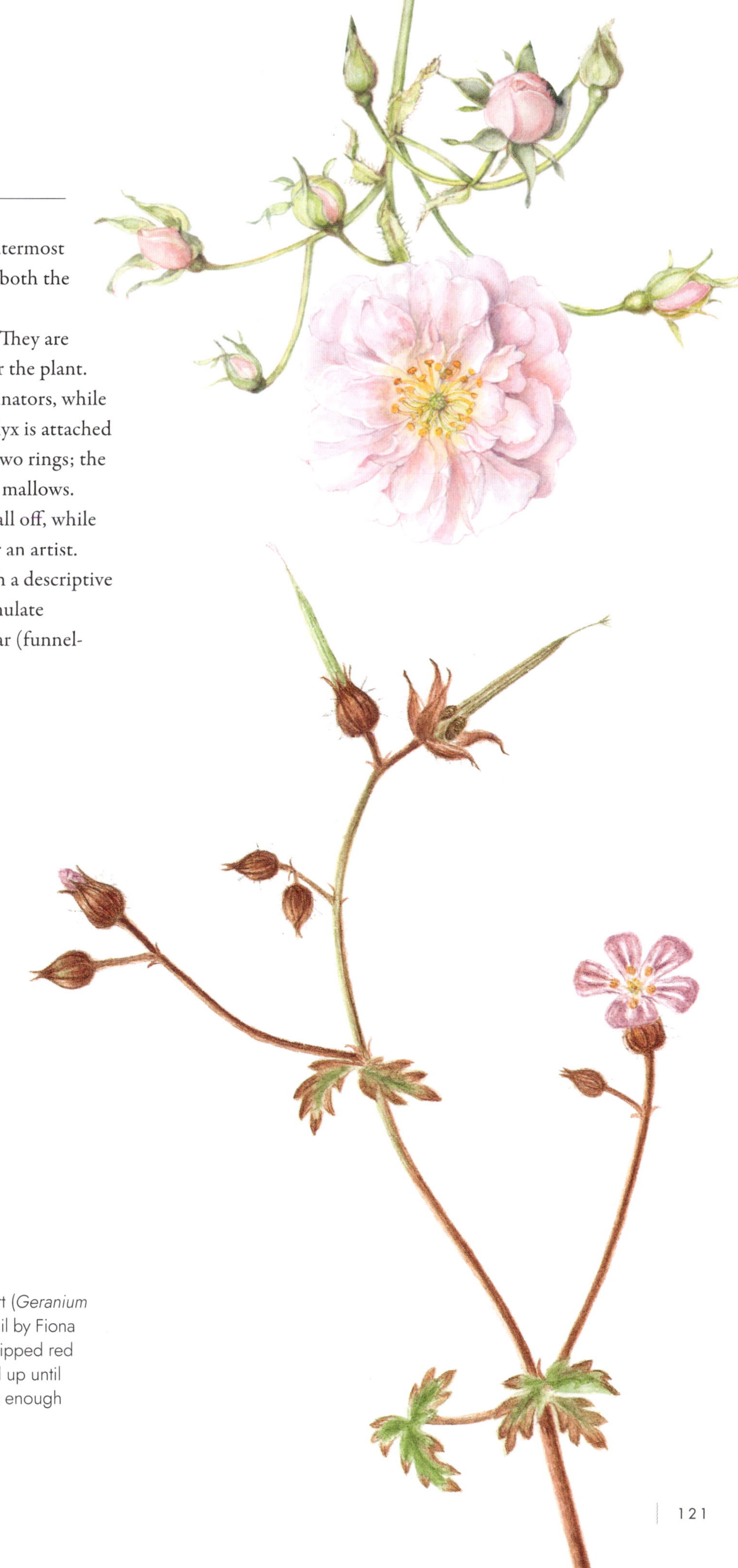

ABOVE RIGHT 'Paul's Himalayan musk' rose detail by Janet Dyer. Sepals are free in this rambling rose, and recurved when the bud opens into a flower.

RIGHT Herb Robert (*Geranium robertianum*) detail by Fiona Ward. The acute-tipped red sepals are closed up until the fruit is mature enough to be released.

Many flowers have petalous sepals that steal the show. They are identified as sepals, rather than petals, because they protect the flower bud and form the outer whorl. With such attractive sepals, some flowers do away with petals entirely.

Sepals provide hellebores with their colour. The actual petals appear as discrete tubes inside the sepals, or are modified into nectaries.

The water lily is one of the most primitive flowers. The yellow water lily (*Nuphar lutea*) has four sepal lobes bent inwards to form a bell shape (see page 82). Inside, there are numerous petals arranged around a domed ovary. Other water lilies vary from white and blue to pinks and purples, sometimes patterned.

Dutchman's pipe (*Aristolochia* spp.) also dispenses with petals, but has a calyx that inflates into a tube called a utricle. This continues into a tubular middle part, which is connected to an expanded limb or lobe. *Aristolochia grandiflora* from Brazil is the largest of them all, with a huge calyx limb about 50cm (20in) wide, and a trailing sepal about 60cm (24in) long. The colours and patterns of the unusual calyx of *Aristolochia* provide exotic subjects for painting.

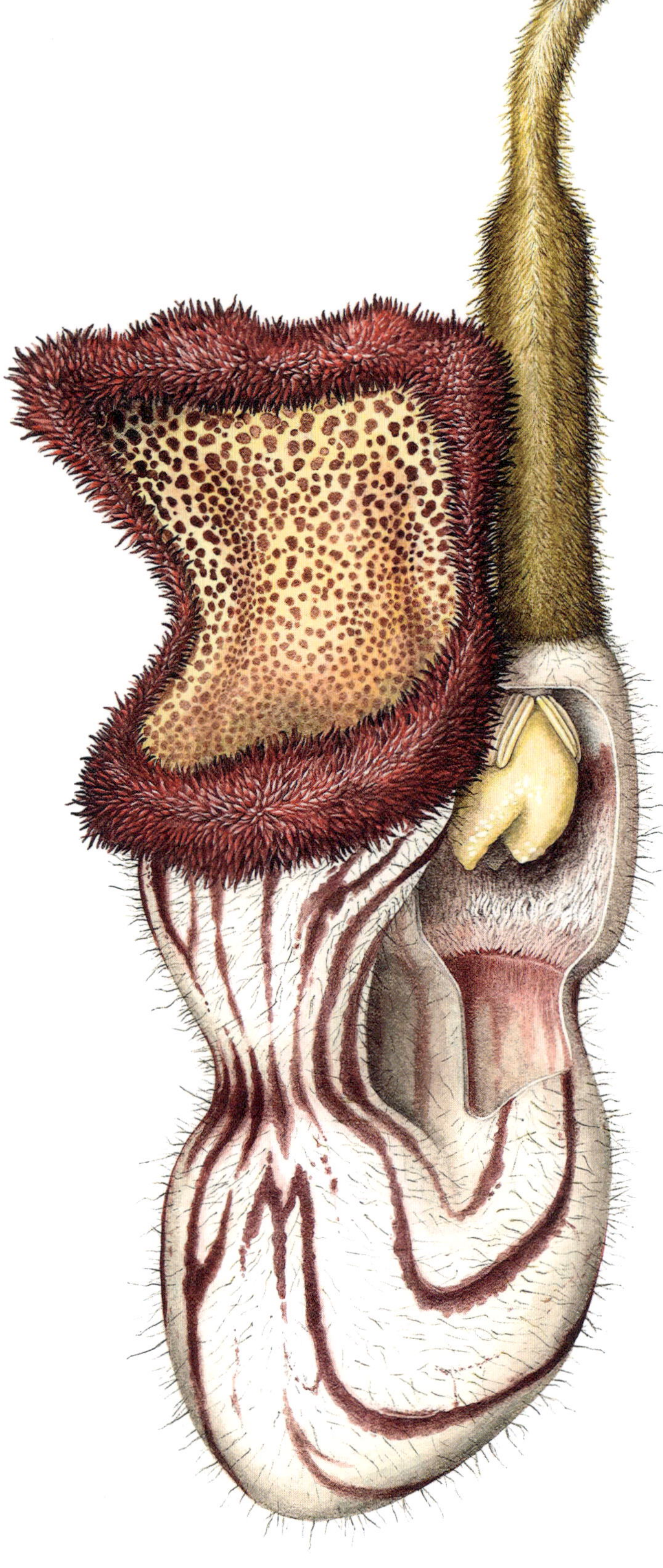

 Dutchman's pipe (*Aristolochia cathcartii*) by Laura Silburn. Its mouth is edged with purple hairs to briefly imprison pollinating insects attracted by the stink of rotting flesh inside.

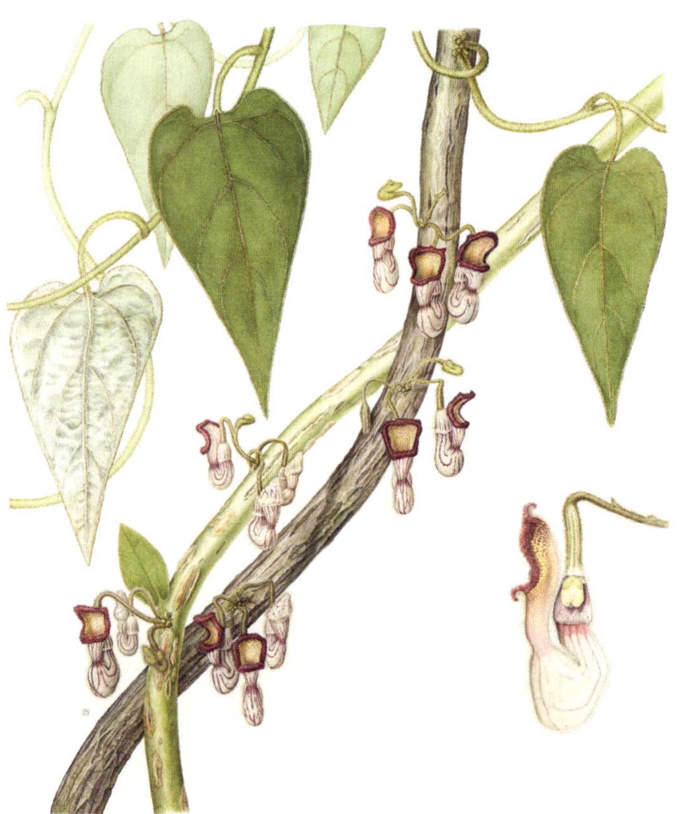

Clematis, which belongs to the buttercup family, has flowers that are particularly attractive for their coloured sepals. There are usually four of them, but cultivated versions can have six. Unlike a buttercup, which has both sepals and petals, clematis has no petals. Like a buttercup, however, it has numerous stamens inside.

The prickly *Acaena novae-zelandiae* from New Zealand, known as red bidibid, is considered invasive in the British Isles. Its many flowers, which form a globular inflorescence, have no petals, but the prickly calyx that surrounds the sexual parts has long spines, aiding the disperals of their seeds. But they are a menacc to remove from clothing and dogs.

Cape gooseberries or the orange Chinese lanterns (*Physalis* spp.) are popular painting subjects because of their persistent calyx. It covers the flowering part until the developing fruit inside forces apart the individual sepals revealing the bright orange fruit inside.

When you paint the papery calyx of a cape gooseberry with its fruit inside, it is best to do the calyx first with a light touch, avoiding the areas where a round fruit will show. Only towards the end should you swish a pale version of the orange, possibly tinged with green, to form the fruit's shape.

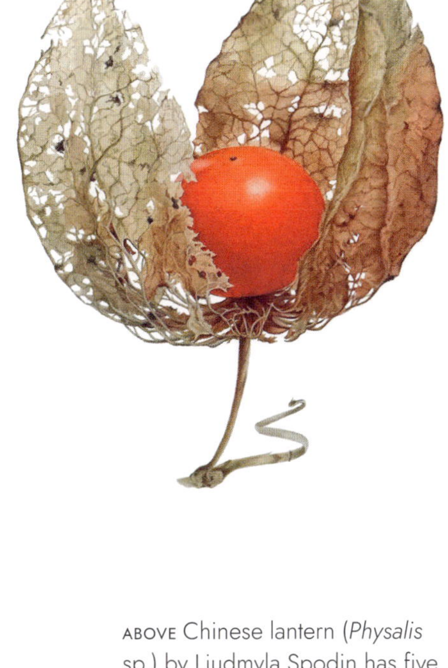

ABOVE Chinese lantern (*Physalis* sp.) by Liudmyla Spodin has five joined-up sepals. They split apart when the fruit is mature.

LEFT *Clematis* 'Princess Diana' by Pamela Richardson. This climber has scarlet petaloid sepals (or tepals), which give it its striking look.

The corolla

The corolla (combined term for petals) is the second whorl of the flower. If the petals are little different from the sepals, they are called tepals. If fused, they may appear as lobes that match the reduced petals. The corolla's main purpose is to attract pollinators, and it is usually colourful, patterned or scented.

Petals are attached to the receptacle at the top of the pedicel. The various ways they overlap, or not, is called aestivation.

A trimerous flower has three parts in each whorl; a tetramerous corolla has four; and a pentamerous has five. There may be multiples or a halving of these numbers, or even a reduction where parts are not seen at all. Eudicots usually have multiples of two or five parts.

A narrowing of the petal towards insertion is a claw. It widens into the limb – the main part. There may be notches or lobes on the outside margin. Some petals have such deep notches they almost look like two petals, for example, ragged robin (*Silene flos-cuculi*). A narcissus flower has a corona, an extra appendage that forms a 'crown' around the stamens and ovary.

Shapes of a radially symmetrical (actinomorphic) corolla in three dimensions can be, broadly speaking, cupuliform (cup-shaped), urceolate (urn-shaped), campanulate (bell-shaped), salver-shaped, tubular or ligulate (strap-shaped).

The backward-facing petals of cyclamen and Turk's cap lilies are strongly reflexed. If the flower is not upright, it may be nutant (nodding).

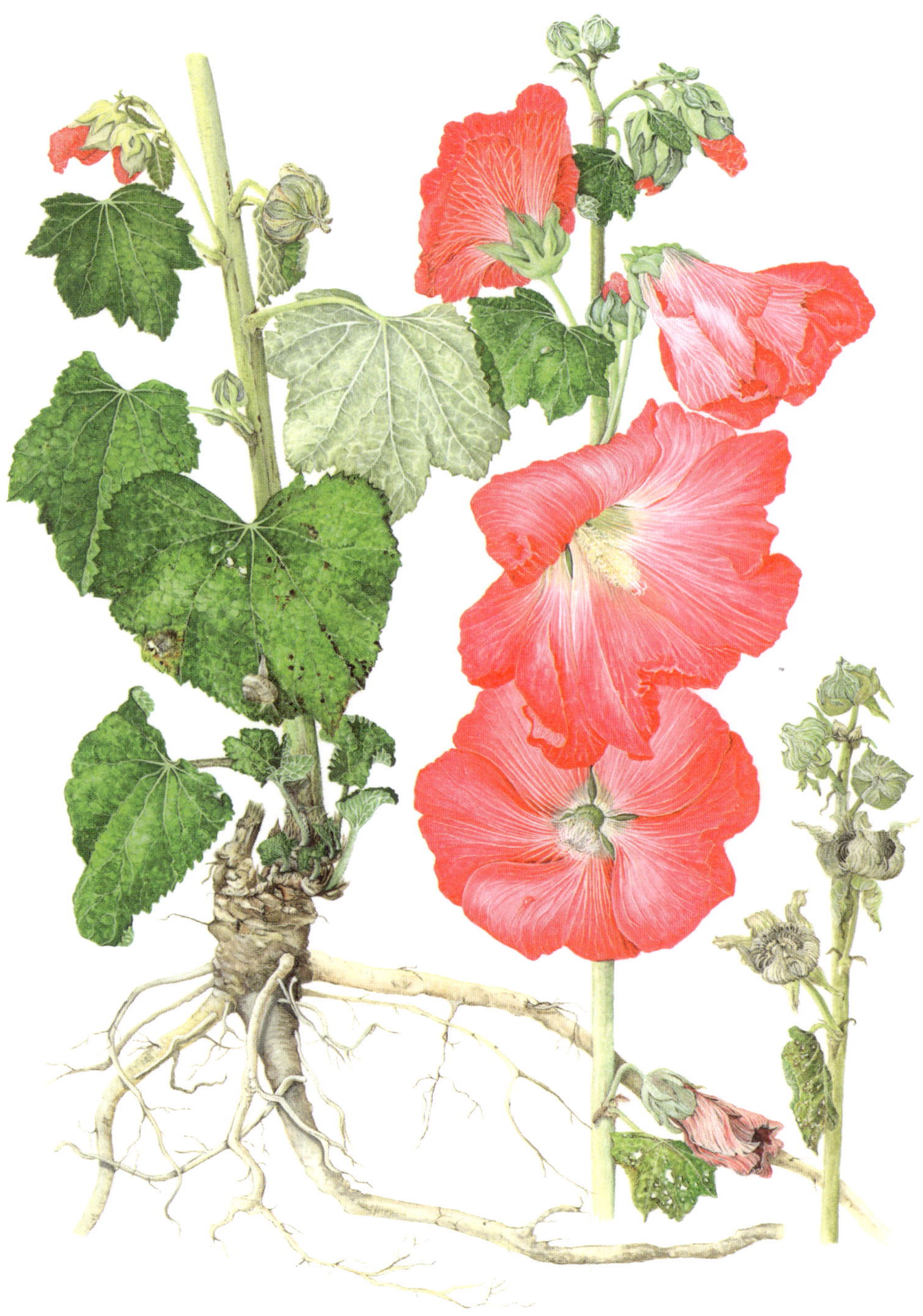

BELOW Hollyhock (*Alcea* cv.) by Tina Bone. The many cultivars of hollyhock have colourful corollas set between the calyx and the male parts inside.

The throat is exactly that, an entrance to the interior of the fused corolla. The throat itself may be marked by patterns, hairs or small appendages. When opened, the inside markings may look different from the outside.

In all cases, any reflexing, or not, of the lobes should be noted. They can spread out as a flat plane, called salverform, like primroses and jasmines, or they can curve inwards.

In many cases, not all petals are fused. Sometimes there may be one or more petals which are free, whereas the rest are fused into a longer tube. This feature is very apparent in bilabiate flowers, but look out for it elsewhere.

Very often, stamens are attached to the inside of a tube, and the attachment point should be determined, whether low down or high up in the tube. Other features, such as hairs, should also be noted.

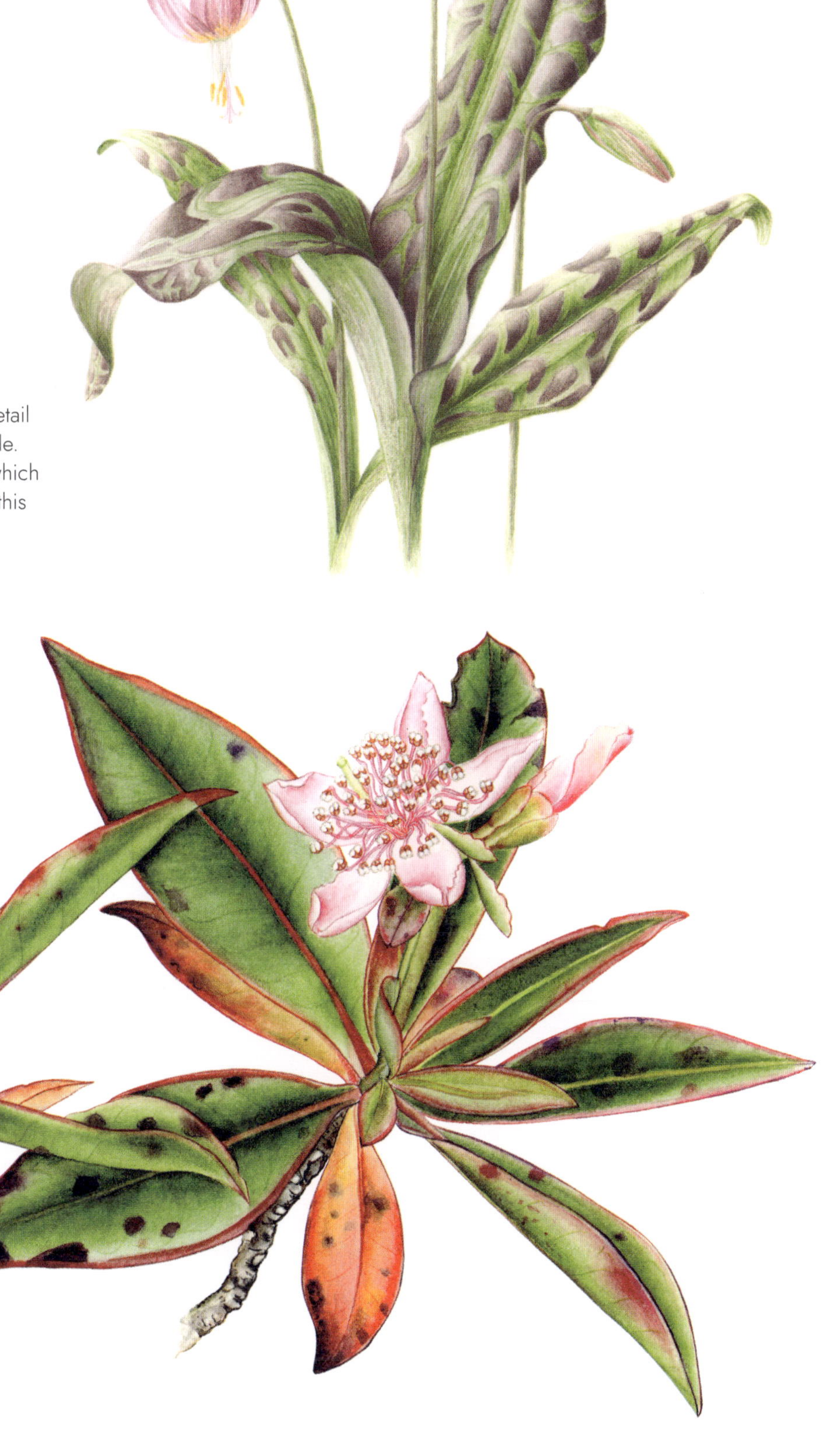

ABOVE RIGHT Fawn lily (*Erythronium revolutum*) by Lyn Campbell. When it is unclear which is tepal or sepal, as in many monocots, they are known as tepals. When painting these nodding North American flowers with strongly reflexed tepals, one threesome should be outside the other.

BELOW *Archytaea triflora* detail by Maria Alice de Rezende. Note the petal margins, which roll inwards (involute) on this South American plant.

Painting an iris

The flower of an iris is structured
as a whole but the petals themselves
are very thin and textured. It is
important to depict both these
elements. The colours in the
specimen shown are light and
therefore it is vital not to over
darken with the shadow colour
and lose some of the light
areas on the petals.

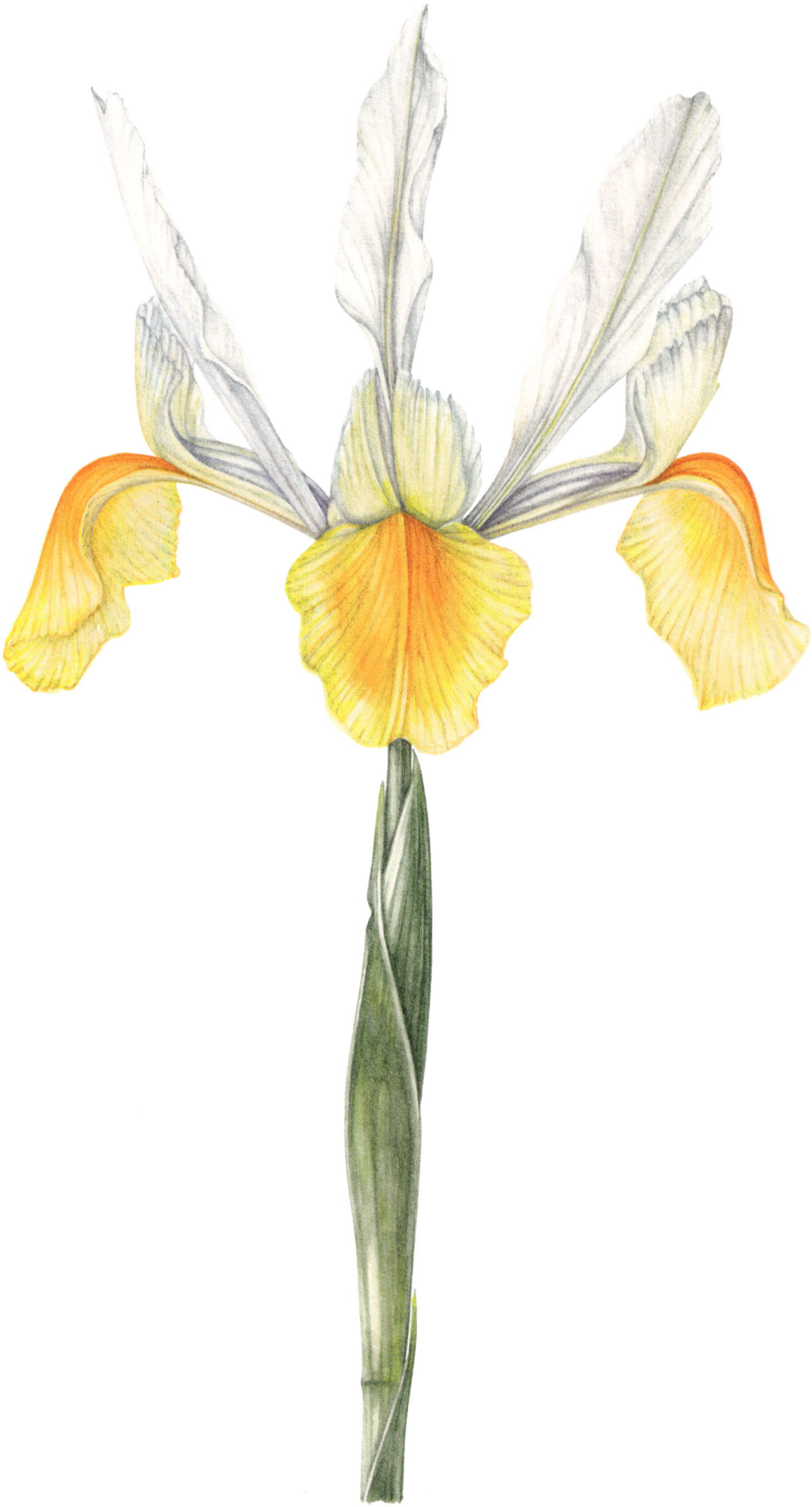

PAINTS USED:

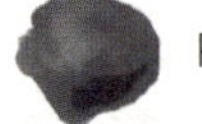 Payne's grey

 Cobalt blue

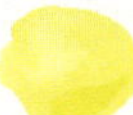 Lemon yellow

 Cadmium yellow

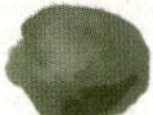 Sap green

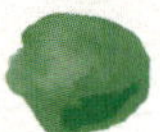 Winsor green
(yellow shade)

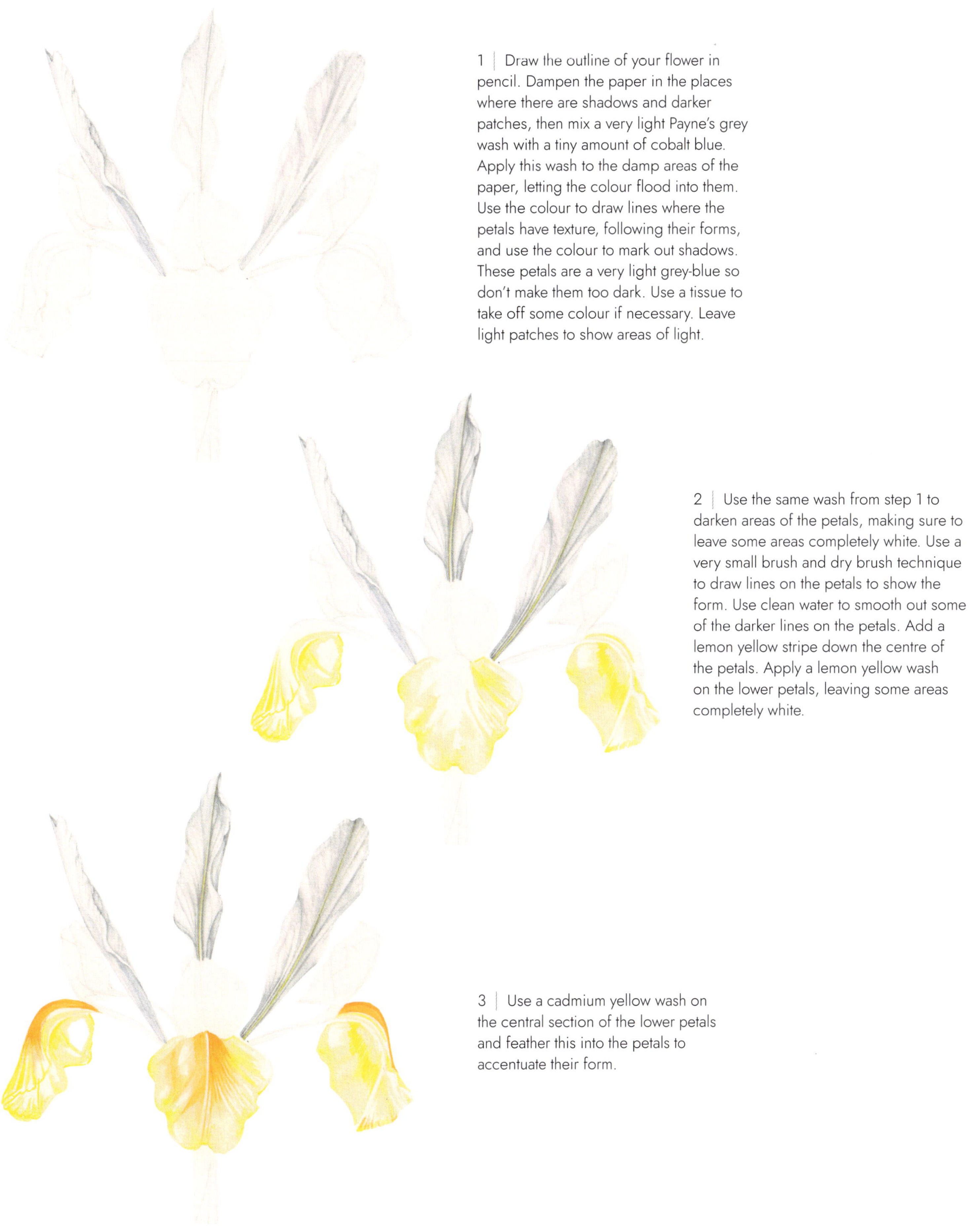

1 | Draw the outline of your flower in pencil. Dampen the paper in the places where there are shadows and darker patches, then mix a very light Payne's grey wash with a tiny amount of cobalt blue. Apply this wash to the damp areas of the paper, letting the colour flood into them. Use the colour to draw lines where the petals have texture, following their forms, and use the colour to mark out shadows. These petals are a very light grey-blue so don't make them too dark. Use a tissue to take off some colour if necessary. Leave light patches to show areas of light.

2 | Use the same wash from step 1 to darken areas of the petals, making sure to leave some areas completely white. Use a very small brush and dry brush technique to draw lines on the petals to show the form. Use clean water to smooth out some of the darker lines on the petals. Add a lemon yellow stripe down the centre of the petals. Apply a lemon yellow wash on the lower petals, leaving some areas completely white.

3 | Use a cadmium yellow wash on the central section of the lower petals and feather this into the petals to accentuate their form.

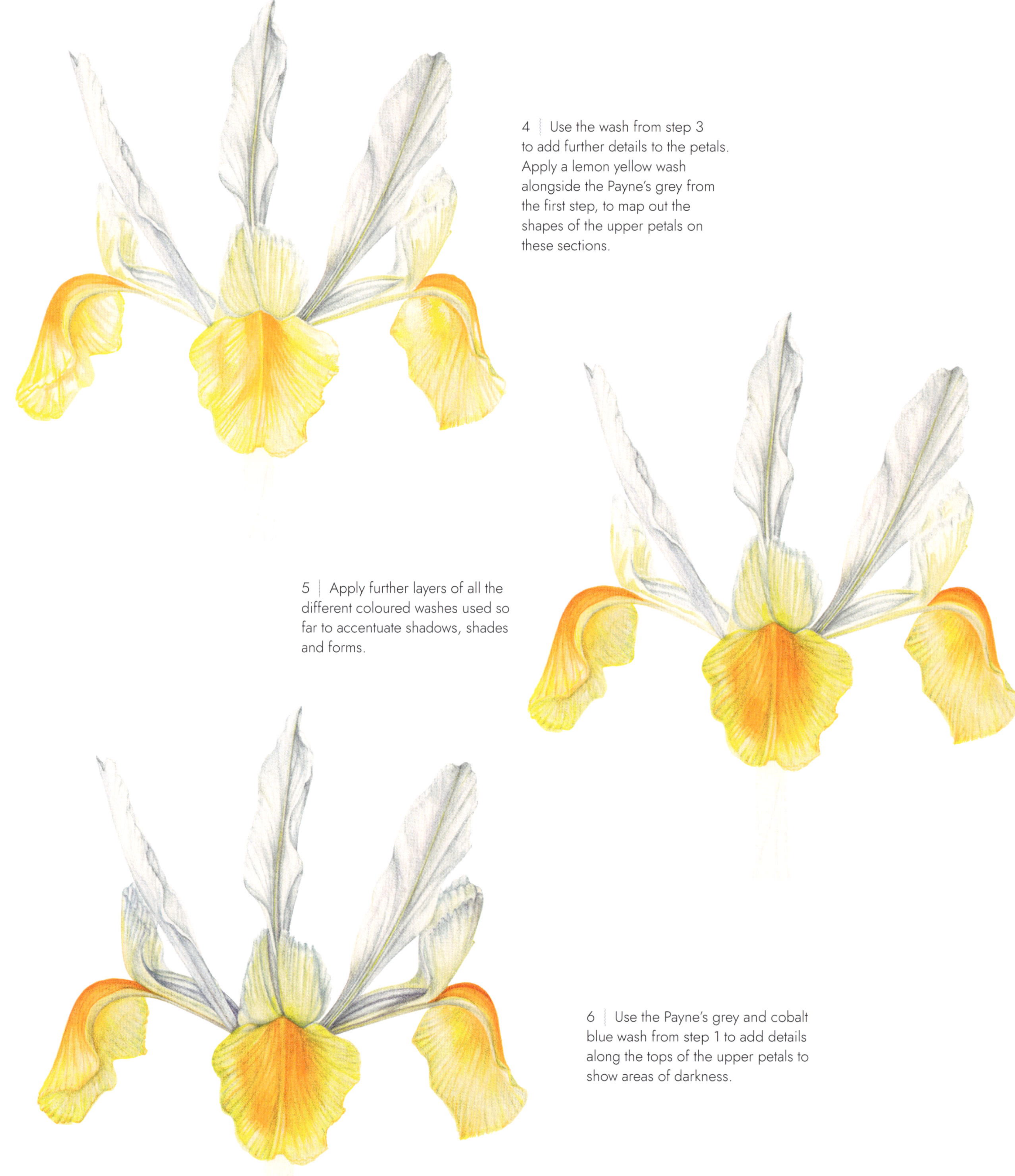

4 | Use the wash from step 3 to add further details to the petals. Apply a lemon yellow wash alongside the Payne's grey from the first step, to map out the shapes of the upper petals on these sections.

5 | Apply further layers of all the different coloured washes used so far to accentuate shadows, shades and forms.

6 | Use the Payne's grey and cobalt blue wash from step 1 to add details along the tops of the upper petals to show areas of darkness.

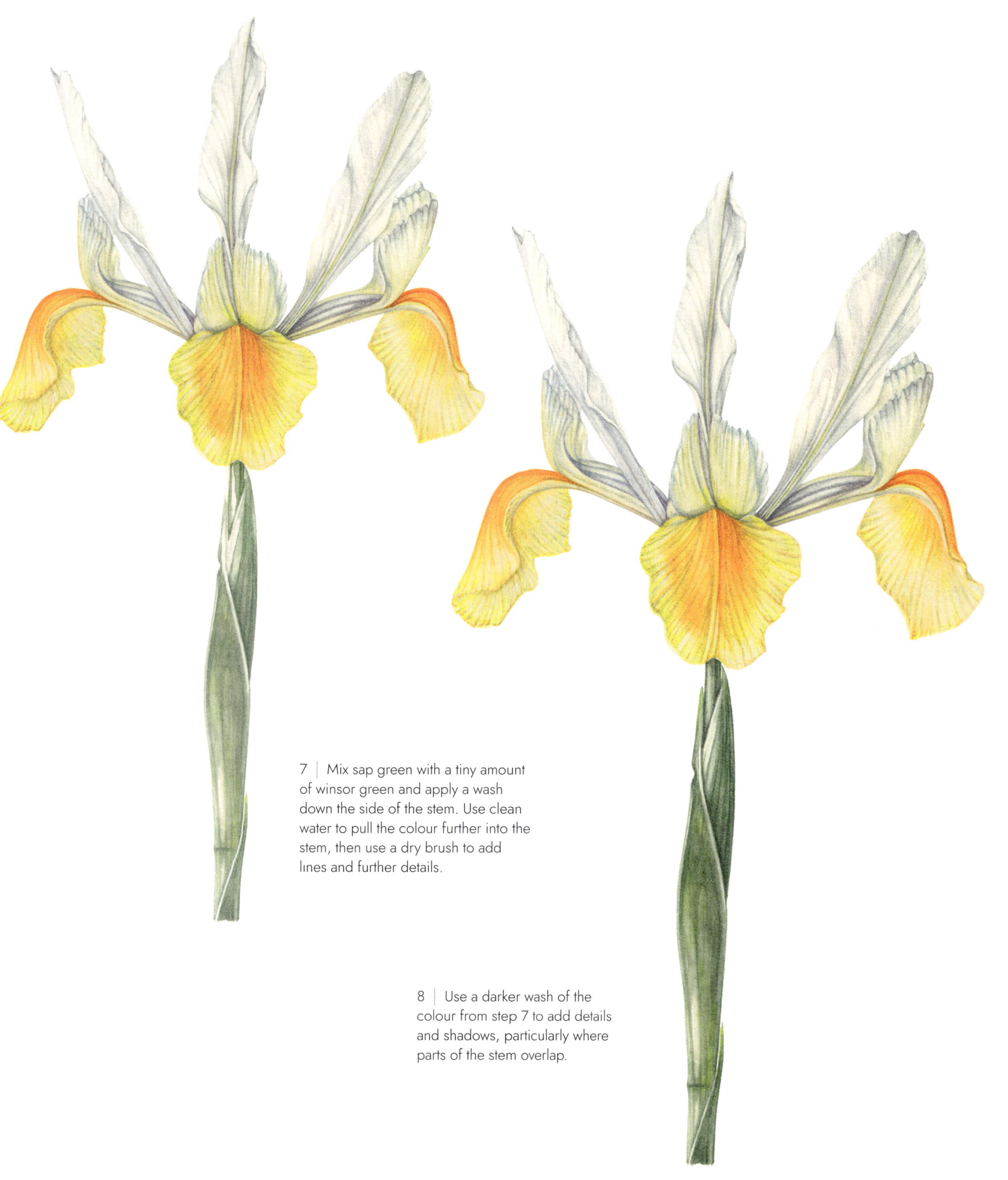

7 | Mix sap green with a tiny amount of winsor green and apply a wash down the side of the stem. Use clean water to pull the colour further into the stem, then use a dry brush to add lines and further details.

8 | Use a darker wash of the colour from step 7 to add details and shadows, particularly where parts of the stem overlap.

New Mexico Grasses

by Vicki Malone

Where grasses are concerned, most people mow them, walk on them and walk past them, but never look at them. Yet grasses can be great fun to paint. Vicki Malone was brought up in Kansas and her appreciation for the prairie grasses of Midwest America has remained, although she now lives on the east coast of North America. As an artist, she delights in the challenge of playing with compositions of New Mexico grasses.

She has brought to our attention one of the most fragile habitats in North America. Much of it has disappeared under urban sprawl or land exploitation. Modern range farming practices, however, are now being deployed to remedy the situation.

ABOUT THE ARTIST

Vicki Malone studied Fine Art at the University of Kansas, and has a certificate of Botanical Art from the Corcoran College of Art and Design in Washington, D.C. She has since worked as an editorial assistant, teacher and artist. She now lives in Maryland.

Her art has been shown at the Corcoran Gallery of Art, Monticello Library, United States Botanic Garden, Delaware Art Museum, and galleries in the Washington, D.C. area. Her work has appeared in *Native Plants of the Mid Atlantic* and *America's Flora*. She exhibits with the American Society of Botanical Artists and is a member of the Guild of Natural Science Illustrators and the Botanical Art Society of the National Capital Region.

The value of grasses

Just as bison, then cattle, roamed the rangelands in search of grassy feed, so Native Americans knew and loved each type of grass for their value as medicine, food or shelter. In New Mexico, grasslands comprise the most extensive habitats of the rangelands, with three or four species of grass dominating a site.

Painting grasses requires a little knowledge of how they are structured. Here, we see culms bent at their nodes, tillers thrown out freely, and a pleasing arc provided by four inflorescences of two different types of grass. We are not given the names of the grasses, but this does not matter at one level, for the picture begs the viewer to take more interest in grasses. The heavy inflorescences are flanked by delicate-looking ones, like supporters on a heraldic coat of arms. The playful curlicues of tillers and sheathing leaves betrays the gravitas of the ripe florets above. Both contrasting types allude to the immense diversity of this plant form.

Vicki painted her grasses mostly in dry brush, a technique that provides ideal control for both long thin structures, and complex detailed ones. On such a big picture, any shake of the hand would immediately be noticeable on a long culm.

RIGHT Grasses provide an elegant and interesting subject, especially when comparing several on one page. Size: 61 × 45.5cm (24 × 18in), watercolour on Arches Aquarelle HP 300gsm.

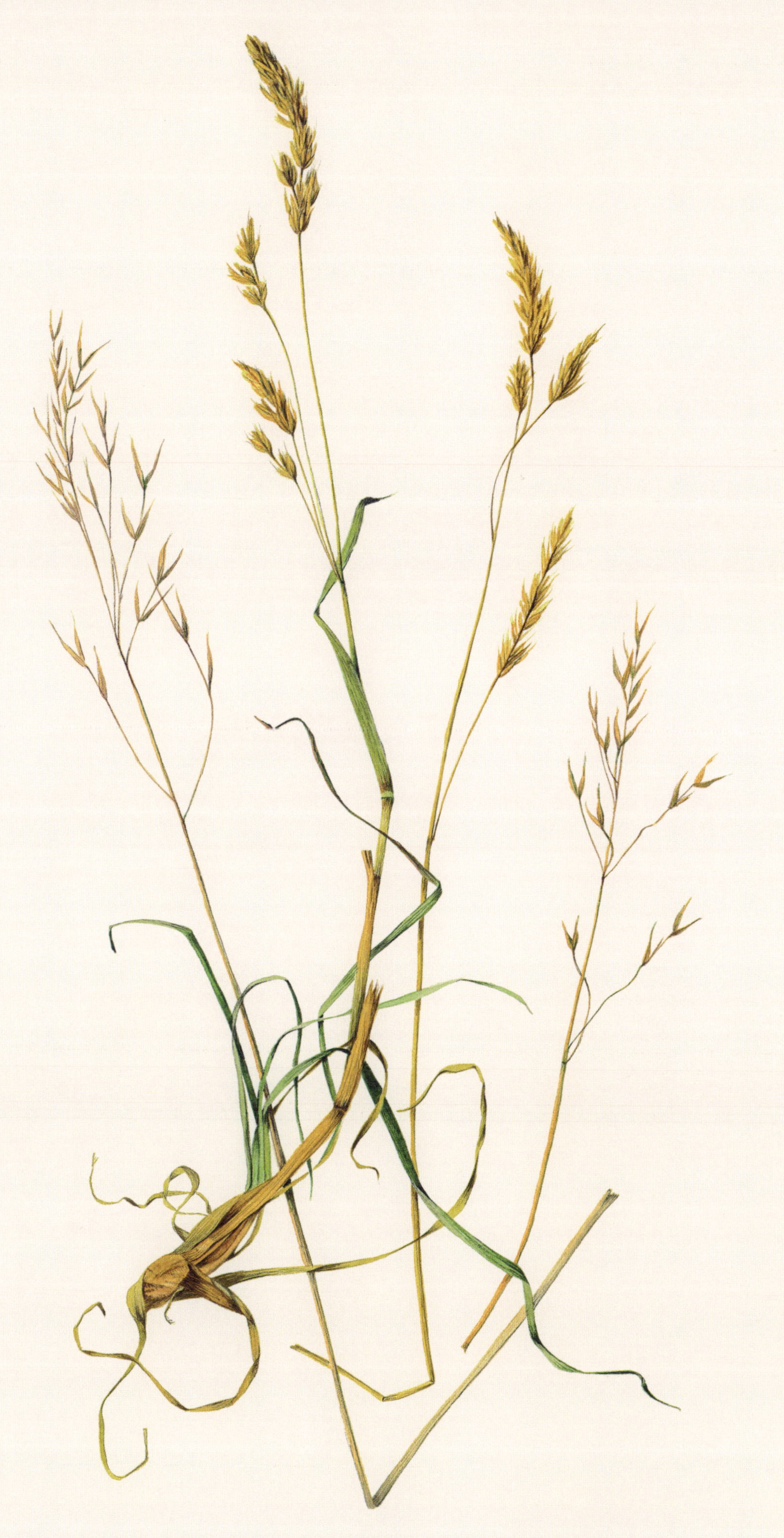

Male flower parts

The androecium is the collective term for the stamens, which produce pollen. They comprise the next, or third whorl.

Each stamen consists of a filament and an anther. An anther consists of pollen sacs, usually paired, called thecae. The number of stamens usually matches the floral parts, or is a multiple of them, though there may be too many to count.

Stamens are often joined up in a tube, or nearly so, to protect the stigma within until it is ready to receive pollen from another plant. Gentians, after pollination, clearly show this feature.

Studying the colour, number, appearance and placement of stamens demonstrates good observational skills. Anther shape, direction and how they open are other features worth noting.

 Dancing lady ginger (*Globba winitii*) by Sunanda Widel belongs to the ginger family, so look for only one fertile stamen (the others are reduced or modified) which, in this case, is long and arching.

- How do the stamens line up with the adjacent sepals and petals? Note that they are spirally arranged in buttercups.

- Are the stamens/anthers extrorse (facing outwards) or introrse (facing inwards)?

- Are they attached to a petal, or at its base?

- Are they of different lengths? Didynamous or tetradynamous?

- Is there more than one whorl?

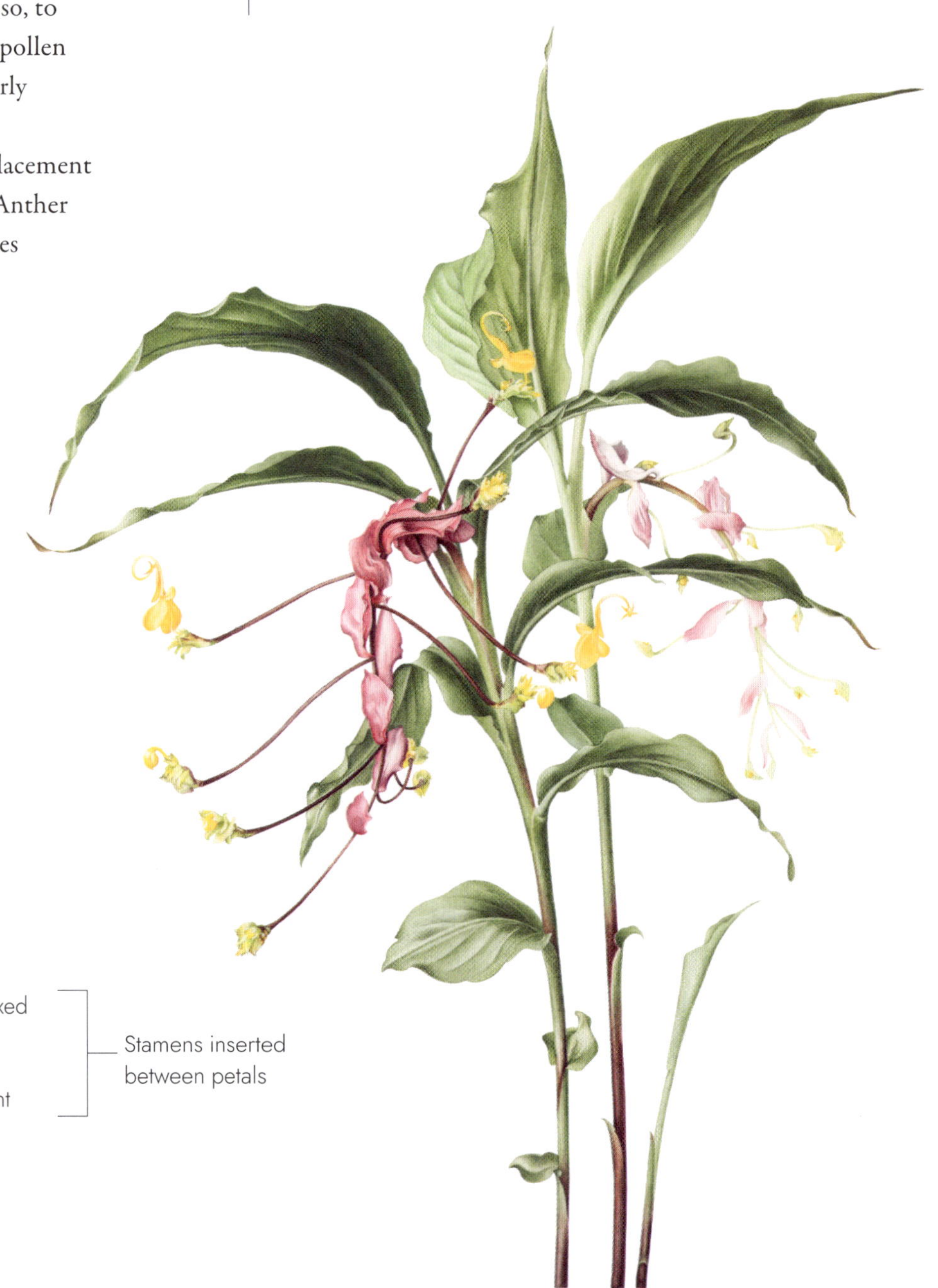

MALE FLOWER PARTS

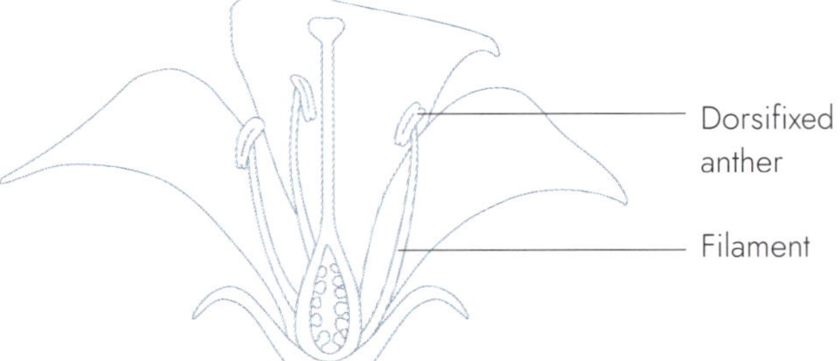

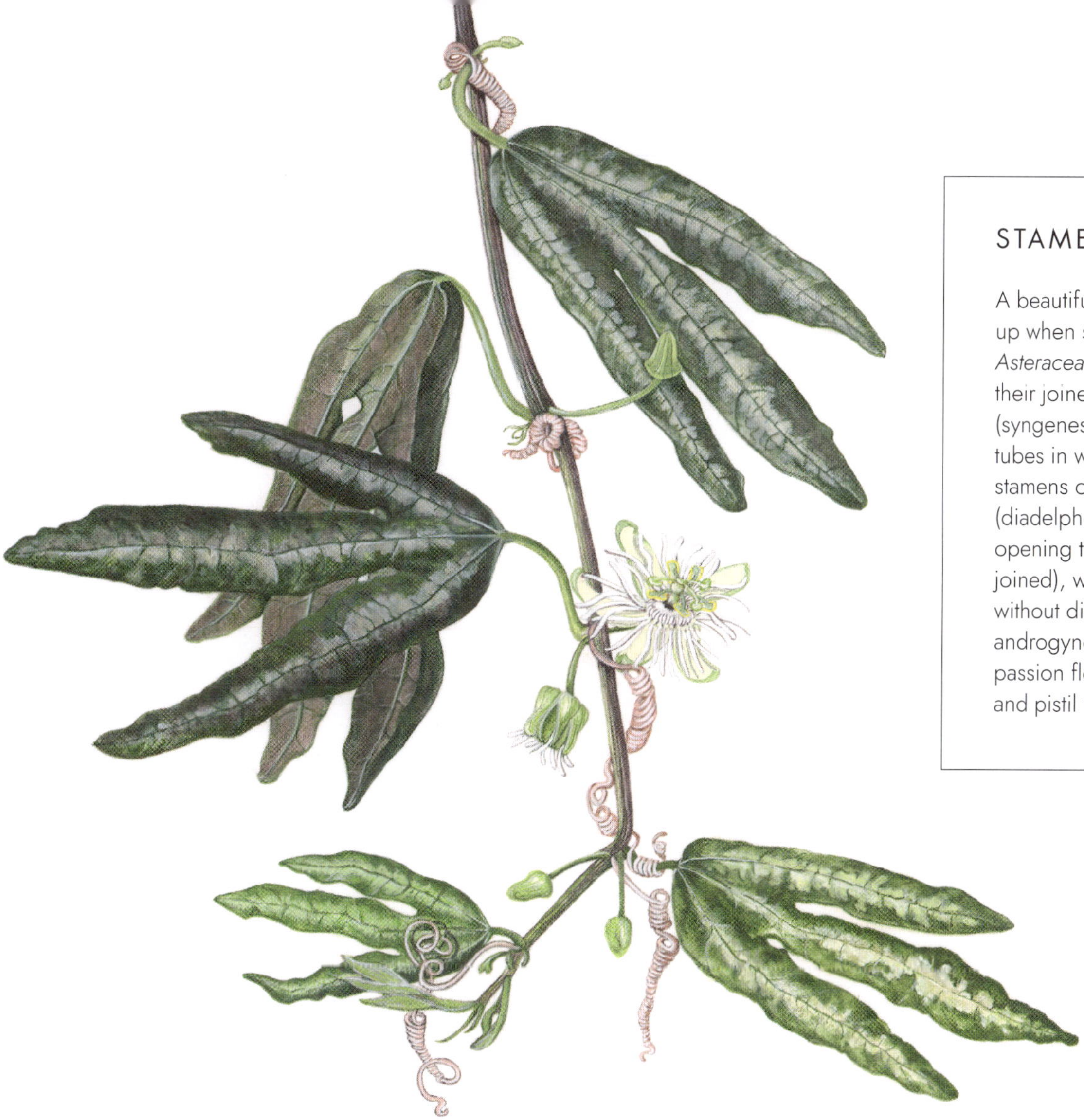

LEFT Three-banded passion
flower (*Passiflora trifasciata*)
by Tina Bone. The towering
androgynophoro of passion
flowers holds a prominent
display of both male and
female parts.

In 'male' only flowers, the anthers function normally, though the pistil does not. In 'female' only flowers, the anthers (or some of them) may be reduced in size, non-functioning, and are called staminodes.

TYPES OF ANTHER ARRANGEMENTS

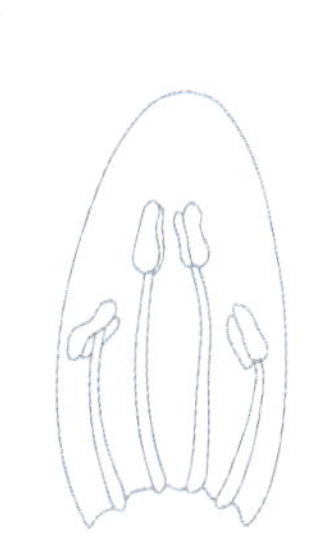

Didynamous anthers

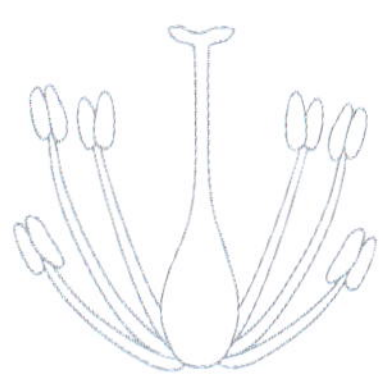

Basifixed anthers

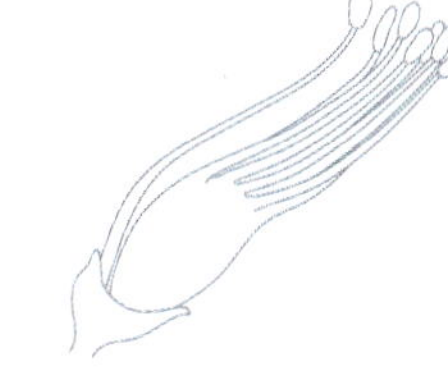

Diadelphous anthers
of a pea

Look closely at the shapes of anthers, their direction and how they dehisce (open). The following terms give a flavour of what to look out for:

- Basifixed: fixed to the filament at the base of the anther.

- Dorsifixed: fixed to the filament in the middle of the two thecae.

- Transverse attachment: fixed to the filament at right angles.

- Poricidal dehiscence: pollen released through pores.

- Valvate dehiscence: pollen released through 'valves'.

- Spurs: appendages to the stamens indicate nectaries or pollination aids.

Female flower parts

The gynoecium (or pistil) forms the innermost whorl of the flower and protects the ovules inside for fertilization and maturation. The word carpel is also used for the basic unit. Since most flowers have more than one carpel, often fused together and also called gynoecium, it is not always easy to discern; the number of styles and compartments (locules) within give a clue. A carpel consists of the ovary, style and stigma. The ovary is the part that develops into fruit. It's important to note whether the ovary is superior (set at the same level as the petals, like lobelia), inferior (set below the petals, like a fuchsia) or inside a hypanthium (typical of roses).

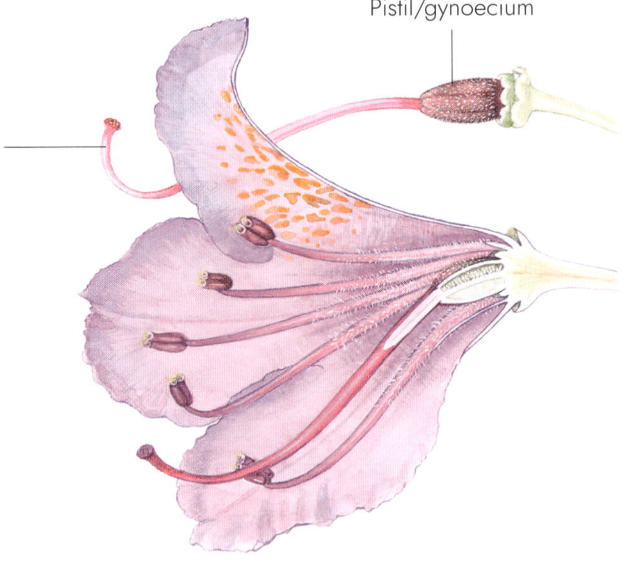

LEFT *Lobelia bridgesii* detail by Hülya Korkmaz. The superior ovary is enclosed by fused stamens and sits at the same level as the petals (not shown).

LEFT *Aconitum carmichaelii* detail by Roger Reynolds, showing apocarpous fruit with carpels that are free from each other, except at the base.

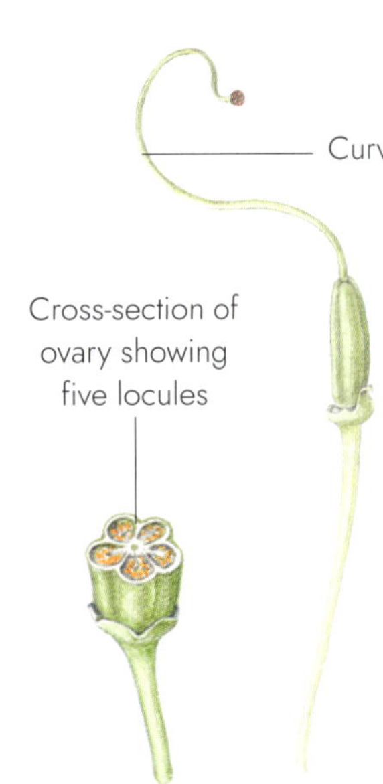

RIGHT *Rhododendron ponticum* detail by Janet Dyer. This has a superior ovary of five fused carpels shown in the five locules containing tiny seeds.

FEMALE FLOWER PARTS

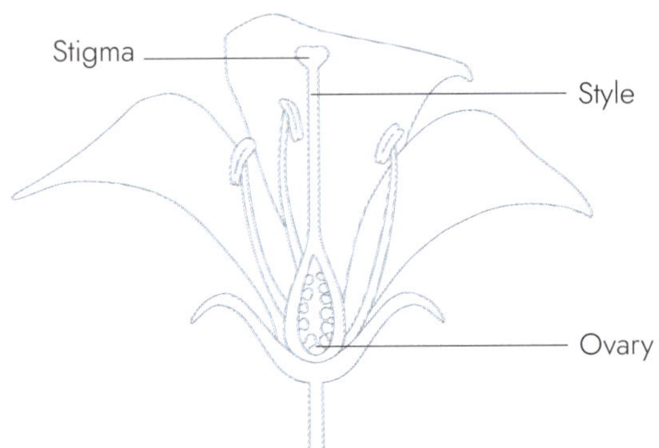

LEFT- OR RIGHT-HANDED FLOWERS

A few plants across a range of families are left- or right-handed, a phenomenon called enantiostyly or 'handedness'. The north American nightshade, *Solanum rostratum*, has flowers with styles deflected either to the left or to the right. The phenomenon can occur on different flowers on one plant, or in separate plants. It is usually associated with unequal stamen length and is linked with efficient cross-pollination.

If there is only one carpel – for example, in sweet peas –
it's monocarpous. If there is more than one free carpel
(separated from each other), it's apocarpous, as seen in
hellebores and buttercups, for example. Most flowers have
fused carpels and so are syncarpous. The number of carpels
follows the floral number, so there may be five, three or two
carpels often united in one gynoecium, or separated.

Inside the ovary there are one or more cavities, called
locules, in which the developing ovules sit. The locules may
represent separate carpels if they are united, and the position
of the ovules relates to each of the carpels.

A transverse section cut through the widest part of the
ovary shows the number of locules (whether syncarpous or
not), along with placentation, the arrangement of the ovules.
Ovules may adhere to a central column (axile), to the
periphery (parietal) or be free.

Where there is only one locule, there may only be a single
ovule positioned at the top (apical), or at the bottom (basal).
In pea pods, which have only one locule, the 'peas' are ovules
attached to one of the margins of the pod.

The ovary is linked to the stigma, which receives the male
pollen by way of the style. The number of style arms relates to
the number of carpels below. There are terms for different
forms of style, for example, flabellate, conduplicate, filiform
and so on. The stigma often have different descriptive terms,
such as bifid (two branches), capitate (like a pin head), lobed
(with lobes), and so on.

Look out for different colours of the different parts, hairs
which might be on some or many of the parts, and take note
of the ovary shape and its bulges, which point to the number
of carpels. For example, there are three bulges on a monocot
ovary, four sides on a euonymus ovary (see page 192).

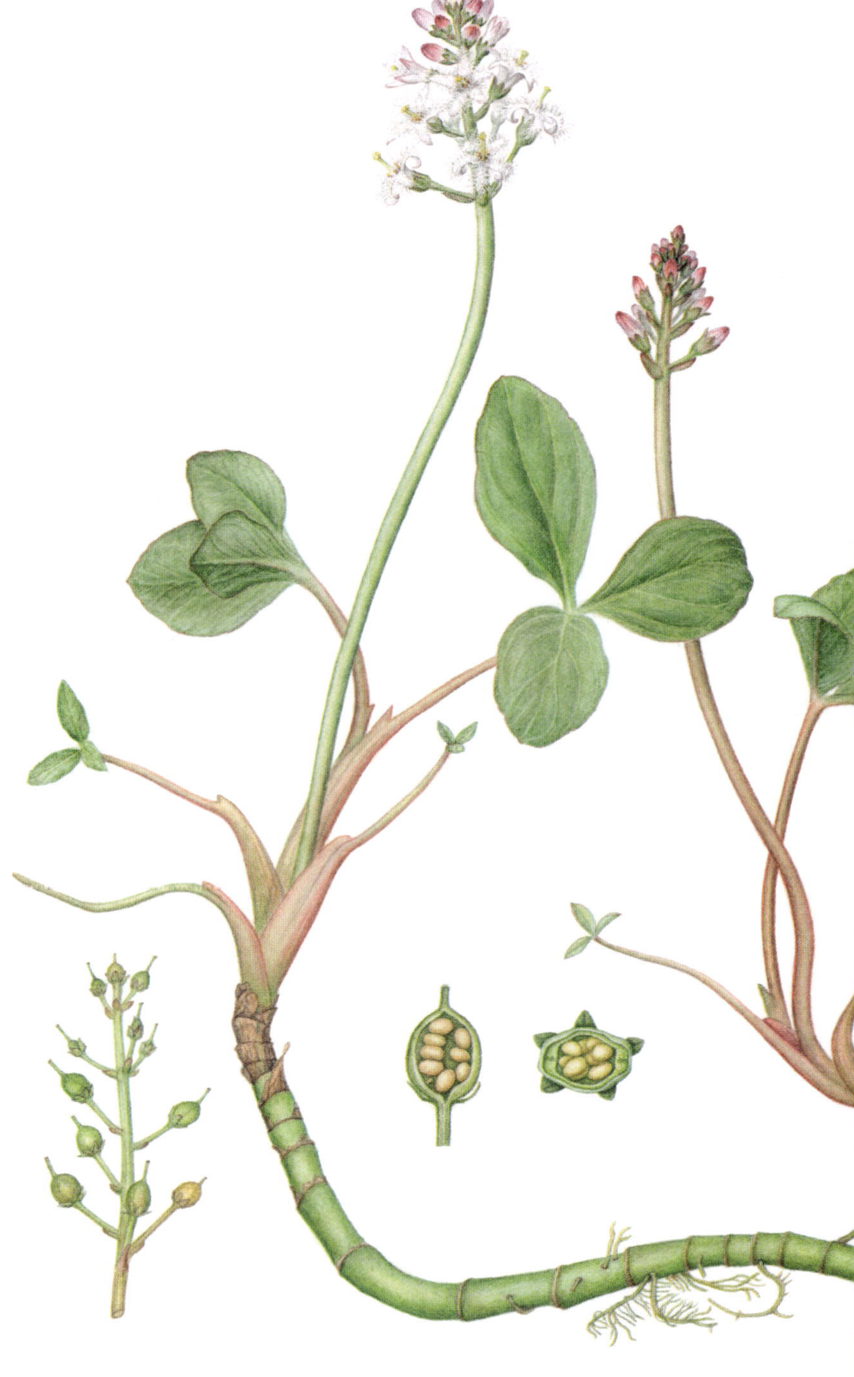

ABOVE Bogbean (*Menyanthes
trifoliata*) by Linda Pitkin. The
bogbean has a single locule
containing ovules, which are
attached to the periphery of the
gynoecium. The five points in
the transverse section provide
a clue that it may be formed
from five fused carpels.

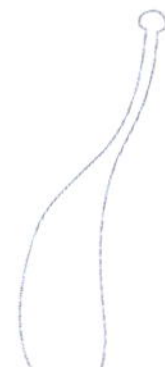
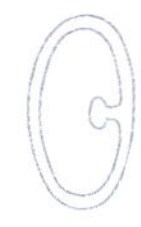

Monocarpous (free)
marginal placenta

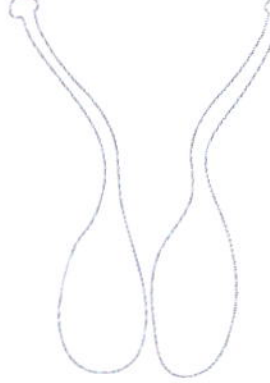
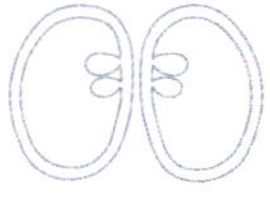

Apocarpous (free)
marginal placenta

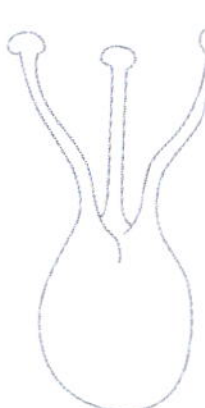
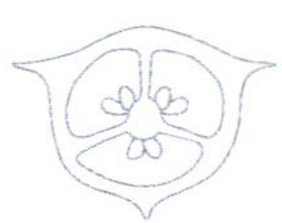

Syncarpous (fused)
axilic placenta

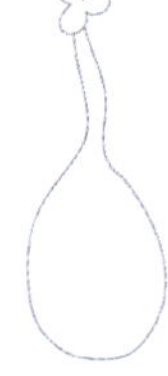
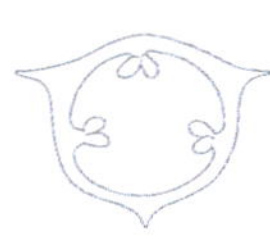

Syncarpous (fused)
parietal placenta

Painting a dahlia

Dahlias are from the *Compositae* family, which means the actual flowers, which are abundant and tiny, are found in the centre of what you might consider 'the flower' — the large petal-like structures are called ray florets. The dahlias in this painting were selected for their magnificent colours, which blend into each other on the ray floret. The colours are vibrant and dahlia petals are thick and heavy, so it is possible to apply the paint in a bold fashion with quite a dark wash from the start, rather than building up with many light washes. In contrast, the centre, which contains the actual flowers, is quite detailed and needs to be painted carefully with a very small brush to try and replicate the details and the pattern, which can be seen when viewed as a whole.

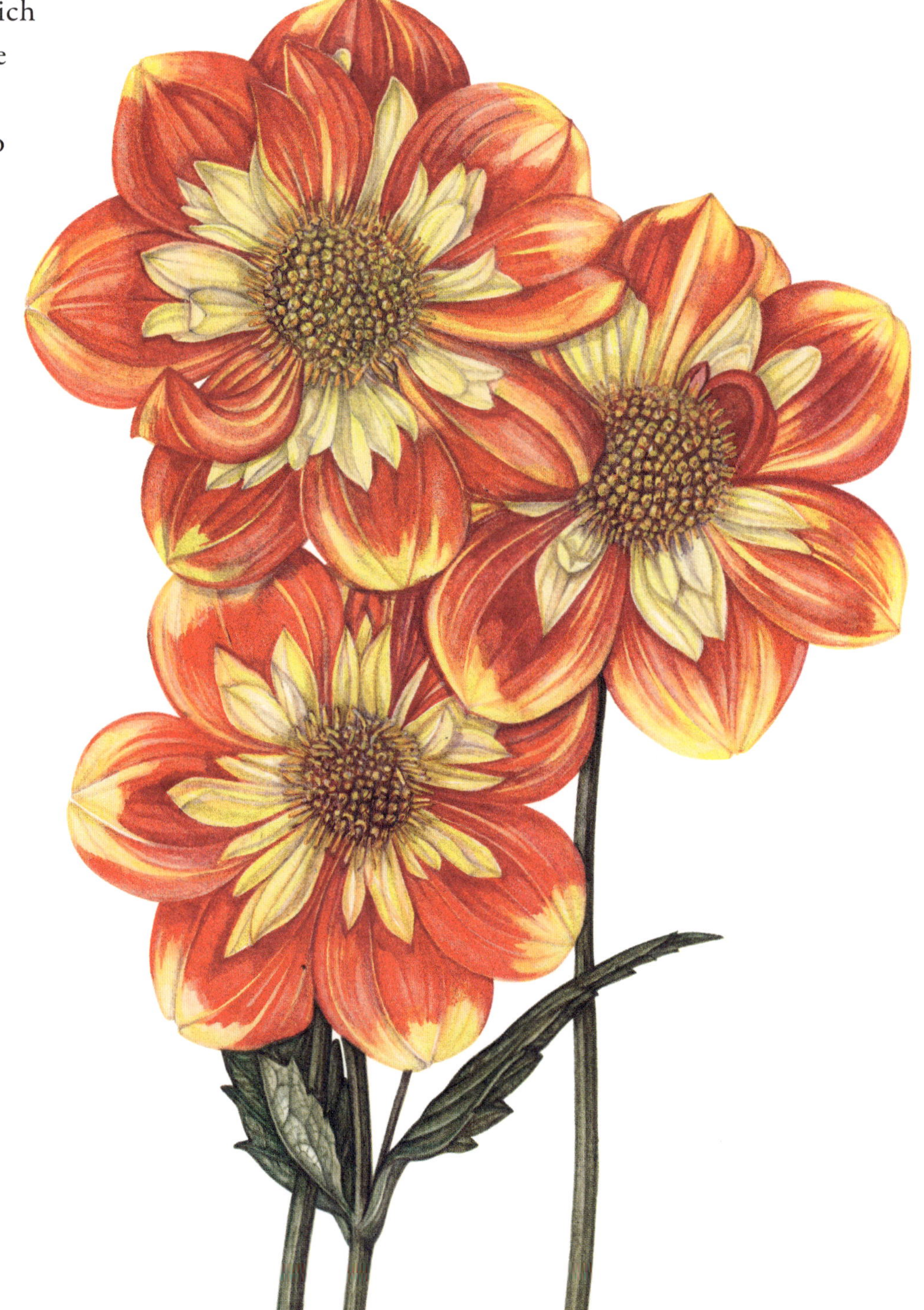

PAINTS USED:

Cadmium red

Lemon yellow

Cadmium yellow

Permanent alizarin crimson

Payne's grey

French ultramarine

Winsor violet (dioxazine)

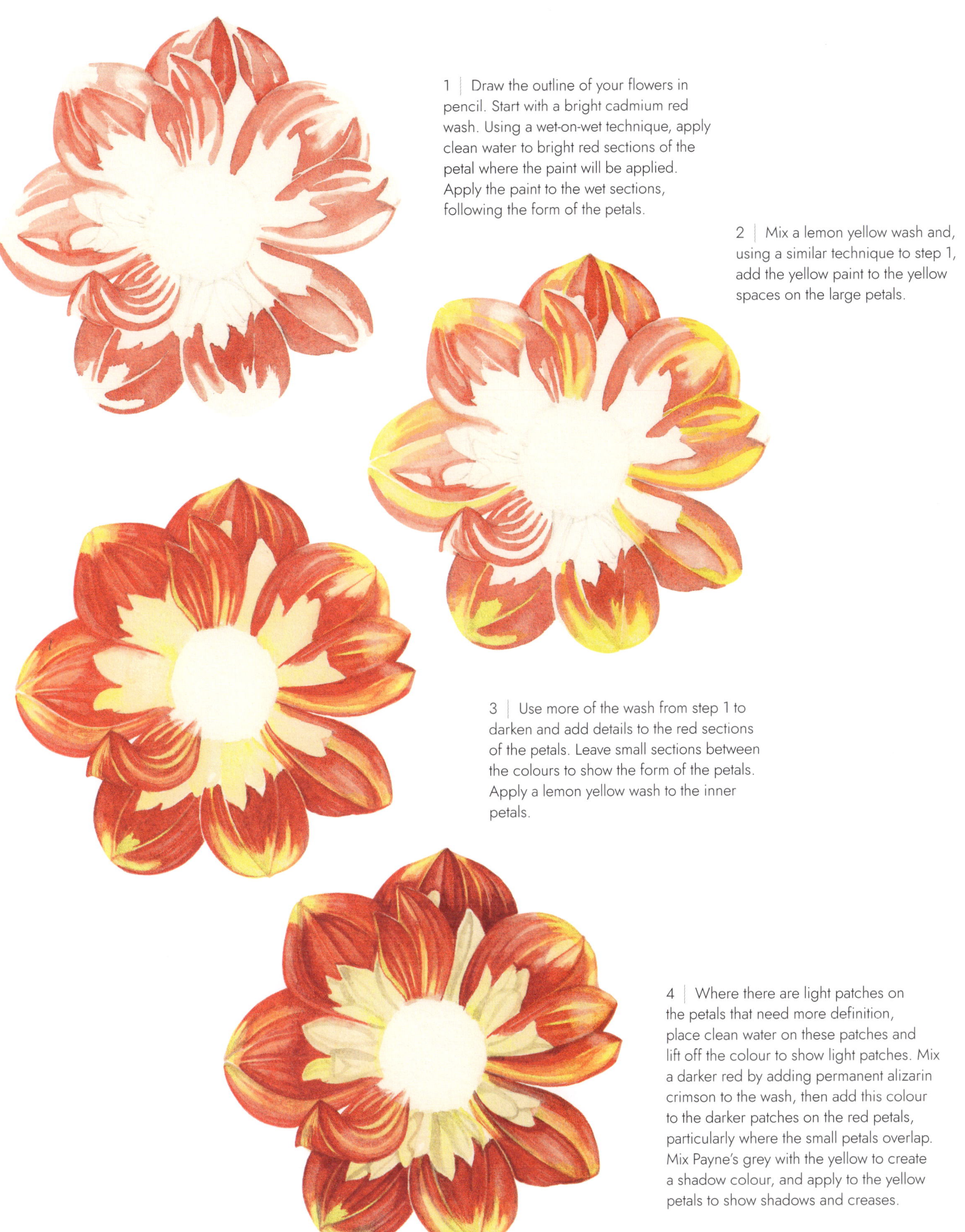

1 | Draw the outline of your flowers in pencil. Start with a bright cadmium red wash. Using a wet-on-wet technique, apply clean water to bright red sections of the petal where the paint will be applied. Apply the paint to the wet sections, following the form of the petals.

2 | Mix a lemon yellow wash and, using a similar technique to step 1, add the yellow paint to the yellow spaces on the large petals.

3 | Use more of the wash from step 1 to darken and add details to the red sections of the petals. Leave small sections between the colours to show the form of the petals. Apply a lemon yellow wash to the inner petals.

4 | Where there are light patches on the petals that need more definition, place clean water on these patches and lift off the colour to show light patches. Mix a darker red by adding permanent alizarin crimson to the wash, then add this colour to the darker patches on the red petals, particularly where the small petals overlap. Mix Payne's grey with the yellow to create a shadow colour, and apply to the yellow petals to show shadows and creases.

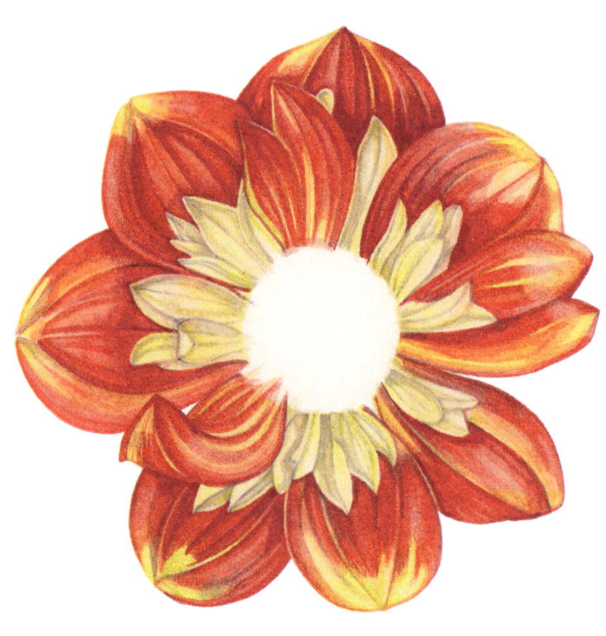

6 | For the centre, use lemon yellow and cadmium yellow, painting lemon yellow circles first then surrounding these with cadmium yellow. Mix a lemon yellow and violet wash for shadows and paint in between each stamen.

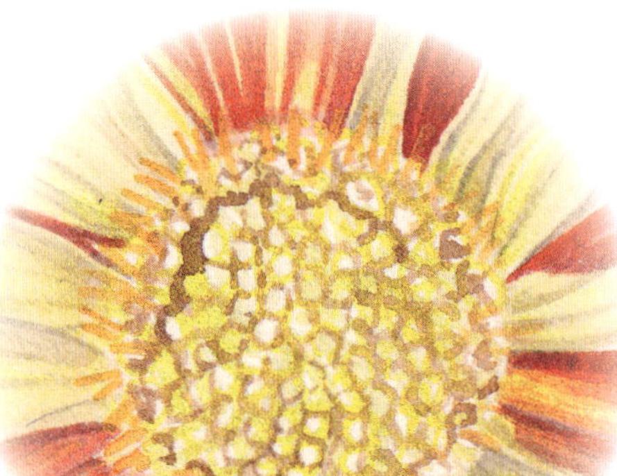

7 | Add further shadow patches around the centre structures, brightening the centres with another wash of lemon yellow. Use cadmium yellow on the other side of the yellow parts and add darker shadows behind the stamens on small petals.

5 | Use the shadow colour on the small yellow petals to enhance the shadows, then use a thin brush to draw small grey lines on the petals to show the central creases. Add lemon yellow patches.

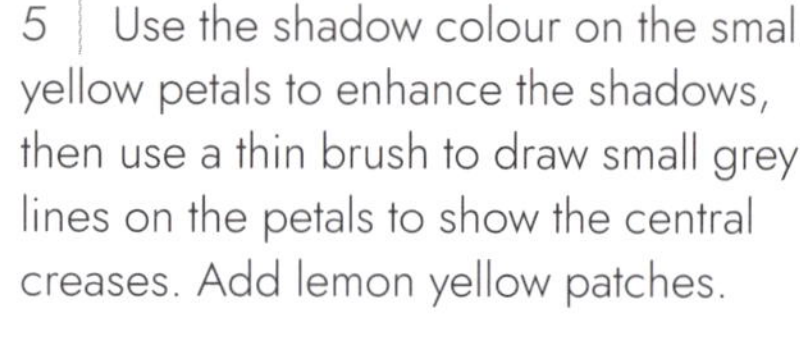

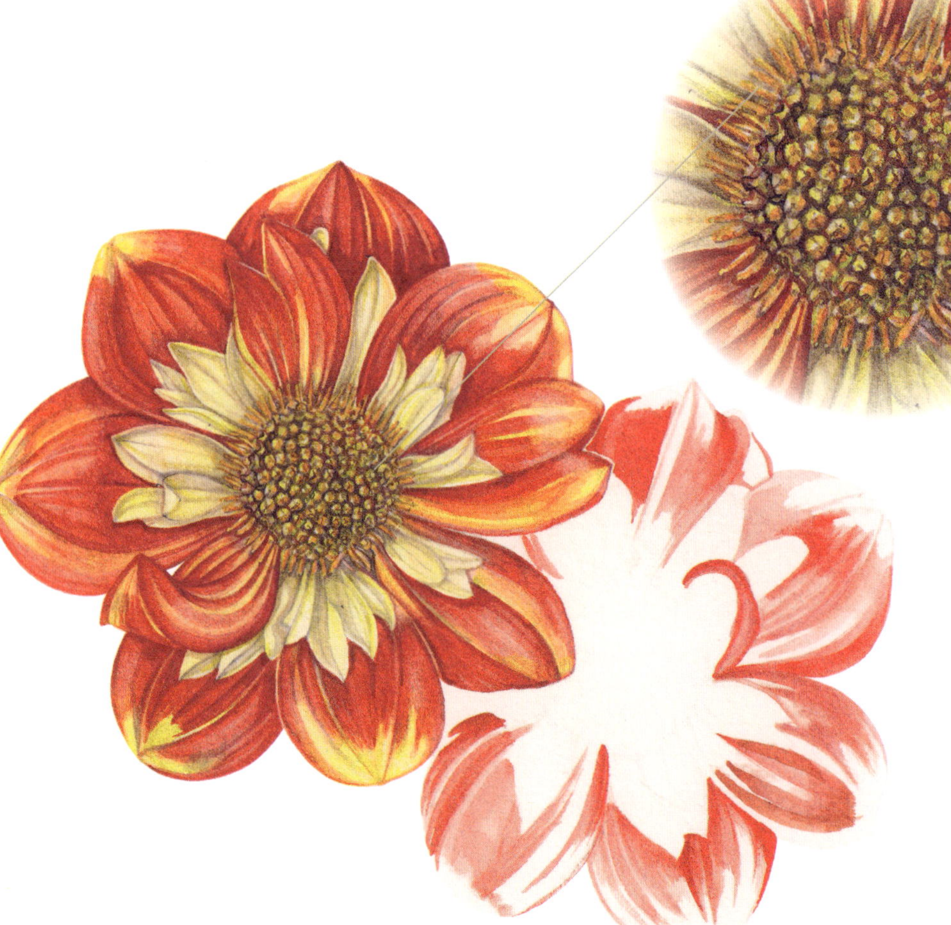

8 | Add alizarin crimson to the shadow colour in the centre to enhance shadows. Start the next flower in the same way using a cadmium red wash and repeat steps 1—8 on both flowers.

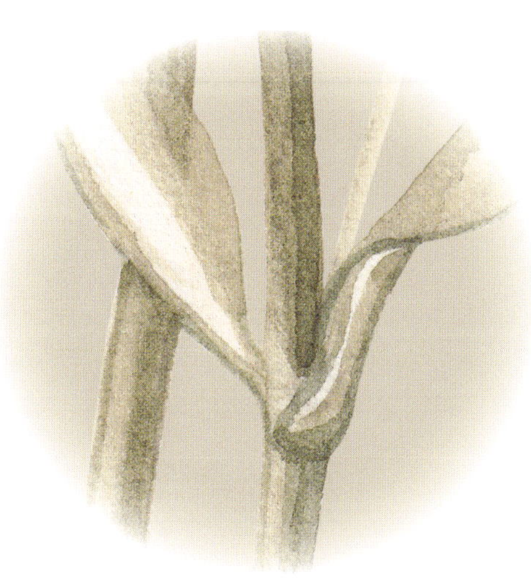

9 | Mix sap green and olive green and apply a first wash to the stem, making one side darker.

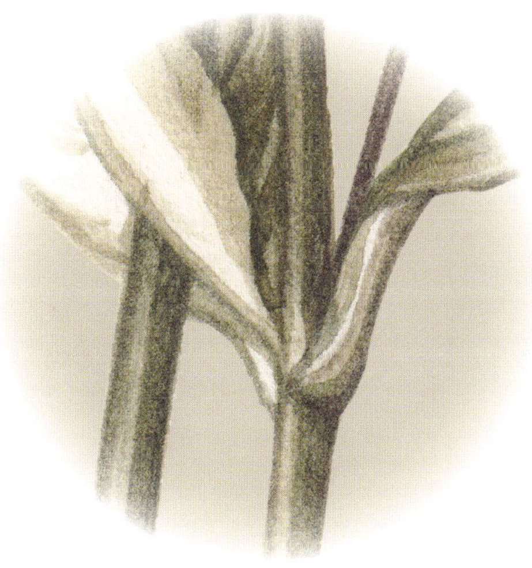

10 | Using the wash from step 9, apply another layer of paint, picking out details such as the veins.

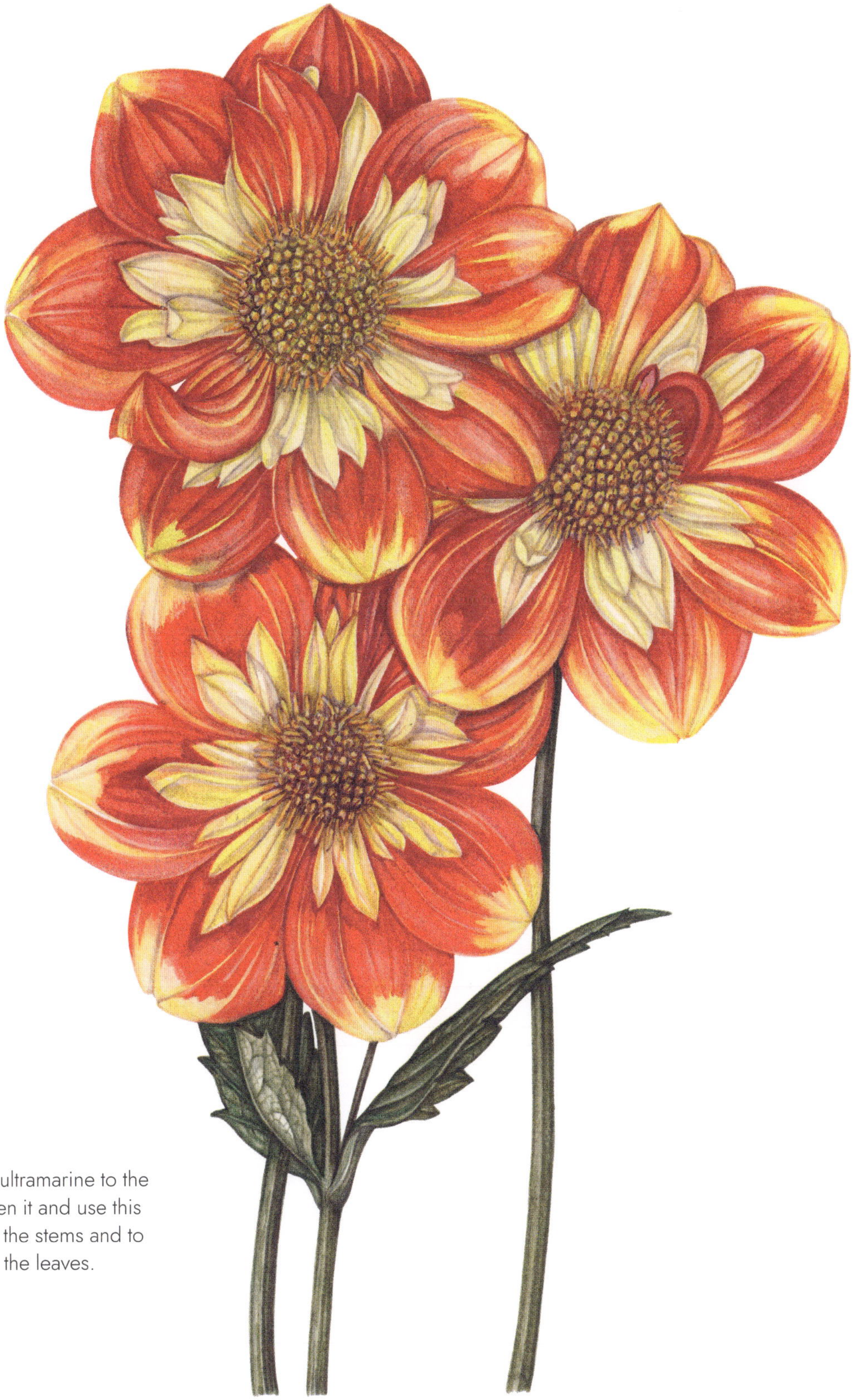

11 | Add French ultramarine to the green mix to darken it and use this down one side of the stems and to highlight veins on the leaves.

Saw banksia (*Banksia serrata*)
by David Reynolds

Saw banksia, or the 'wiriyagan' tree was one of the original banksias collected in Botany Bay in 1770 by Joseph Banks during Captain James Cook's expedition to Australia. It was likely sketched by Sydney Parkinson, the expedition artist, who died on the voyage. Banksias are members of the protea family and the adult trees are fairly fire resistant. The long, saw-like dark green leaves, clustered on the upper branches of the tree, are topped by spikes comprising hundreds of flowers. The fruit, called a 'cone', has up to 30 follicles, each with one or two seeds. Today, it is a familiar sight in many gardens and public parks in Australia.

Exquisite complexity

The viewer is drawn to the sheer complexity of this plant at every level — as was David Reynolds. The banksia's huge flowers and strong shapes were an irresistible attraction to him at one level. At another level, as an experienced botanical artist, he wanted to tackle the myriad flowers on the spike, and the many serrated leaves crowded underneath. David's background in design and storytelling shows, for this is what we get: a five-part story of a symbolic Australian plant from genesis to decay. He has given us several flower spikes, but only one dominates. This is the virtuoso piece. The grey, yellowish flowers are each carefully painted against judicious points of contrast behind them. The flowers are a tubular set of four tepals, each containing four anthers. What one mainly sees is a style that extends beyond the other parts. David is not afraid to spend hours on one small section, painting in small details as clearly and as accurately as possible, as 'accurate realism' is important to him. But precision on a painting like this would be lost if it were not for a balance in the overall colour. Only one feature leads the eye in — the other features, the dried old flower, the two fruiting cones and a small spike of buds take their place discreetly behind. All the parts are contained within an oval, which is slightly, and appealingly, tilted.

RIGHT The intricate flower heads and architectural cones make a rewarding subject. Size: 63 × 59cm (24³/₄ × 23¹/₂in), watercolour on 300gsm HP paper.

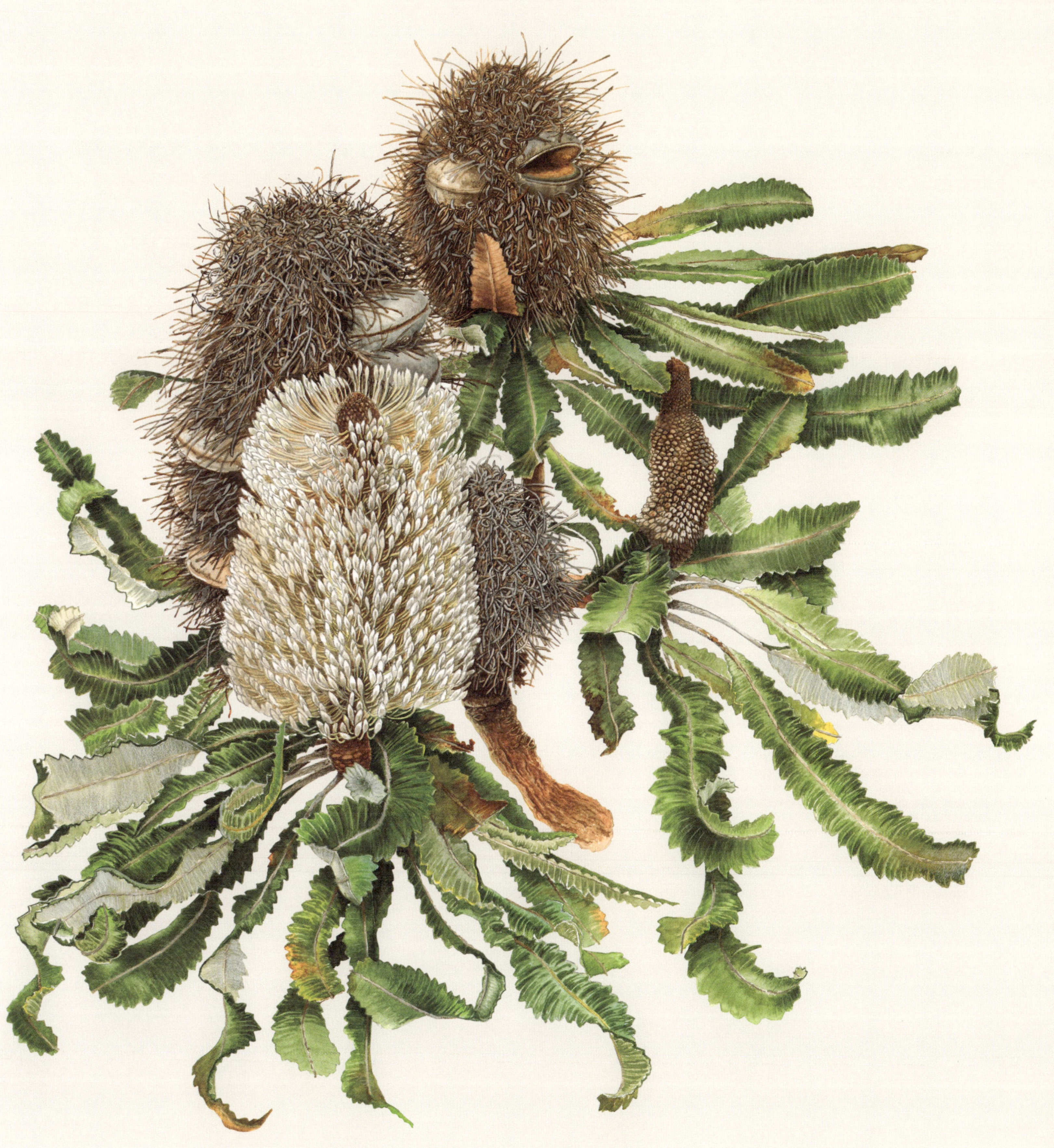

Nectaries

Where they exist, nectaries hold a sweet liquid with which
to reward pollinators or animals that protect the plant.
They can look like another floral part, be modified from an
existing structure, or simply be very well hidden. Nectaries
are usually positioned where a pollinator can best pick up
pollen whilst taking its reward. Often this means the nectary
is deep in the flower's structure.

Hellebore nectaries are modified petals and as a result
are visible as green rhomboid structures between the
petaloid calyx and the whorl of stamens. Buttercups, in the
same family, secrete their nectar from a small pocket at the
base of each of their five petals. Nectaries on other plants
may be much more discreet, or not needed.

Nectary spurs are often modified petal parts, such as
the hollow spurs of many orchid genera. Many violas are
defined by their spurs. *Viola tricolor* has a spur that is longer
than the sepals. It extends backwards and two stamens
secrete nectar into it. Columbines (*Aquilegia* spp.) have
five spurs formed by each petal, which face upwards.
They make an interesting flower study.

The spurs in many delphiniums are formed from
both sepals and petals. One of the sepals forms a spur.
Inside the sepals are four small petals, two of which
also form a nectary spur that extends deep into the
sepal spur above and is protected by it.

Impatiens spp. have spurs derived from the calyx,
where two sepals are modified into a nectary.

Nectaries
(modified petals)

TOP RIGHT *Helleborus × hybridus*
by Carolyn Jenkins. The tubular
green structures on the left
and right are modified petals
containing nectar.

RIGHT *Delphinium wellbyi* by
Sarah Howard, a scented
Ethiopian delphinium which,
like other delphiniums, has
long nectary spurs adding
drama to the corolla.

The nectary spur of garden nasturtiums (*Tropaeolum majus*) is formed from the hypanthium (the floral cup extending from the stem tip to surround the ovary). Fleshy, disc-like structures under the floral parts, the ovary or the stamens are most likely a nectary. They are called nectariferous discs.

A nectary not on a floral part – for example, on leaves, stems and bracts – is called an extrafloral nectary. From their position, extrafloral nectaries do not help with cross-pollination, but they may reward fauna that protect the plant.

DARWIN'S ORCHID

Darwin's orchid (*Angraecum sesquipedale*) from Madagascar is famous for its long nectar spur, which runs up to 35cm (14in). It was named for Charles Darwin because he predicted that the orchid could only be pollinated by a moth with a proboscis as long as the nectar spur, then unknown. Forty years later, a moth with an unusually long proboscis was found in the vicinity, and, indeed, proved to be the orchid's pollinator.

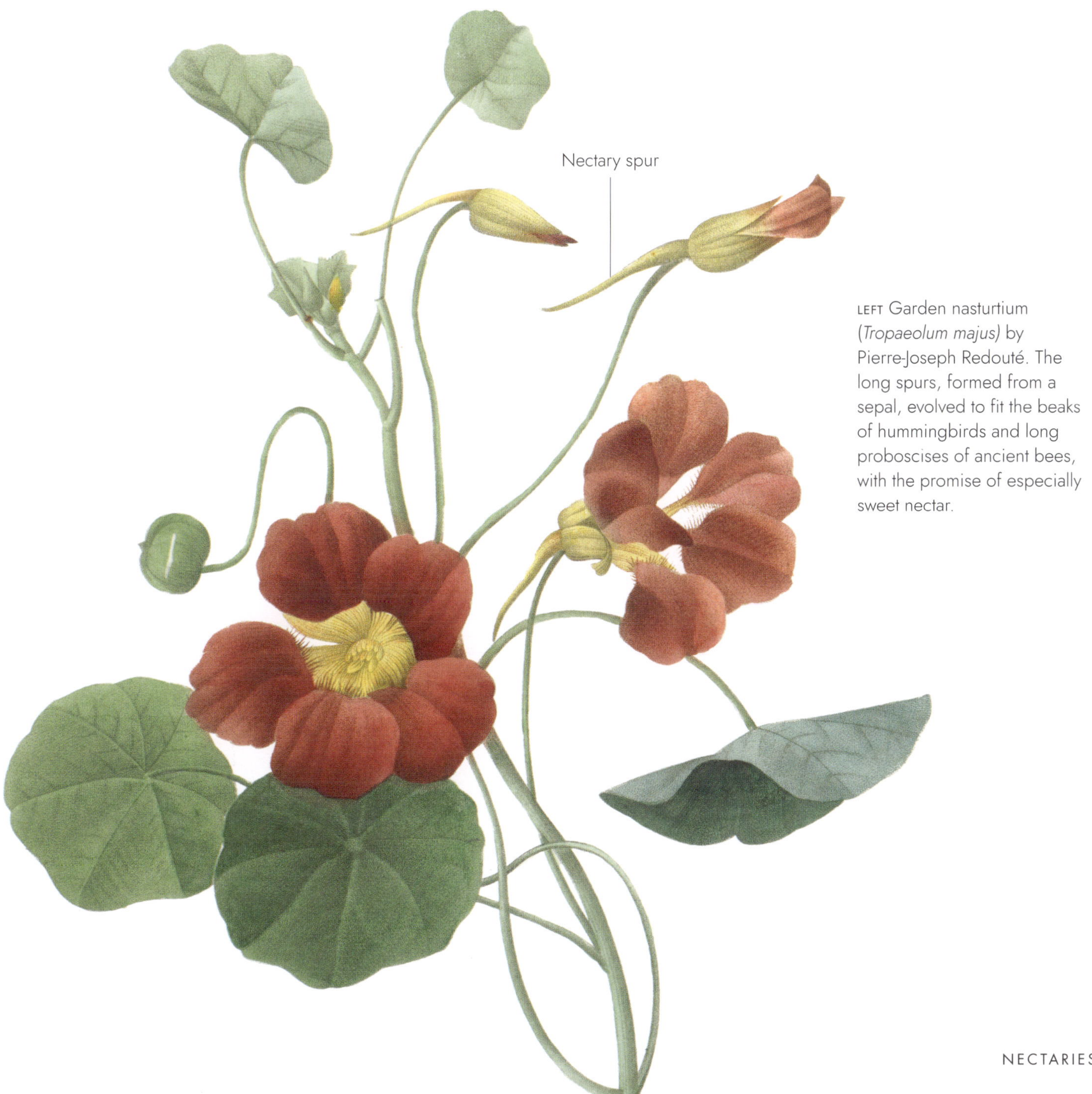

LEFT Garden nasturtium (*Tropaeolum majus*) by Pierre-Joseph Redouté. The long spurs, formed from a sepal, evolved to fit the beaks of hummingbirds and long proboscises of ancient bees, with the promise of especially sweet nectar.

Pollination

The anatomy of both animals and plants has co-evolved to suit each other's needs. Aids to pollination include devices such as colour and shape, patterns and guides, hairs and triggers. It is wise to research the pollination method of the plant you are painting in case you miss a structure that is important.

BELOW Butterfly bush (*Buddleja davidii*) detail by Janet Dyer. This is a seductive source of nectar for butterflies, but the plant is seriously invasive outside its native China.

ABOVE RIGHT European beech (*Fagus sylvatica*) detail by Ingrid Arthur. Many wind-pollinated flowers have separate male and female flowers that face in different directions.

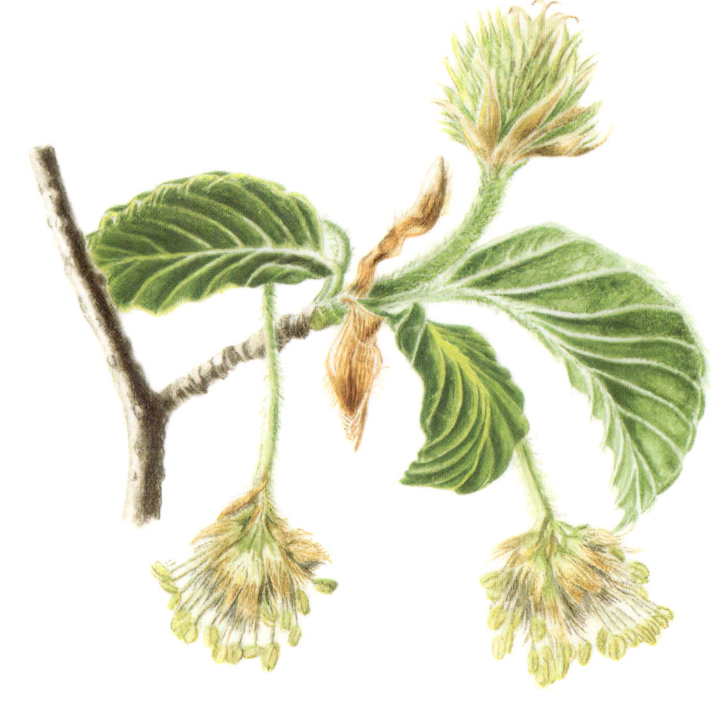

WHAT TO LOOK FOR

- Identify the pollinator, or method of pollination, and you will have a better idea of what structural forms to look out for. If wind pollinated, for example, do the anthers and stigmas protrude, as grass stamens do?

- Be aware of parts and colours that change during the flower's maturation before, during and after pollination.

- Study the number, length and placement of stamens carefully: they do not always produce pollen. They can be different lengths, close together, or wide apart.

- Look for any mechanisms that prevent self-fertilization, such as male catkins, which rely on the wind and usually dangle on the tree, whereas female catkins face upwards.

- Note patterns on the petals that help pollinators to find their reward.

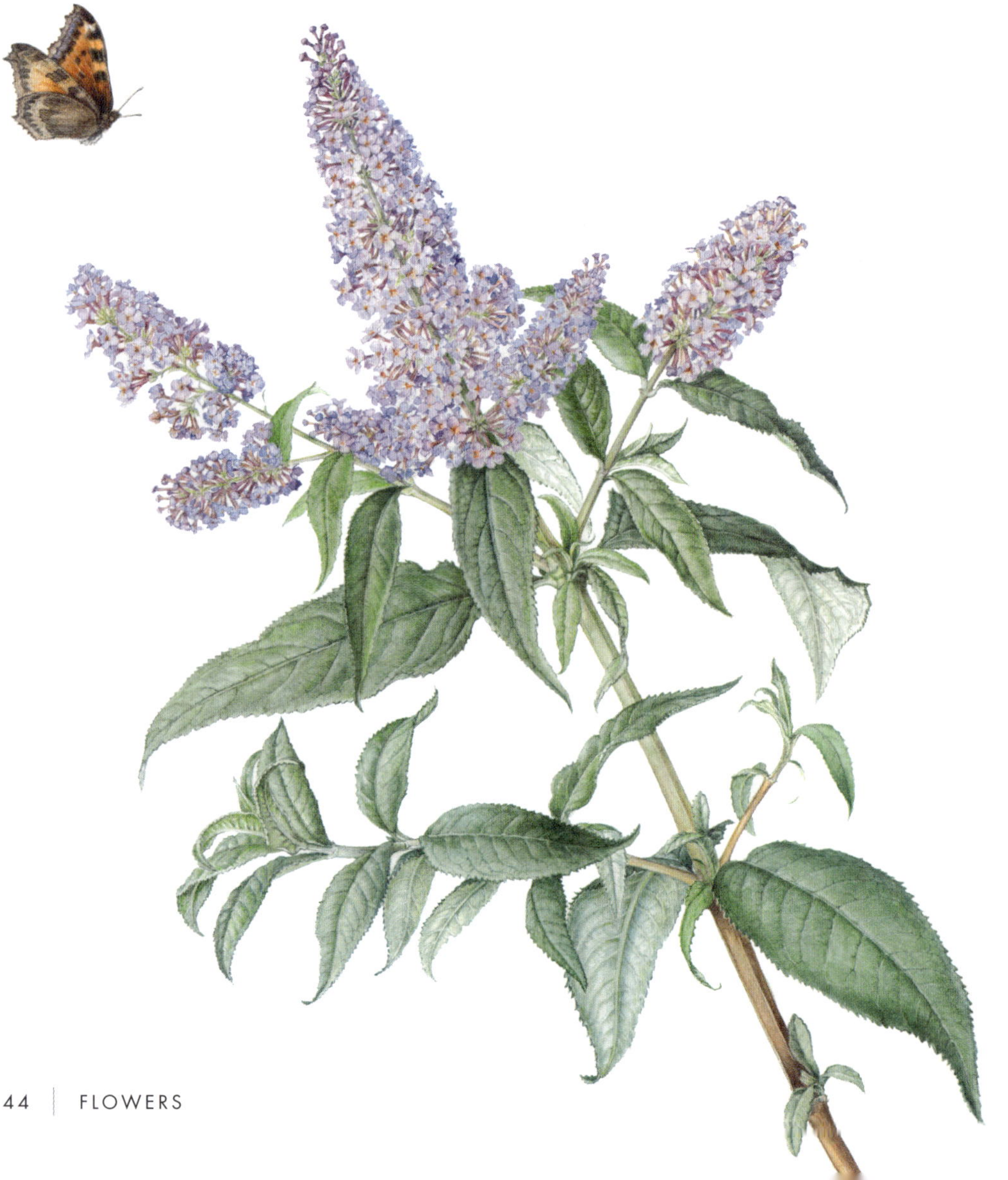

There are also mechanisms to prevent self-pollination, such as covering the stigma with stamen tubes, allowing pollen to be collected, but not dropped onto its own stigma until it, too, is receptive – for example, *Gentiana* and *Solanum* species, or the bump (rostellum) on the orchid column that separates pollen from stigma (see page 156).

PAINTING NECTAR GUIDES

Painting insect guide markings, or patterns, on a plant can be challenging if the markings are paler than the surrounding colour. Jacqueline Pestell painted the foxglove (*Digitalis purpurea*) on this page. She explains how to paint the nectar guides on this popular subject:

- Mark the centre of each white blotch and the shape in pencil.

- Begin by mapping the veins in the darkest magenta of the flowers.

- Build up layers to create the tubular flower forms, ensuring all highlights are retained.

- When dampening the paper for painting, avoid the areas of the splodges, though allow bleed from the surrounding area.

- Finish off the centre of the blotches with a dry brush and more concentrated pigment.

There are also preventative measures to keep unwanted animals out. Hairs in the funnel of a corolla can stop certain insects from crawling into flowers such as foxgloves and pansies.

To mitigate self-pollination, primroses have two types of flower on the one plant in equal numbers: 'pin' flowers have a prominent style, while 'thrum' flowers have prominent stamens. Have a look when you next collect one.

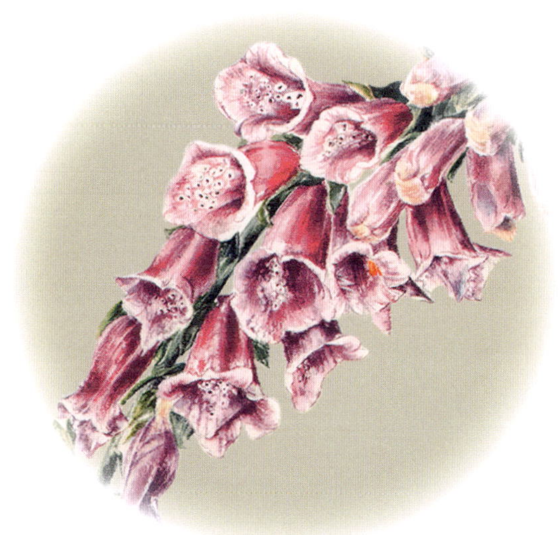

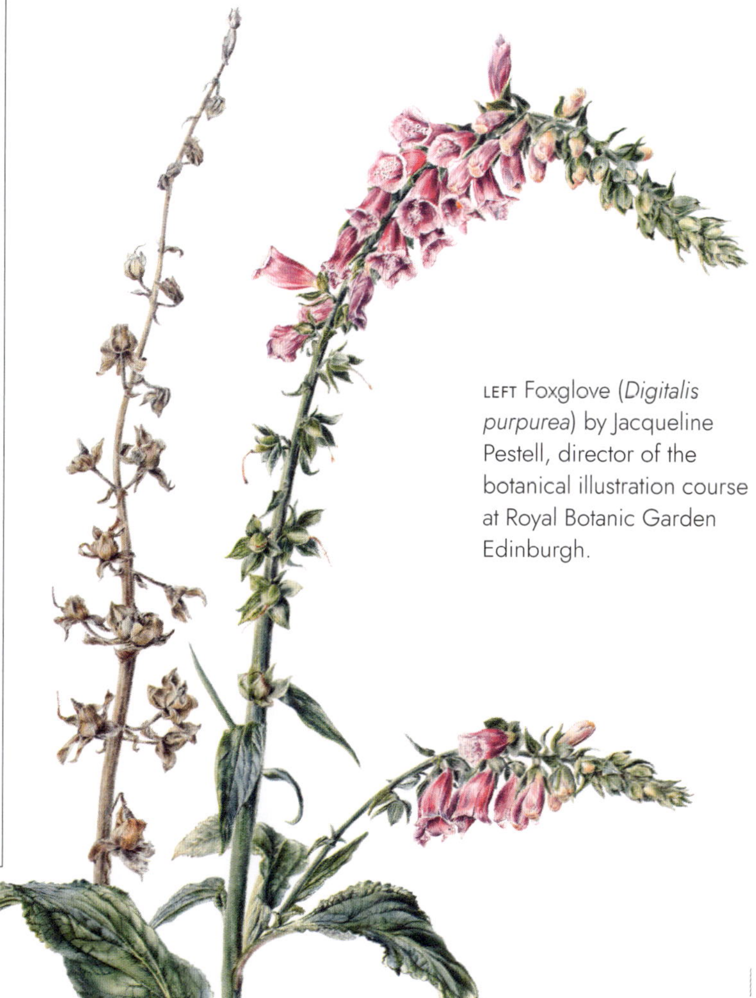

LEFT Foxglove (*Digitalis purpurea*) by Jacqueline Pestell, director of the botanical illustration course at Royal Botanic Garden Edinburgh.

Painting a hollyhock

Hollyhocks are large showy flowers, and there is a crinkled texture to the petals that can be depicted using light and dark colours. They are quite often two-toned and this can be achieved by putting different colours on different sections of the petal and overlapping them in places.

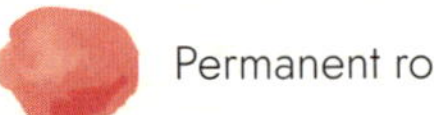

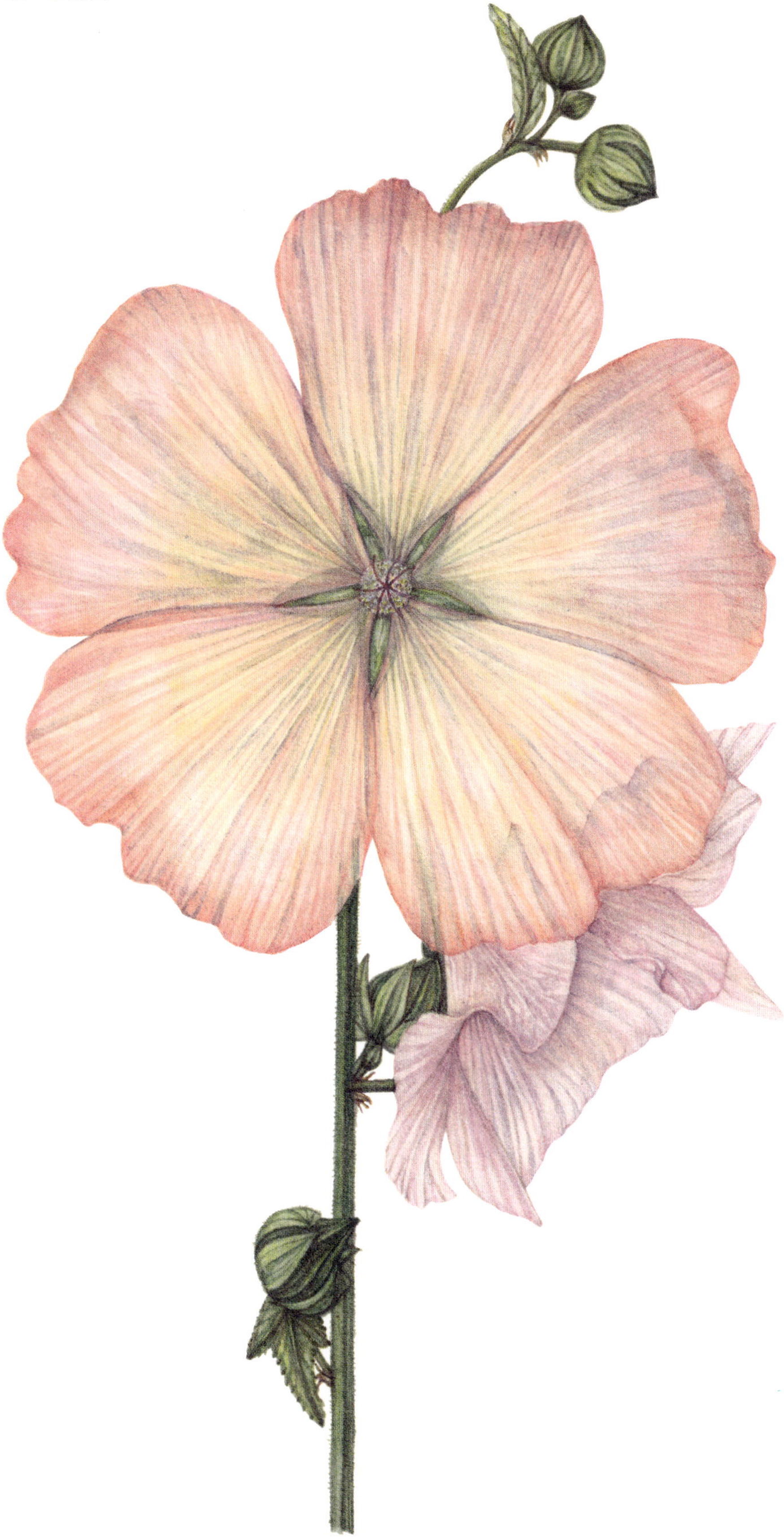

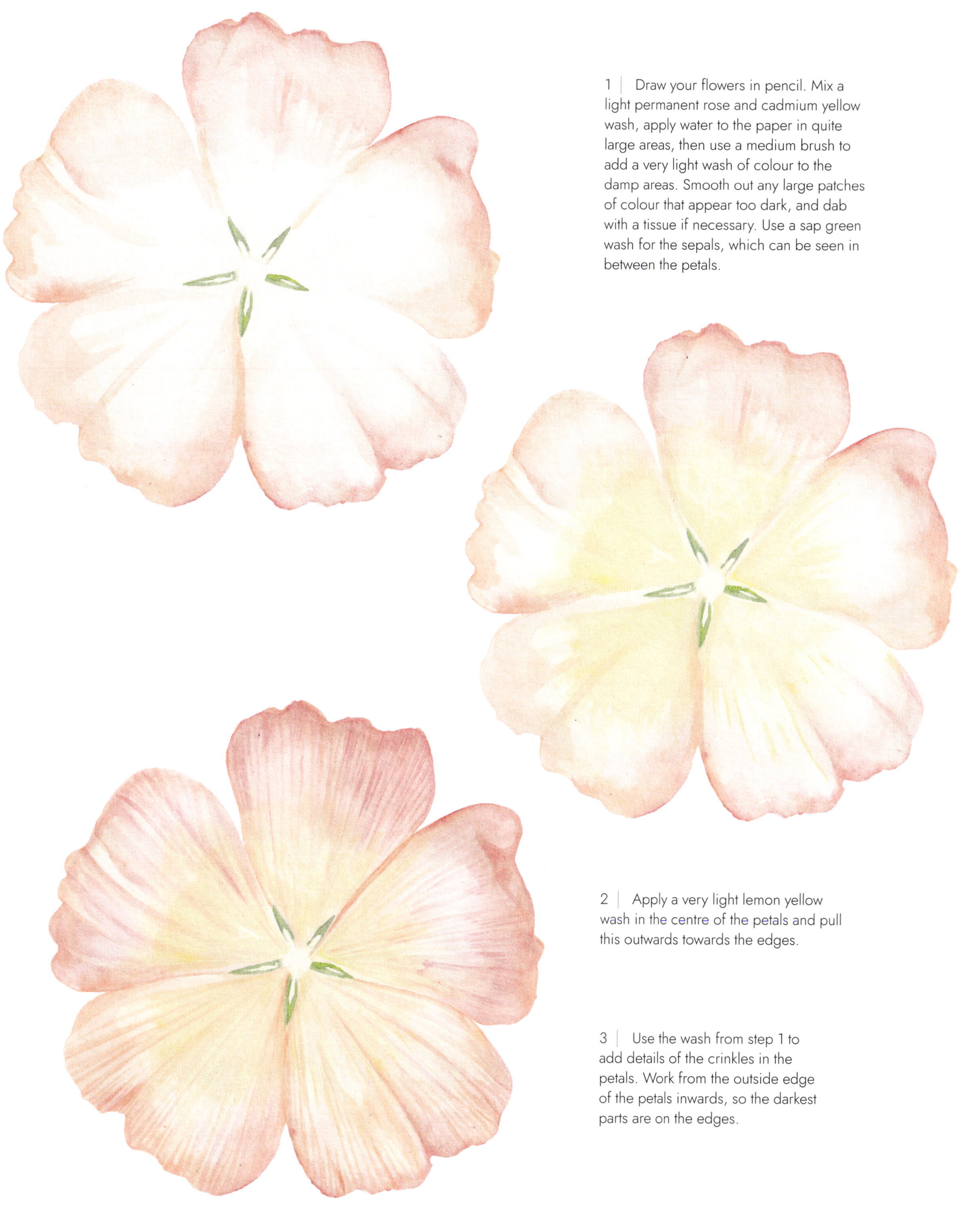

1 | Draw your flowers in pencil. Mix a light permanent rose and cadmium yellow wash, apply water to the paper in quite large areas, then use a medium brush to add a very light wash of colour to the damp areas. Smooth out any large patches of colour that appear too dark, and dab with a tissue if necessary. Use a sap green wash for the sepals, which can be seen in between the petals.

2 | Apply a very light lemon yellow wash in the centre of the petals and pull this outwards towards the edges.

3 | Use the wash from step 1 to add details of the crinkles in the petals. Work from the outside edge of the petals inwards, so the darkest parts are on the edges.

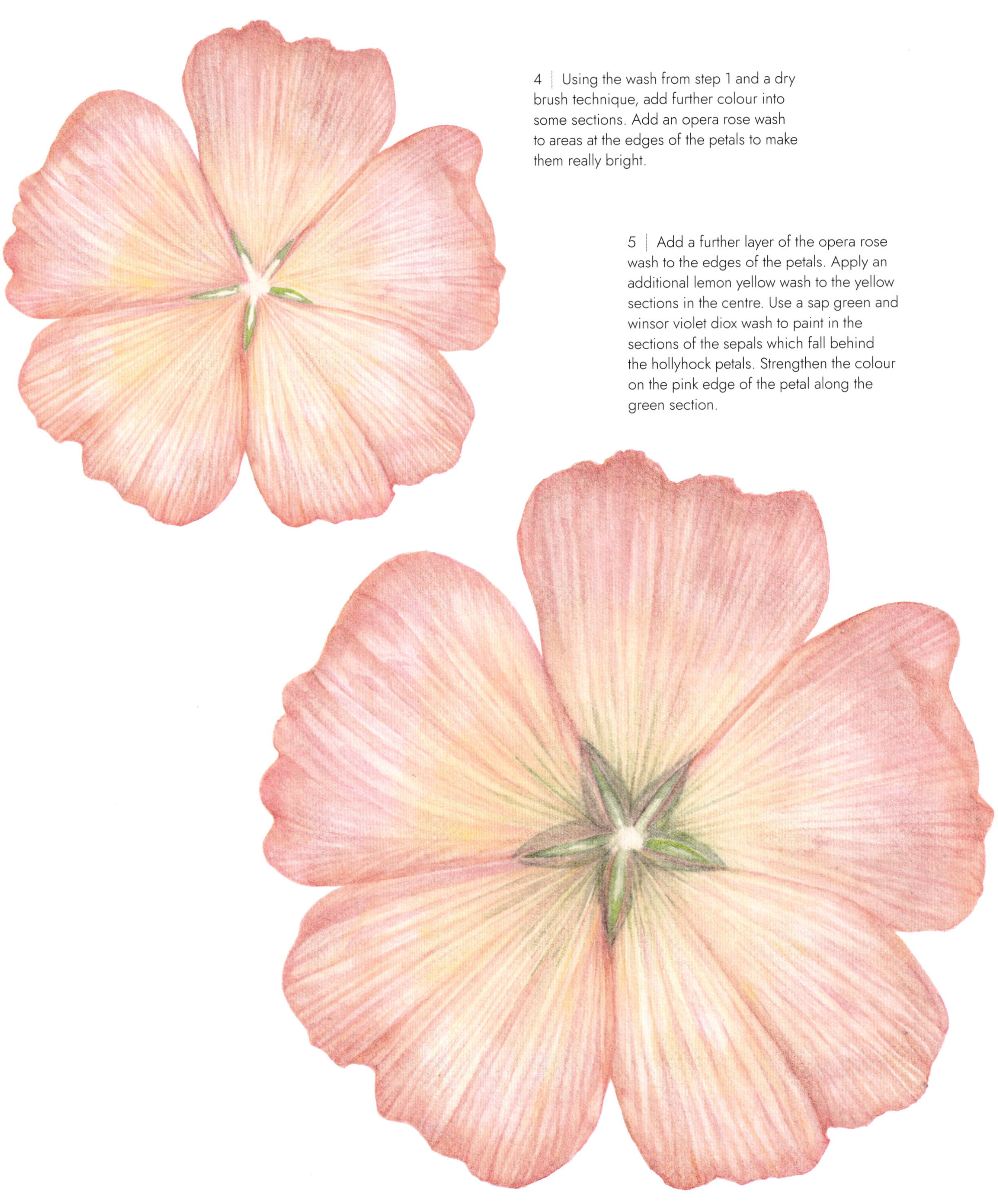

4 | Using the wash from step 1 and a dry brush technique, add further colour into some sections. Add an opera rose wash to areas at the edges of the petals to make them really bright.

5 | Add a further layer of the opera rose wash to the edges of the petals. Apply an additional lemon yellow wash to the yellow sections in the centre. Use a sap green and winsor violet diox wash to paint in the sections of the sepals which fall behind the hollyhock petals. Strengthen the colour on the pink edge of the petal along the green section.

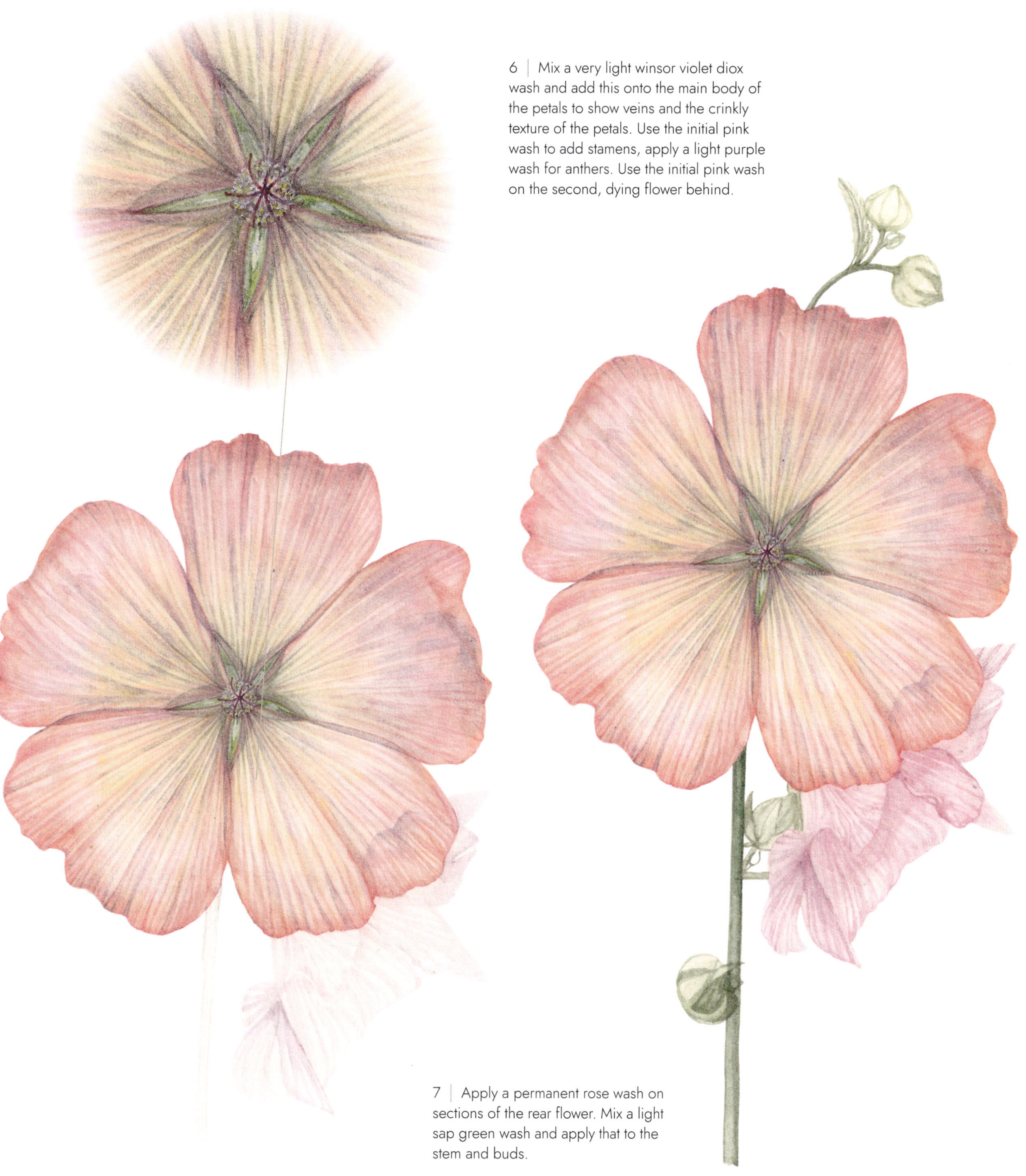

6 | Mix a very light winsor violet diox wash and add this onto the main body of the petals to show veins and the crinkly texture of the petals. Use the initial pink wash to add stamens, apply a light purple wash for anthers. Use the initial pink wash on the second, dying flower behind.

7 | Apply a permanent rose wash on sections of the rear flower. Mix a light sap green wash and apply that to the stem and buds.

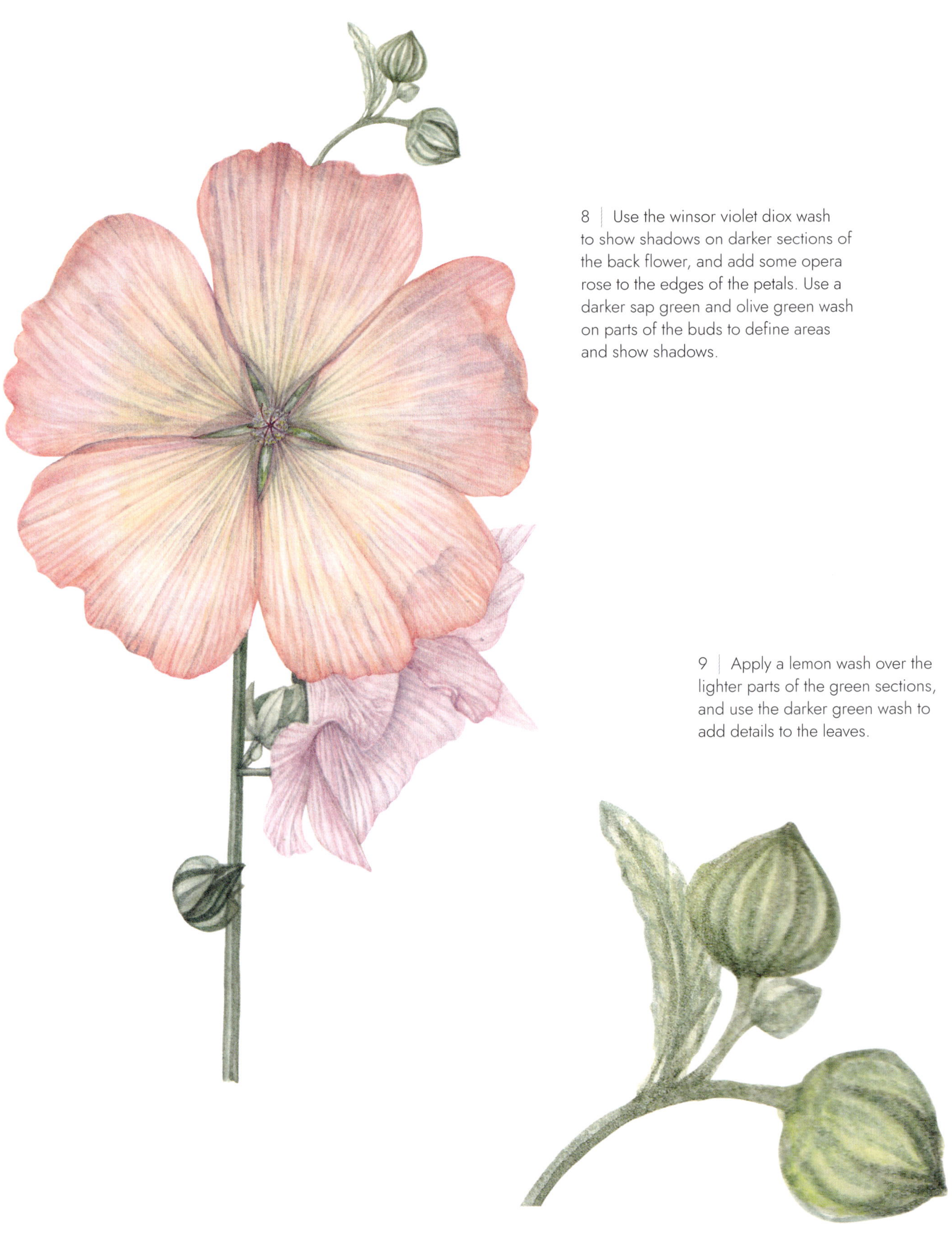

8 | Use the winsor violet diox wash to show shadows on darker sections of the back flower, and add some opera rose to the edges of the petals. Use a darker sap green and olive green wash on parts of the buds to define areas and show shadows.

9 | Apply a lemon wash over the lighter parts of the green sections, and use the darker green wash to add details to the leaves.

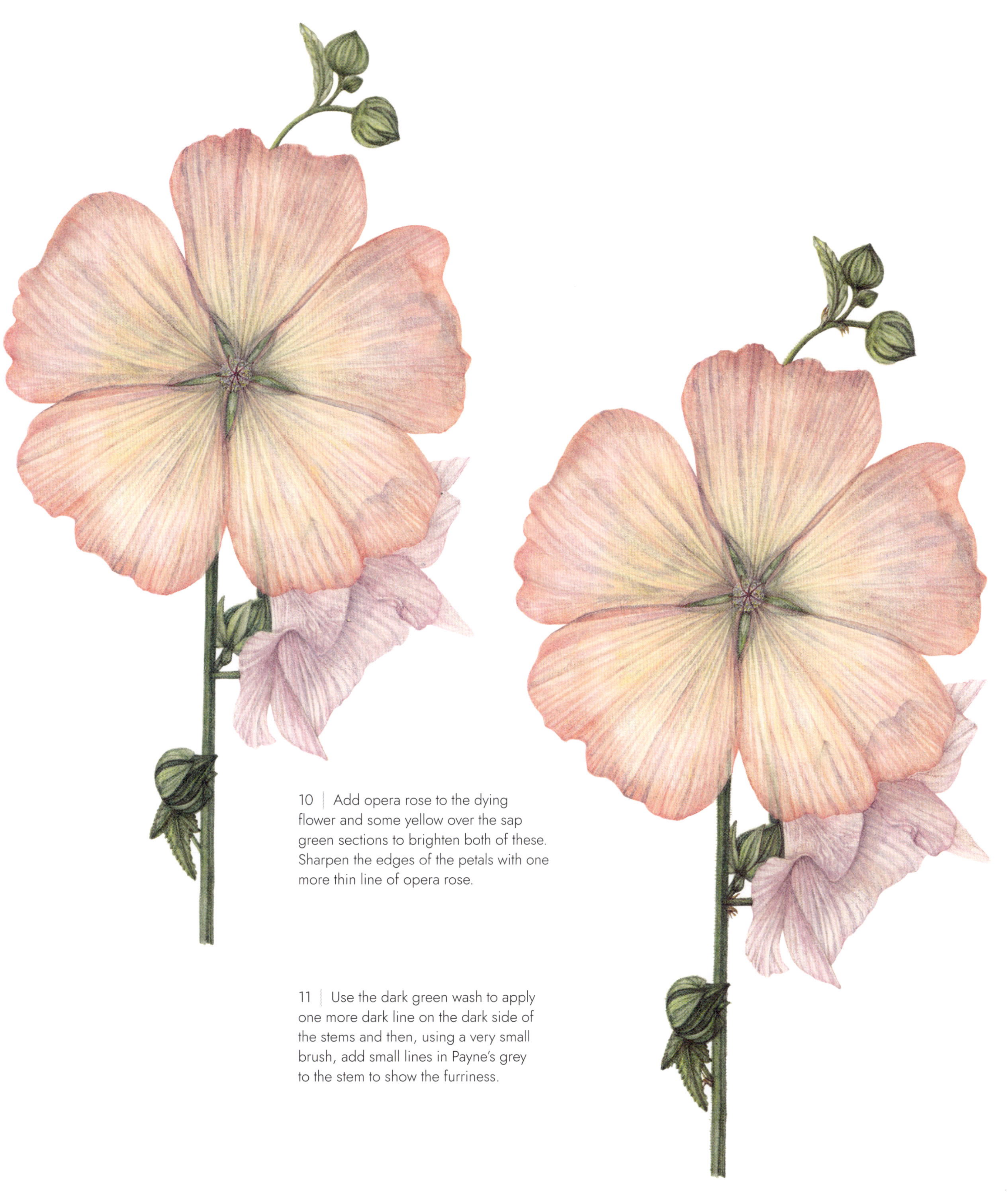

10 | Add opera rose to the dying flower and some yellow over the sap green sections to brighten both of these. Sharpen the edges of the petals with one more thin line of opera rose.

11 | Use the dark green wash to apply one more dark line on the dark side of the stems and then, using a very small brush, add small lines in Payne's grey to the stem to show the furriness.

Himalayan cobra lily (*Arisaema consanguineum*)

by Marianne Hazlewood

Every year the Himalayan cobra lily rises from the ground, displays its hooded spathe enclosing a spadix, and each female part then ripens into mesmerizingly bright-coloured globose fruit.

Arisaema are also irresistibly fascinating to artists, and Marianne is a specialist in painting this genus, growing many in her garden in the Pentland Hills near Edinburgh, Scotland. *Arisaema consanguineum* is a relatively tall plant, up to a metre (3ft) high, and grows wild in the Himalaya mountains and other hilly parts in that region. It bears a single, deciduous leaf with radial leaflets, each with a long, tendril-like tip. A green and purple striped hood appears under its leafy umbrella, also with an elongated tip, which provides a ladder for crawling insects to get to the spadix. In cultivation it can be grown in damp woodland gardens in temperate regions.

Focusing on detail

Marianne shows us this *Arisaema* through the eyes of an insect. You can see the detail of the stem, but you cannot see the whole plant. From this perspective you notice the detailed grooves running down the stem, not one line crossing another, the slight bump of the leaf's attachment to the stem, then the overhang at the base of the spathe. Crawl further and you see the tantalizing emergent spadix, which is what attracts the insect.

A perfectionist who focuses on the fine detail, Marianne displays her virtuosity in the colouring of the plant, where different colours dive in and out of parallel grooves on stem, leaf and spathe, never once losing sight of the shadows cast by light or a part of the plant.

As much care has gone into the composition as in the painted detail. Much has been made of the parallel lines of the flowering and fruiting stems. Artfully, Marianne has taken the edge off the parallels by curving them slightly to make them look natural. The drooping fruit, with the trailing male part, is positioned to make a third parallel feature. The palette is limited to greens, maroons and the brighter red of the fruit, all placed at different horizontal levels.

RIGHT The attractive reds and greens reflect the differing maturing times of the fruit. Size: 45 × 32cm (17^3/4 × 12^1/2in), watercolour on Fabriano 5 paper.

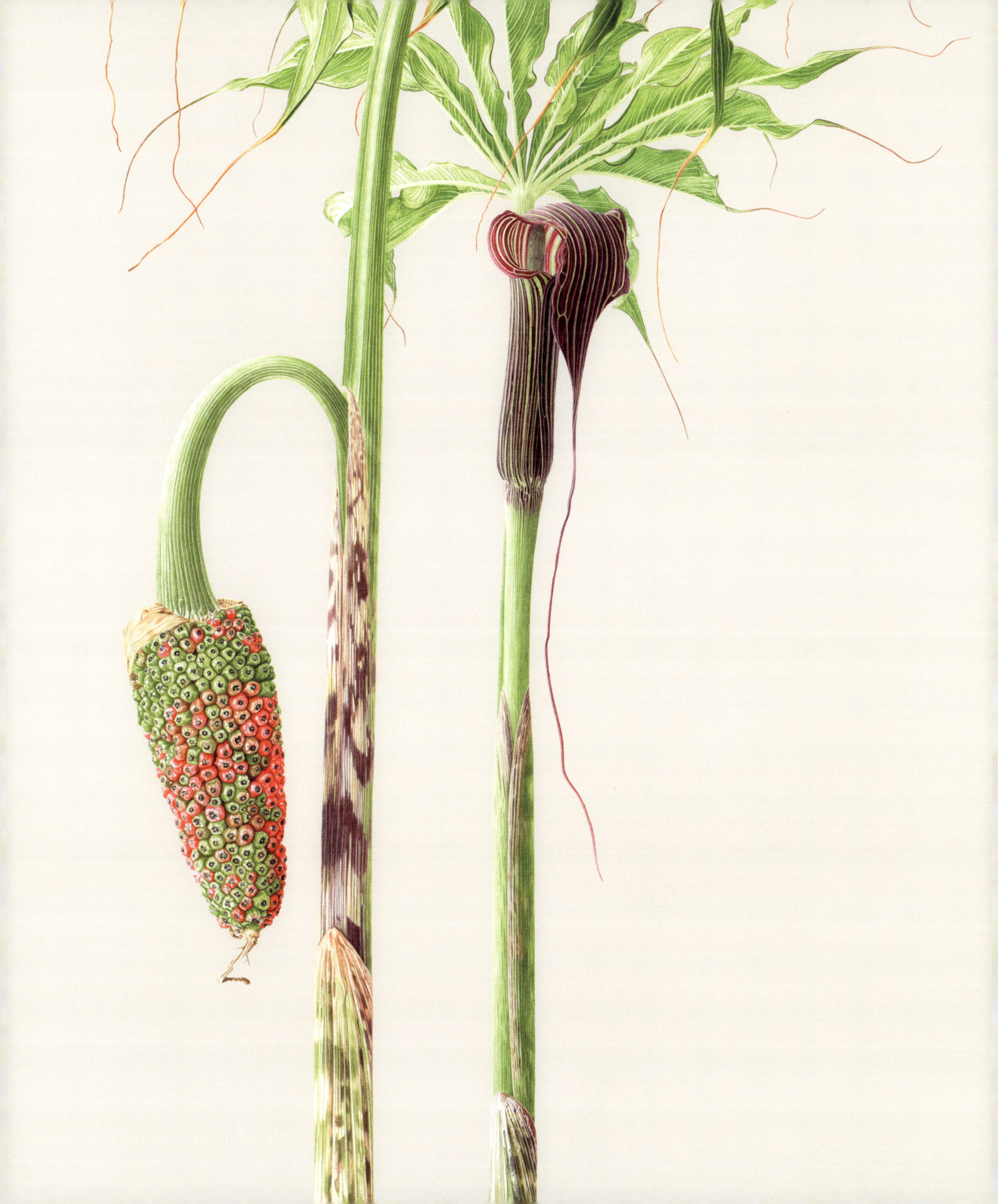

Plant families with specialist terms

Some notable plant families have unique structures with terminology to match. It's worth trying to grasp the terms, as they point to structures that can be difficult to understand.

Asteraceae

The *Asteraceae* family has thousands of species, including daisies and thistles, which all have in common a stem that holds a dense flower head called a capitulum.

Asteraceae flowers are easily recognized by their composite, often flat, sometimes globose inflorescences set above tightly packed sets of bracts. Inside, where many identifying features are found, the structures become even more esoteric. Working from the outside of the inflorescence to the inside:

Receptacle: the swollen tip of the pedicel holding all the flower parts; the longitudinal shape may be flat, convex, conical, columnar and so on.

Capitulum: the head of flowers.

Involucre: the system of bracts surrounding the flower head.

Phyllary: a bract (involucral bract), several of which form an involucre.

Ray floret: the outer, often sterile, showy flowers on the periphery with one extended lobe (ligule).

Disc floret: the inner, usually fertile, tubular flowers in the centre, with five equal, very short lobes.

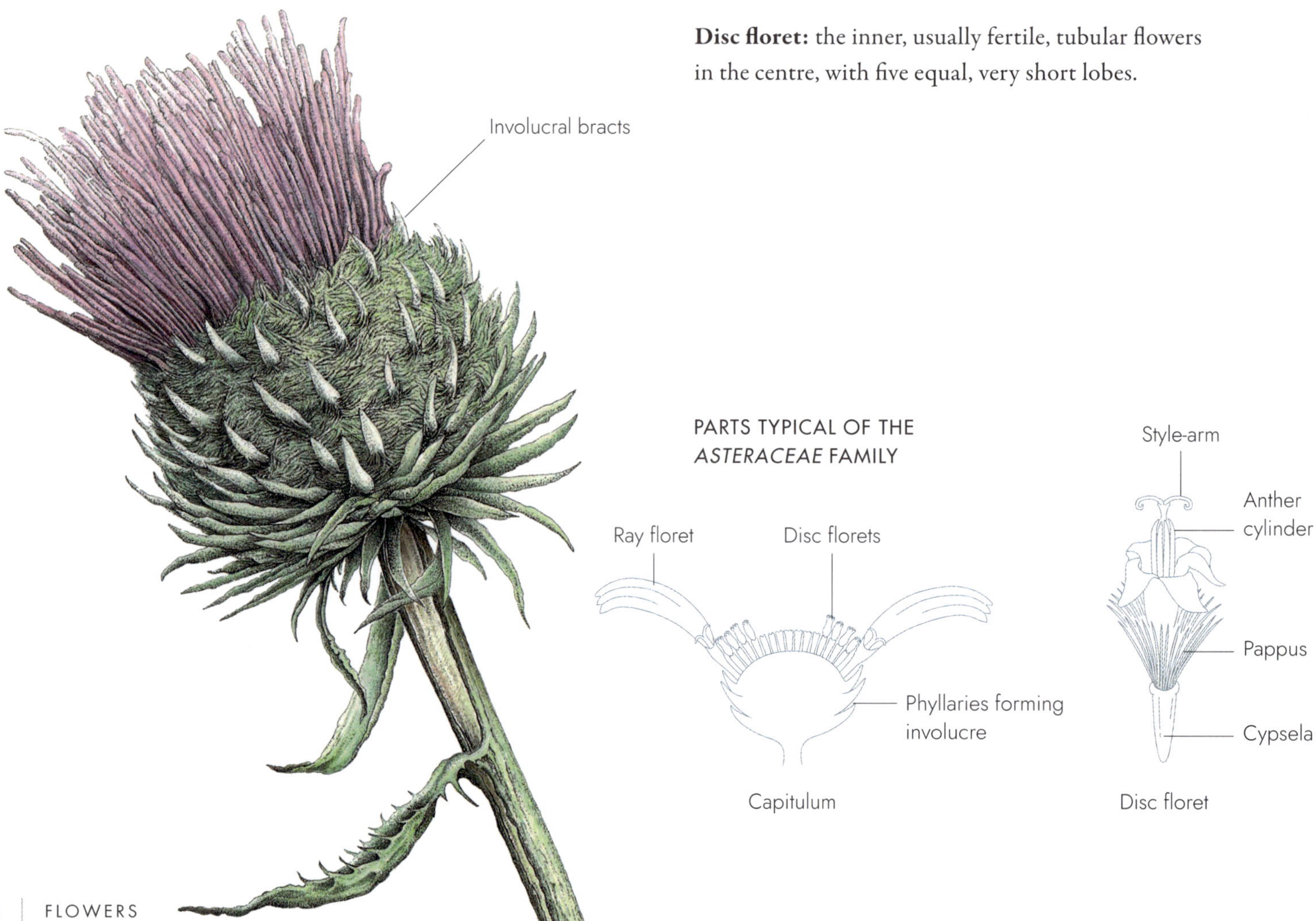

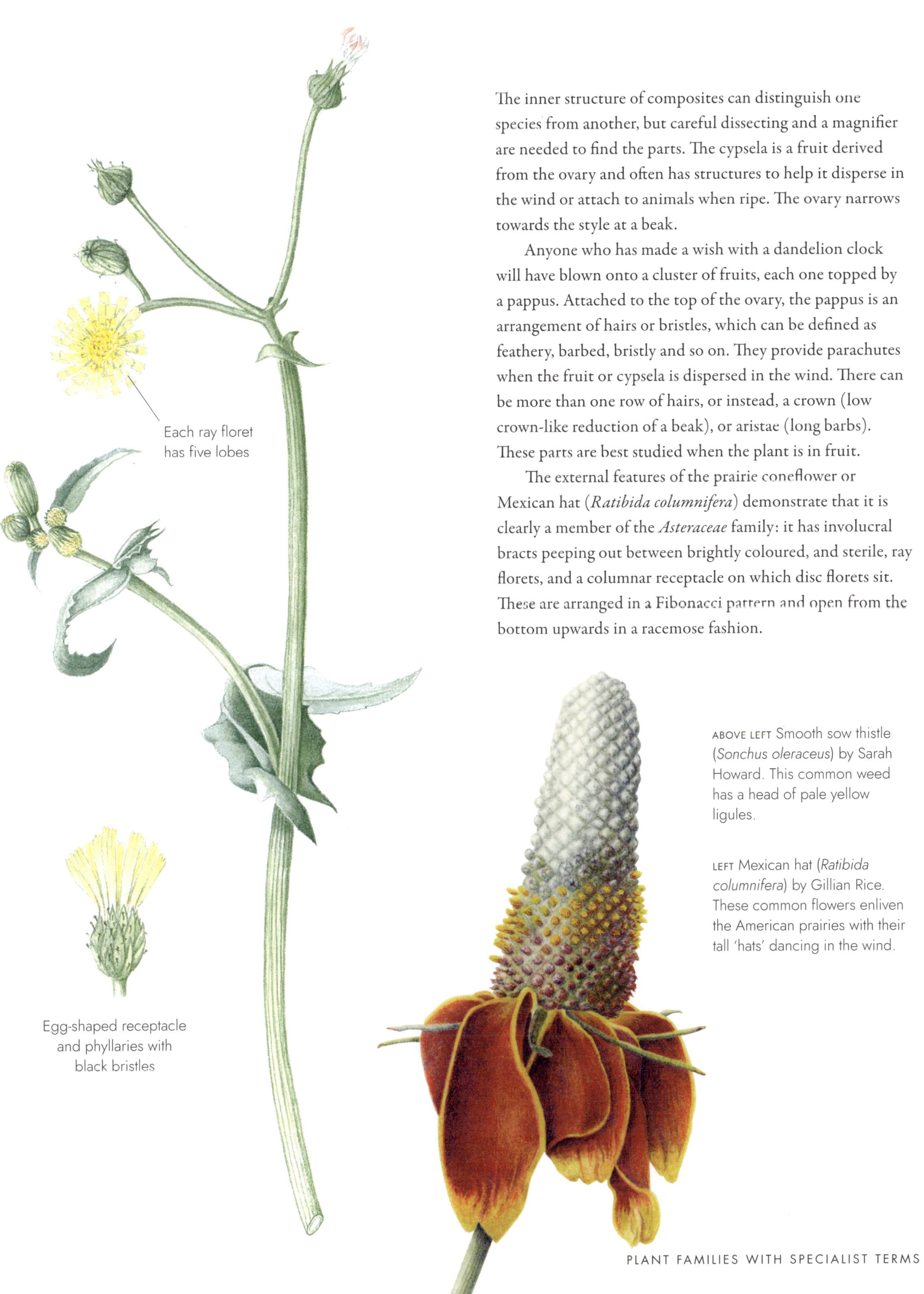

The inner structure of composites can distinguish one species from another, but careful dissecting and a magnifier are needed to find the parts. The cypsela is a fruit derived from the ovary and often has structures to help it disperse in the wind or attach to animals when ripe. The ovary narrows towards the style at a beak.

Anyone who has made a wish with a dandelion clock will have blown onto a cluster of fruits, each one topped by a pappus. Attached to the top of the ovary, the pappus is an arrangement of hairs or bristles, which can be defined as feathery, barbed, bristly and so on. They provide parachutes when the fruit or cypsela is dispersed in the wind. There can be more than one row of hairs, or instead, a crown (low crown-like reduction of a beak), or aristae (long barbs). These parts are best studied when the plant is in fruit.

The external features of the prairie coneflower or Mexican hat (*Ratibida columnifera*) demonstrate that it is clearly a member of the *Asteraceae* family: it has involucral bracts peeping out between brightly coloured, and sterile, ray florets, and a columnar receptacle on which disc florets sit. These are arranged in a Fibonacci pattern and open from the bottom upwards in a racemose fashion.

Each ray floret
has five lobes

Egg-shaped receptacle
and phyllaries with
black bristles

ABOVE LEFT Smooth sow thistle (*Sonchus oleraceus*) by Sarah Howard. This common weed has a head of pale yellow ligules.

LEFT Mexican hat (*Ratibida columnifera*) by Gillian Rice. These common flowers enliven the American prairies with their tall 'hats' dancing in the wind.

Orchids (monocots)

Orchids are the second largest seed plant family and rely on specialist pollinators. Their structures and the names for them reflect this aspect of their life cycle. Definitions are from *The Manual of Cultivated Orchid Species* by Helmut Bechtel et al:

Sympodial and monopodial: the two main groups based on their growth form. In sympodial growth each new shoot is determinate and terminates in a potential inflorescence or solitary flower. Monopodial growth continues from a terminal bud from season to season.

Velamen: a parchment-like sheath or layer of spiral-coated air-cells on the root which may act as protective insulation.

Scape: a leafless peduncle arising directly from a rosette of basal leaves.

Pseudobulb: a swollen aerial stem.

Lip or labellum: a lip, referring to the enlarged, often highly modified abaxial petal of the orchid flower.

Epichile or epichilium: the terminal part of the lip when it is distant from the basal portion.

Column: an advanced structure composed of a continuation of the flower stalk, together with the upper part of the female reproductive organ (pistil) and the lower part of the male reproductive organ (stamen).

Rostellum: the often beak-like sterile third stigma lying between the functional stigmas and stamen.

Pollinium: a pollen mass.

RIGHT Spider orchid (*Brassia cochleata*) by Betty Gadsby. This American epiphytic sympodial orchid shows off its pseudobulb from which new growth appears.

PARTS OF AN ORCHID

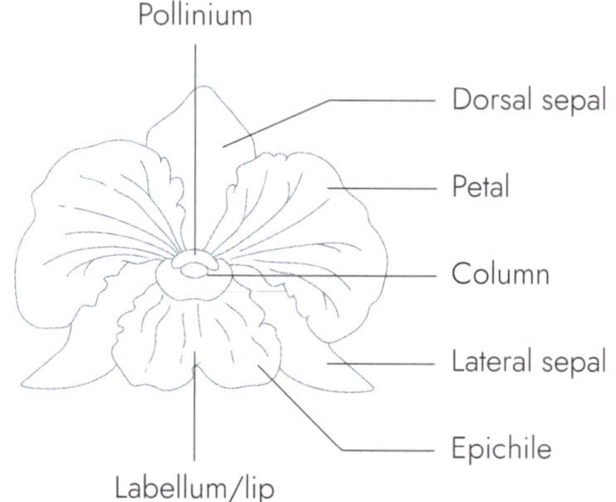

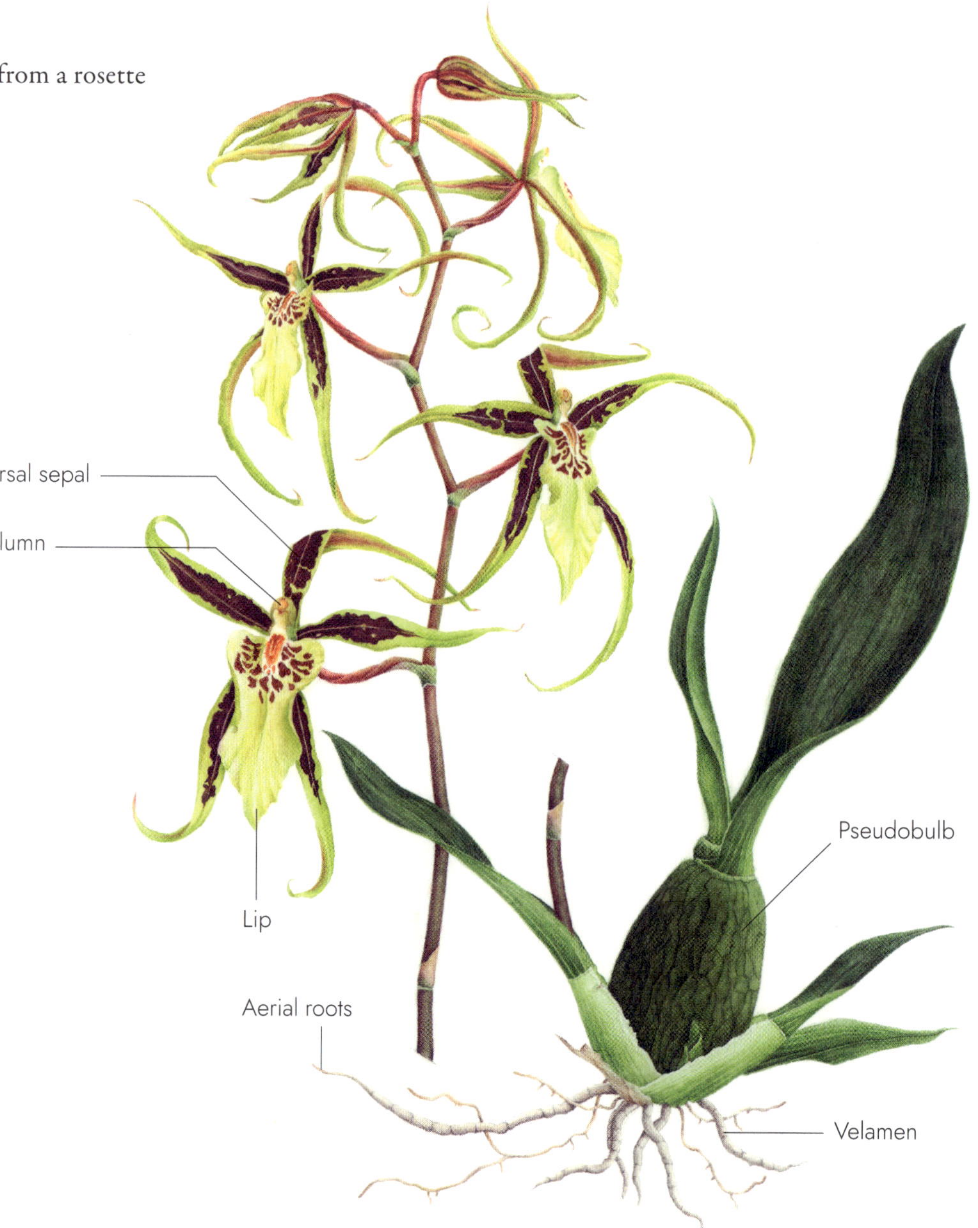

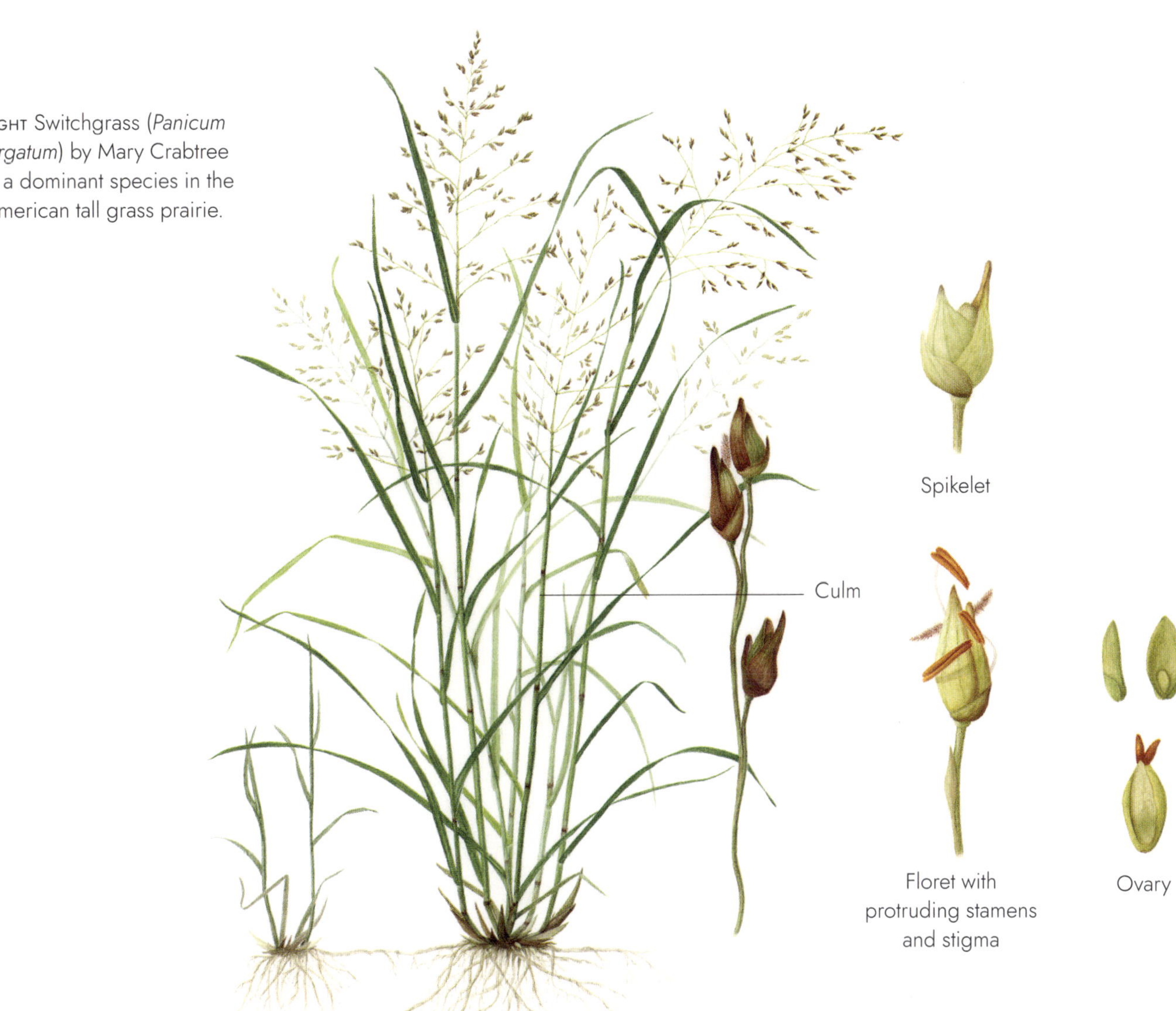

RIGHT Switchgrass (*Panicum virgatum*) by Mary Crabtree is a dominant species in the American tall grass prairie.

Grasses (monocots)

Humanity is reliant on the *Poaceae* family for food. It provides our cereals, and grass and hay for animal feed. Grasses present a lovely, yet challenging, subject for artists because of their intricate small flower parts, but the attempt to understand them is rewarding:

Tillers: sucker or branch emerging from the base of the plant.

Culm: the stem.

Sheath: tubular part of the leaf which envelops the stem.

Node: where a new leaf is attached and enables bending.

Ligule: a membranous projection of the sheath at the node, which often has defining characteristics for identification.

Auricle: leaf margin lobes at the node.

Spikelet: the small clusters of florets that make up the inflorescence.

Rachilla: axis of the spikelet.

Glume: equivalent to bracts at the base of a spikelet. There are upper and lower glumes.

Floret: a single flower.

Lemma: the outer bract.

Palea: the inner bract.

Awn: extension of the lemma, sometimes very long and thin.

Lodicule: small scales below the ovary, possibly a reduced perianth.

Coast coral tree (*Erythrina afra*)
by Daleen Roodt

Splendid orange-scarlet inflorescences cover the bare branches of *Erythrina* trees in the Southern African spring. They entice numerous bird and insect pollinators, such as the Cape white-eye, seen here. The name 'erythrina' comes from the Greek word for red, hence they are often and ambiguously known as 'flame trees' or 'coral trees'. Artists, too, find them irresistible with their keels and standards of pea-like flowers, whiskery stamens and bold brown new buds on the ends of racemose inflorescences.

Achieving a sense of scale

Outdoors, Daleen Roodt always does quick sketches to compare natural light with indoors, where she finishes her paintings. Her drawings are fully tonal, so that the shadows and highlights are not lost later. Her palette is simple, and outdoors, where paint can dry quickly, her painting is watery. Indoors, she prefers good control of the brush to work her way round the elements.

This inflorescence is complex, with numerous flowers at different stages, facing different ways and buds inserted at the tip. Added to this, Daleen needs to allow for the protruding stamens, cleverly shown to best advantage against the light. The vibrancy of the red is maintained by tonal shifts from pale to intense. The diagonal line of the composition adds drama to this tightly knit study. Her emphasis is on the top inflorescence, but Daleen artfully aligns the bird with the inflorescence below and gives the plant a sense of scale (Cape white-eyes are about 12cm/4^3/4in long). Neither feature detracts from the other. Rather, they tell the story of the bird's rapturous enjoyment of the nectariferous flowers.

RIGHT Each scarlet keel leaves little doubt the *Erythrina* is a member of the pea (*Fabaceae*) family. Size: 35 × 24.8cm (13^3/4 × 9^3/4in), watercolour on paper, 2016.

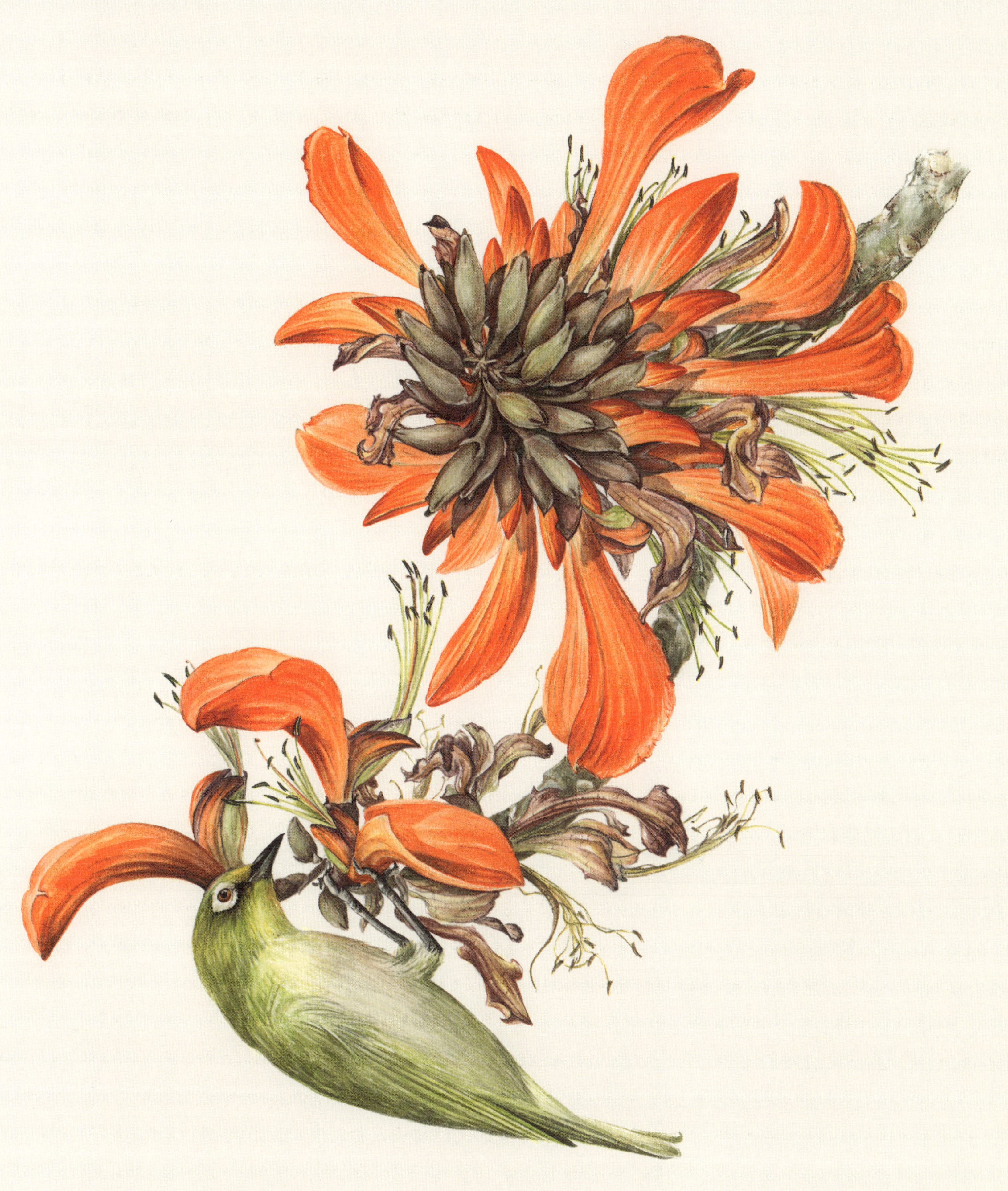

Inflorescences

An inflorescence is a flower head consisting of a structure bearing the flowers and their protective bracts and bracteoles. It may be very simple with only a few flowers, or it may have many branches carrying millions of flowers.

Patterns and branching

It is important to get the pattern of branching right, if you can see it. But do not despair if you find patterns impossible to discern; artists rely on a botanist's help when they run into difficulties, to clarify the specific botanical meanings many general terms have.

If the inflorescence is loose or lax, it is more likely the branching pattern can be seen. It means looking at the pattern on both the main and any lateral stems, as well as the flowering order. Most botanists distinguish between racemose or cymose; indeterminate or determinate; and flowering from the bottom or from the top.

LEFT *Aloe* aff. *percrassa* by Sarah Howard. The spiral arrangement of the florets, together with their bracts and flowering order, clearly indicates a racemose type

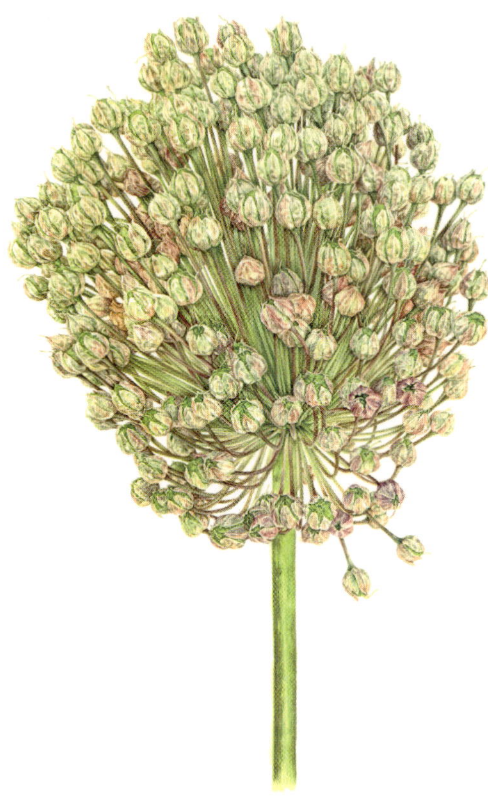

The variations are numerous and some elaborate flower heads show a combination of both patterns. Tomatoes, for example, are described by gardeners as indeterminate or determinate in their growth but the pattern of branching is always cymose. It is best to keep in mind the two basic types, racemose and cymose. Variants of both are given on the following pages.

It can be difficult to distinguish between racemose and cymose inflorescences on superficially similar shapes. The carrot family, *Apiaceae*, has distinctive flat-topped inflorescences which are racemose. But, yarrow, also flat-topped, is cymose.

The carrot family is also identified by its umbels (meaning parasol in Latin), where the pedicels arise from the same point, like spokes on a wheel. Many in the monocot *Amaryllidaceae* family, such as alliums and agapanthus, also have an umbellate form held on a scape, though it is thought these are not true racemose umbels. Look carefully at the flowering order in both cases.

ABOVE Fading allium (*Allium* sp.) by Toni Dade. Once the flowers are finished it is easier to see that the inflorescence is determinate and arises from the top of a scape in an umbellate form

RIGHT Himalayan balsam (*Impatiens glandulifera*) detail by Janet Dyer. The seed pods at the base of the inflorescence indicate an indeterminate racemose structure. They explode when ripe, giving its common name, 'Touch me not'.

DETERMINATE GROWTH

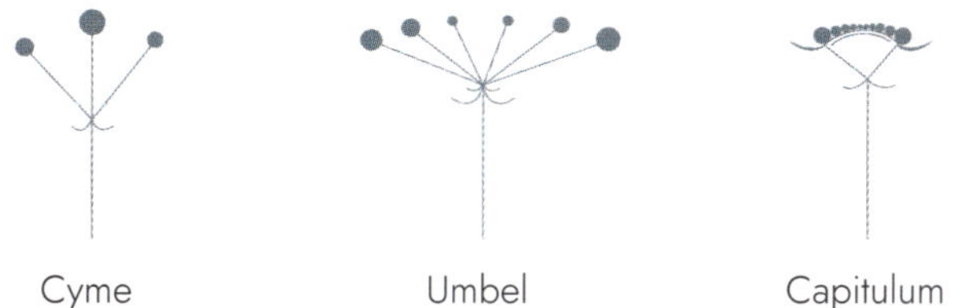

Cyme   Umbel   Capitulum

INDETERMINATE GROWTH

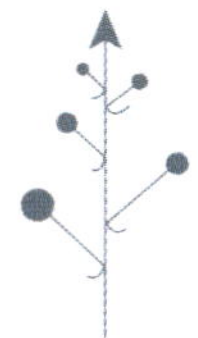

Raceme

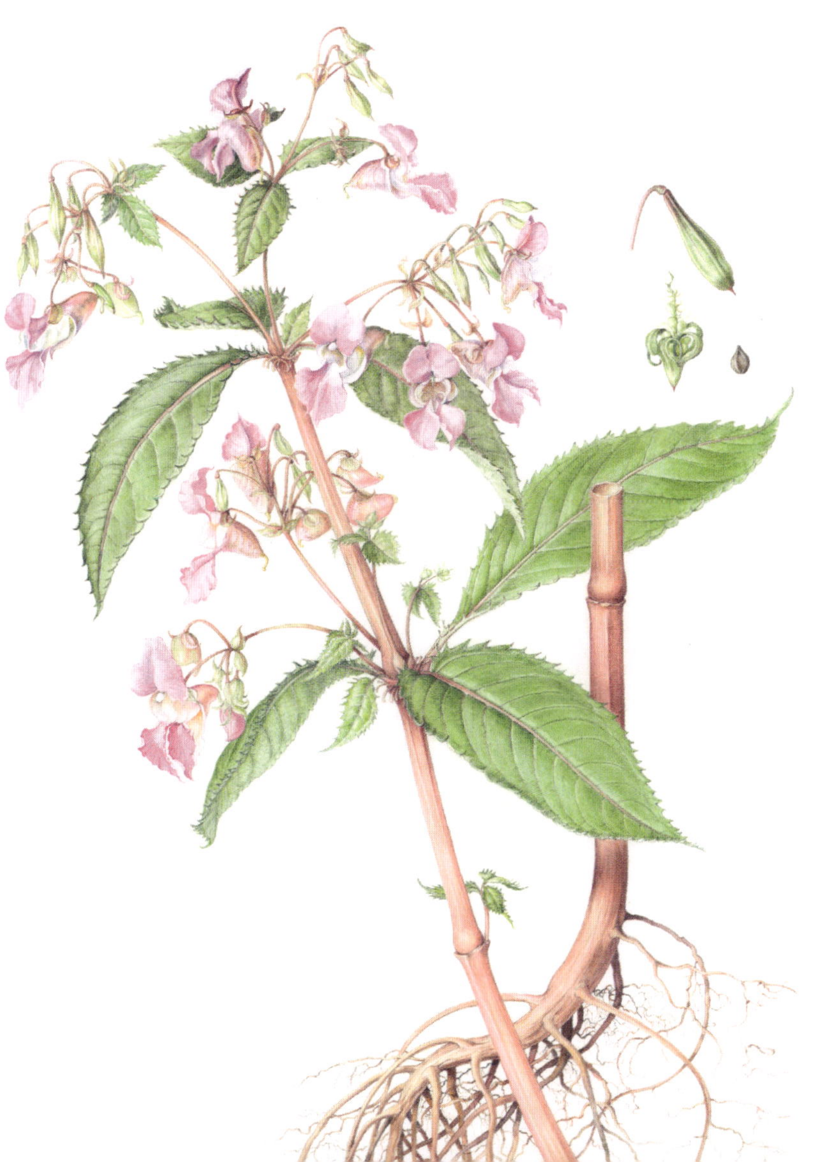

Inflorescence features

The main stem (axis) of the inflorescence is the peduncle. After the point where flowering begins the peduncle is called the rachis. Any branching of the rachis is guided by genetically determined patterns (see page 160). Each individual flower (floret) is usually borne on its own stalk, a pedicel. Flowers lacking a pedicel are termed sessile. There is often a leafy bract at the base and sometimes a tiny pair of bracteoles further up the axis.

In some (mostly tropical) woody species, the flowers arise directly from the main stems, rather than from leaf axils, a condition called cauliflory. Cocoa (*Theobroma cacao*) is a classic example. A cluster of flowers arising from a leaf axil, as in some cherry trees, is called a fascicle. Where clusters of flowers arise around nodes, they are whorls.

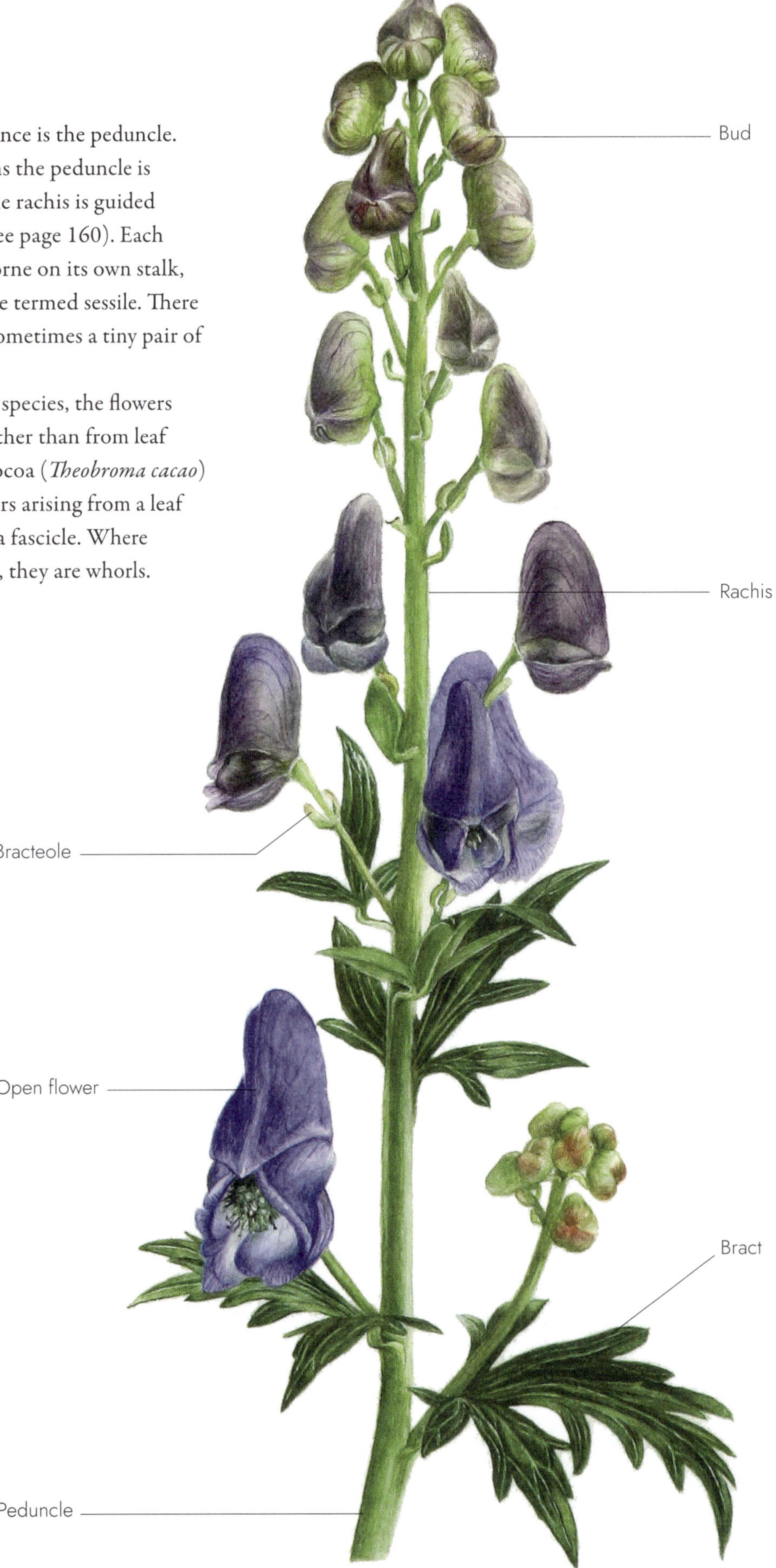

RIGHT Carmichael's monk's hood (*Aconitum carmichaelii*) detail by Roger Reynolds. A simple racemose inflorescence which flowers from the bottom up.

WHAT TO LOOK FOR

- Decide what needs to be shown and how much detail is needed.

- Decide what the flowering order is because that will help to discern whether the pattern is racemose or cymose.

- Find out where any bracts or bracteoles are. They will guide your thinking about the hierarchy of the flowering pattern in a compound inflorescence.

- If you cannot see any pedicels, bracteoles or the phyllotaxis along the axis, it is fine not to include them, but details could be useful in another part of the picture.

- The overall shape of a dense inflorescence may be important: triangular, cylindrical, globular and so on.

- Include a few faded flowers along with buds and open flowers to show flowering order. Ensure the rachis narrows towards the tip if necessary.

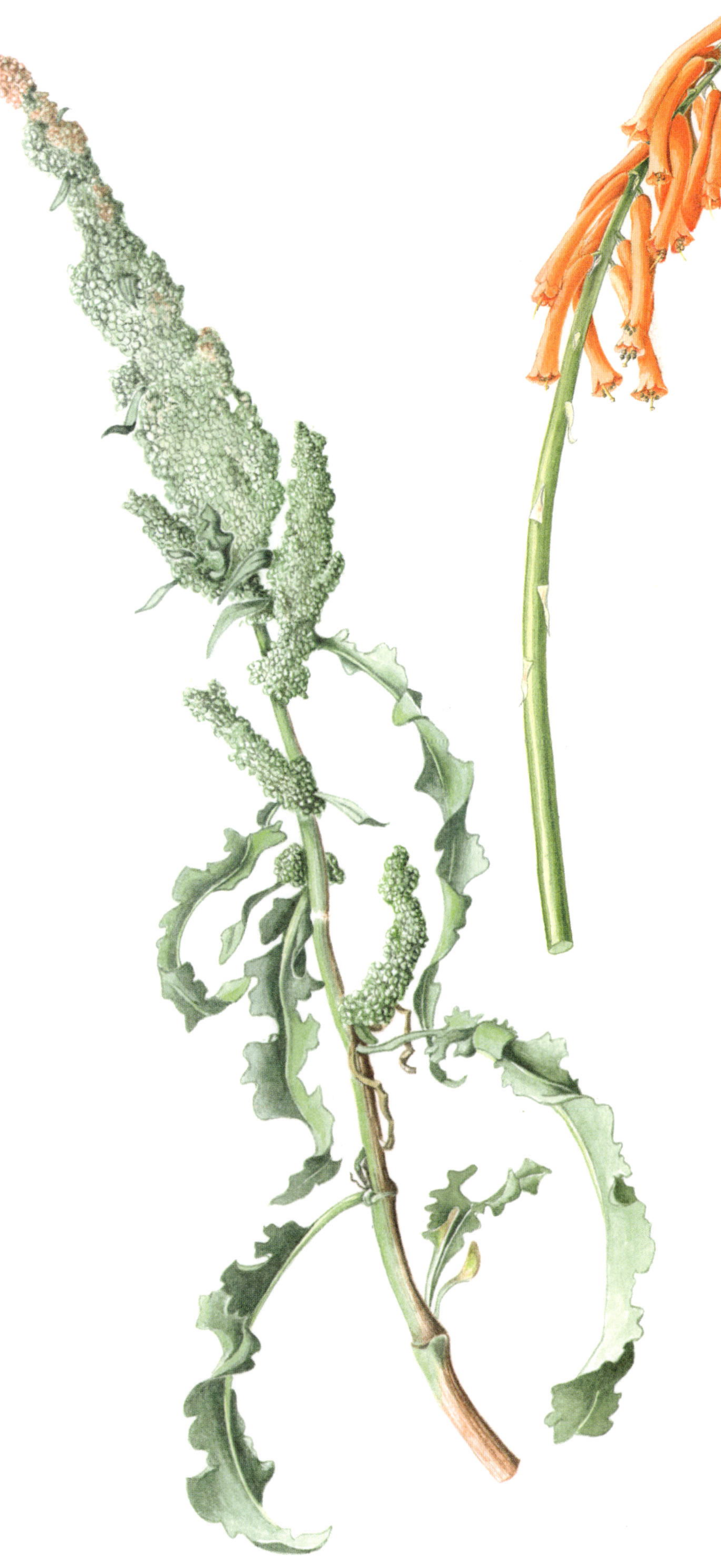

ABOVE RIGHT *Kniphofia schimperi* detail by Sarah Howard, a lax inflorescence revealing bracteoles, pedicels, flower arrangement and order of flowering.

RIGHT Curled dock (*Rumex crispus*) by Sarah Howard. This is a dense inflorescence but take a look inside to note the spiral arrangement of the tiny flowers on the axis indicating a raceme.

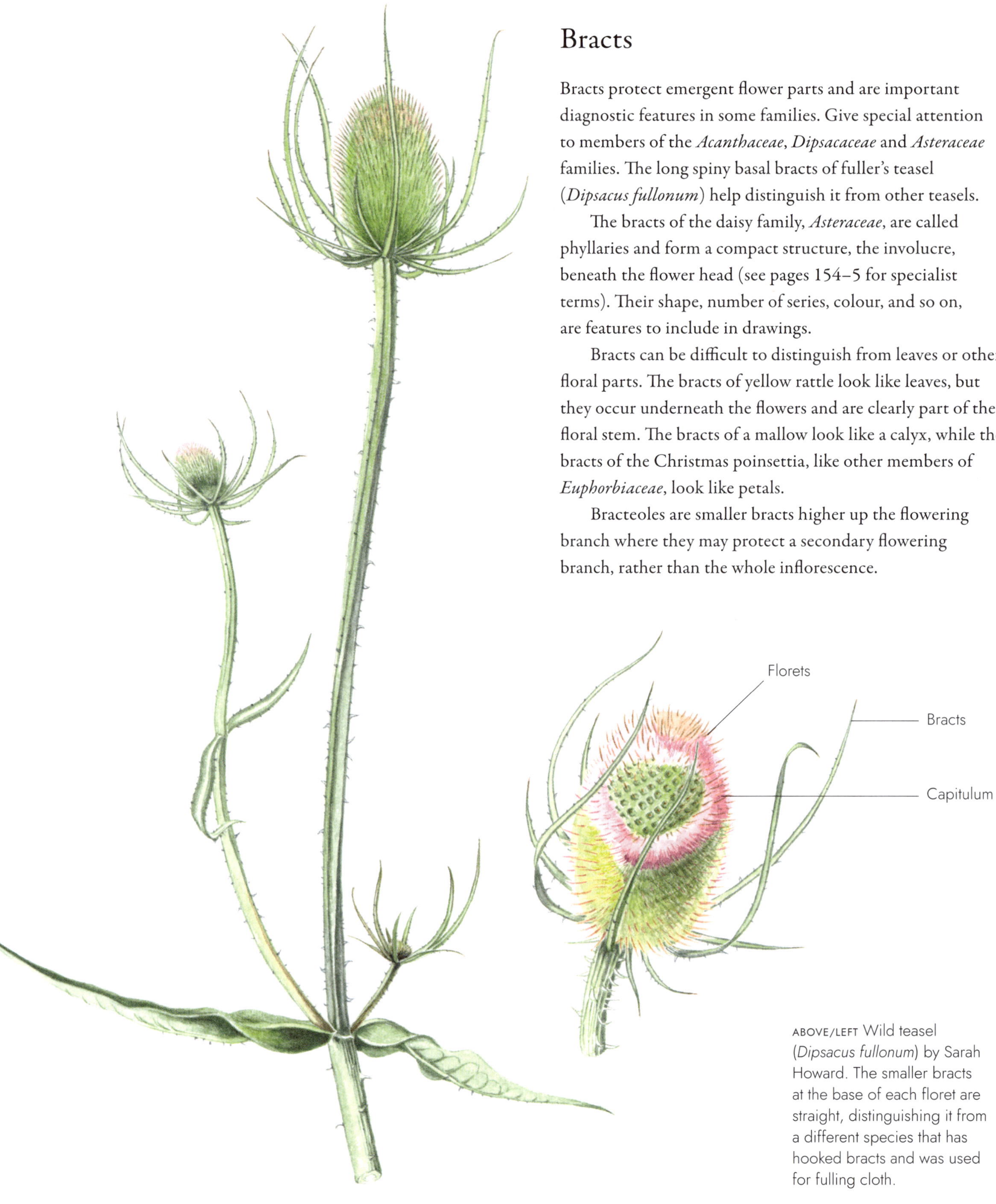

Bracts

Bracts protect emergent flower parts and are important diagnostic features in some families. Give special attention to members of the *Acanthaceae*, *Dipsacaceae* and *Asteraceae* families. The long spiny basal bracts of fuller's teasel (*Dipsacus fullonum*) help distinguish it from other teasels.

The bracts of the daisy family, *Asteraceae*, are called phyllaries and form a compact structure, the involucre, beneath the flower head (see pages 154–5 for specialist terms). Their shape, number of series, colour, and so on, are features to include in drawings.

Bracts can be difficult to distinguish from leaves or other floral parts. The bracts of yellow rattle look like leaves, but they occur underneath the flowers and are clearly part of the floral stem. The bracts of a mallow look like a calyx, while the bracts of the Christmas poinsettia, like other members of *Euphorbiaceae*, look like petals.

Bracteoles are smaller bracts higher up the flowering branch where they may protect a secondary flowering branch, rather than the whole inflorescence.

ABOVE/LEFT Wild teasel (*Dipsacus fullonum*) by Sarah Howard. The smaller bracts at the base of each floret are straight, distinguishing it from a different species that has hooked bracts and was used for fulling cloth.

Bracts come in many guises:

- Petaloid and coloured on poinsettia and other euphorbia (called cyathophyll by specialists), or dogwood *Cornus florida*.

- Scaly bracts on 'catkins' and on conifers (see page 220).

- Fused on an acorn cup, forming a cupule.

- In clusters like those beneath a thistle head where they are called involucral bracts.

- Hooded to cover the whole inflorescence, often in monocots, where they are called spathes; for example, alliums and arisaema (see page 152).

- Like a calyx in some mallows and in strawberries where they have two, the outer one called an epicalyx.

- In grasses, the glume and lemma are bracts, and the palea is the bracteole (see page 157 where these parts can be seen).

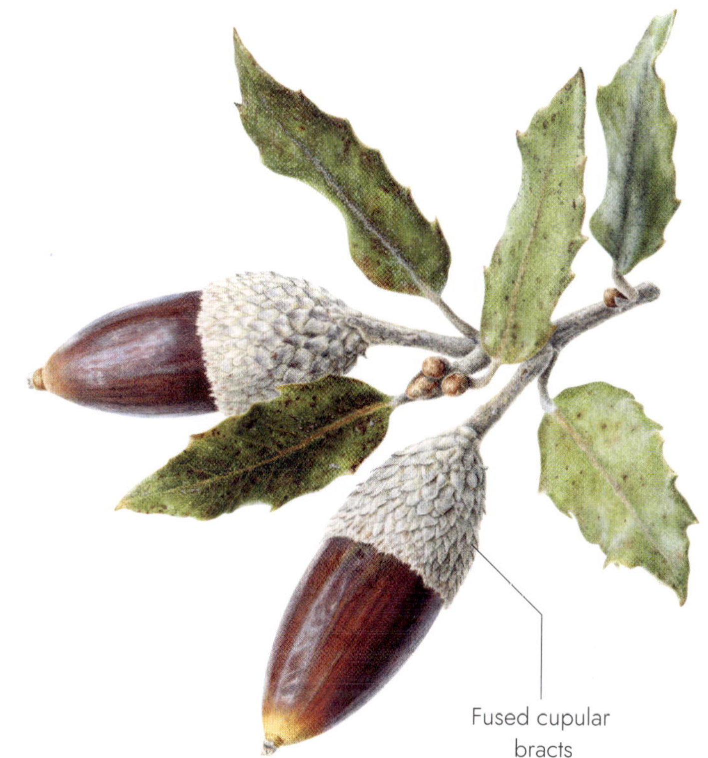

ABOVE Holm oak (*Quercus ilex*) by Carolyn Jenkins shows the oak's characteristic cupules which 'cup' the fruit, the acorn.

LEFT Globe artichoke (*Cynara cardunculus*) by Carolyn Jenkins. Known for its edible involucral bracts, this Mediterranean garden thistle was referred to by Homer in the 8th century BCE.

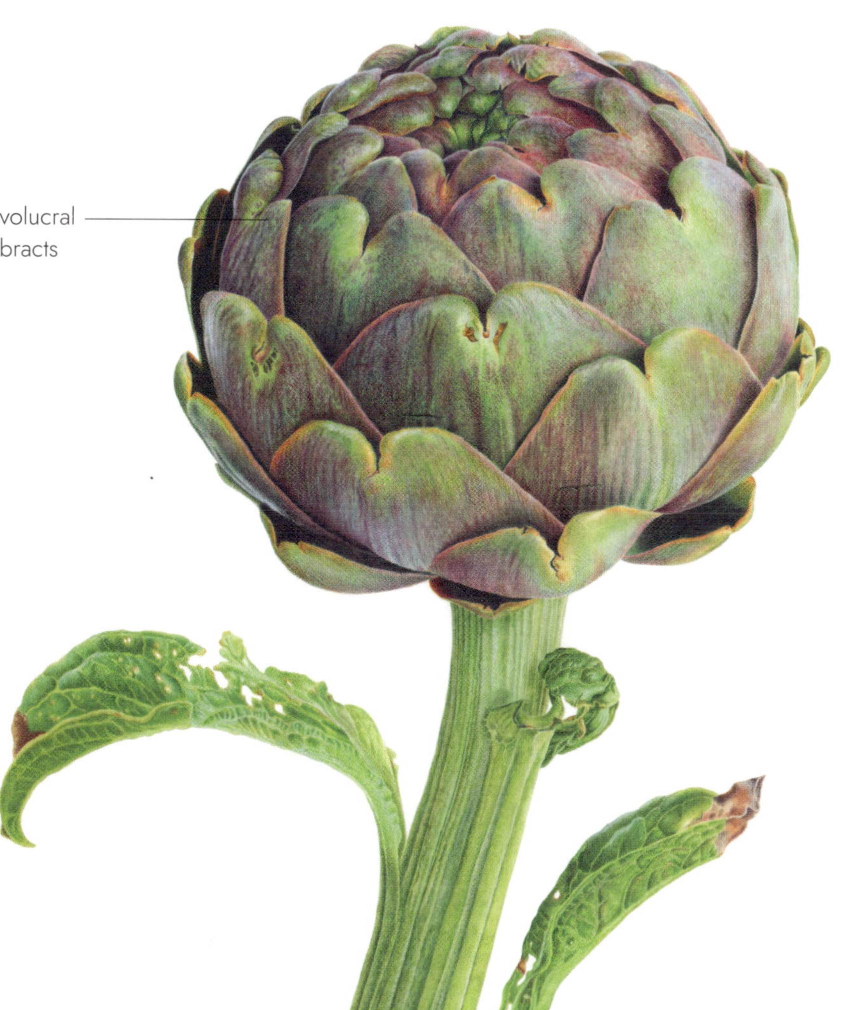

Racemes

A raceme is an inflorescence with flowers, or florets, arranged on a main axis that keeps growing and extending, so the newest buds are at the top. In flat-topped racemose inflorescences, flowers begin opening at the sides and continue into the middle.

A raceme's florets are attached to the stem, often in spirals, by a pedicel. There may be a bract, or bracteole at the base of the peduncle or pedicel.

If sessile with no pedicel, they are spikes, for example, foxgloves or hyacinth. An often-drooping, cylindrical spike with crowded flowers is a catkin; for example, oaks, hazel and birches.

Corymbs are flat topped because their stems, set at intervals along the main axis, are of different lengths, which brings their flowers to the same level; for example, rowans (below).

Also flat topped, or globose, are the flowers of *Apiaceae* such as parsley and carrots, which are characterized by their umbels, usually compound, and emanate from a single point at the stem tip. A composite grouping of flowers sitting on a domed or flat receptacle is a capitulum, characteristic of the daisy family, for example, *Gaillardia* (see page 37). Notice the flowering is from the sides to the centre.

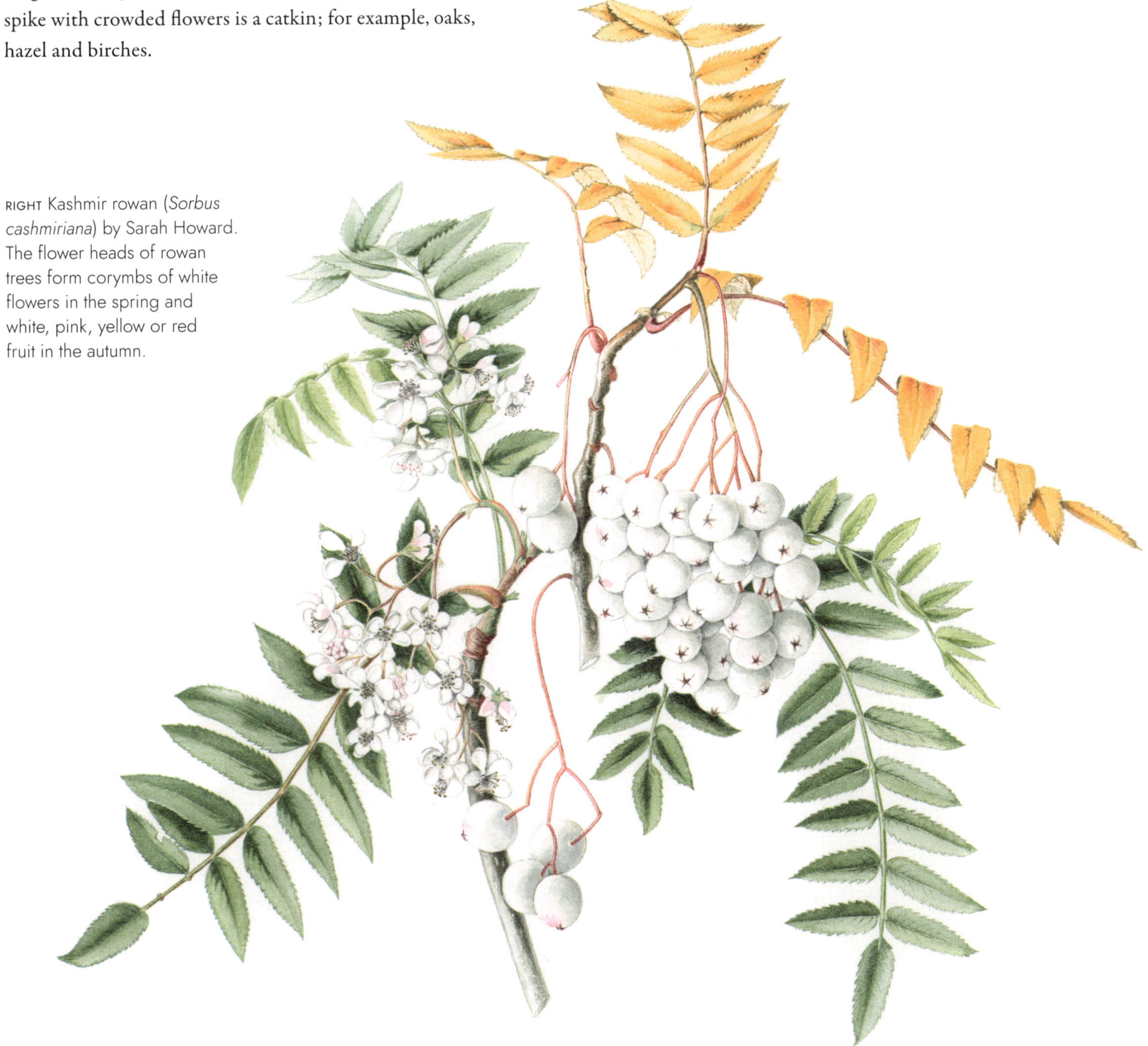

RIGHT Kashmir rowan (*Sorbus cashmiriana*) by Sarah Howard. The flower heads of rowan trees form corymbs of white flowers in the spring and white, pink, yellow or red fruit in the autumn.

Yarrow produces its flowers in a capitulum, like others
of the daisy family, though the inflorescence is in the form
of a corymb.

A secund raceme has florets on only one side, like
foxgloves or Solomon's seal (below).

A thyrse is a mixed form; the main axis is racemose
and keeps growing. Its lateral branches, however, are cymose,
for example *Paulownia* species.

THE TINY WORLD OF THE SPADIX

Arums and anthuriums have specialized spikes called
a spadix. In their case, the inflorescence stalk, the
peduncle, is thick, long and fleshy. Tiny stalkless
flowers sit on it, protected by a large colourful spathe.
Often, male and female flowers sit on the same spadix
with female flowers towards the bottom and male
towards the top, preventing self-fertilization. It is an
imaginative leap to think such tiny flowers can produce
the brightly coloured and relatively large globose fruit
you see in *Arisaema* (see page 152).

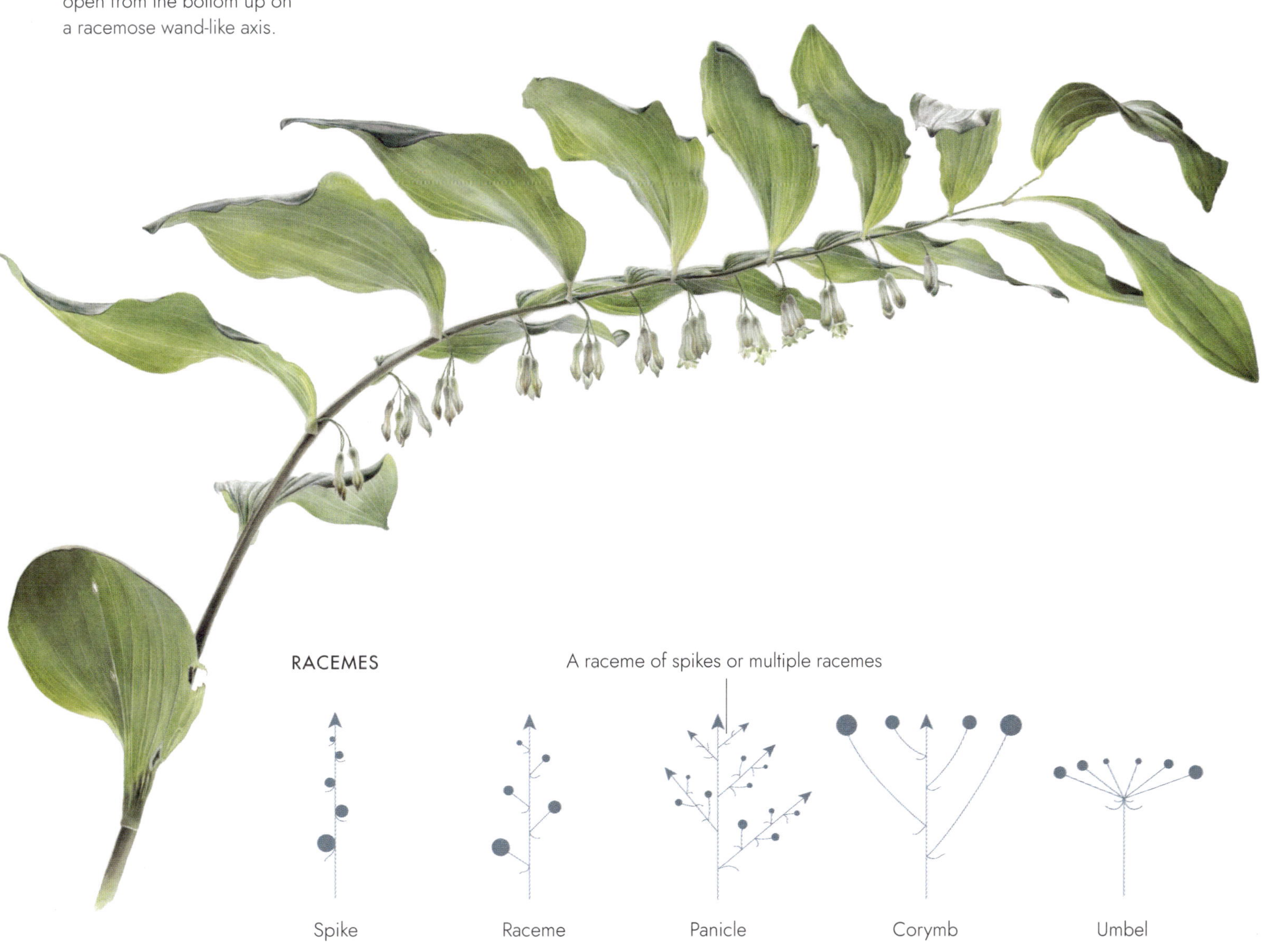

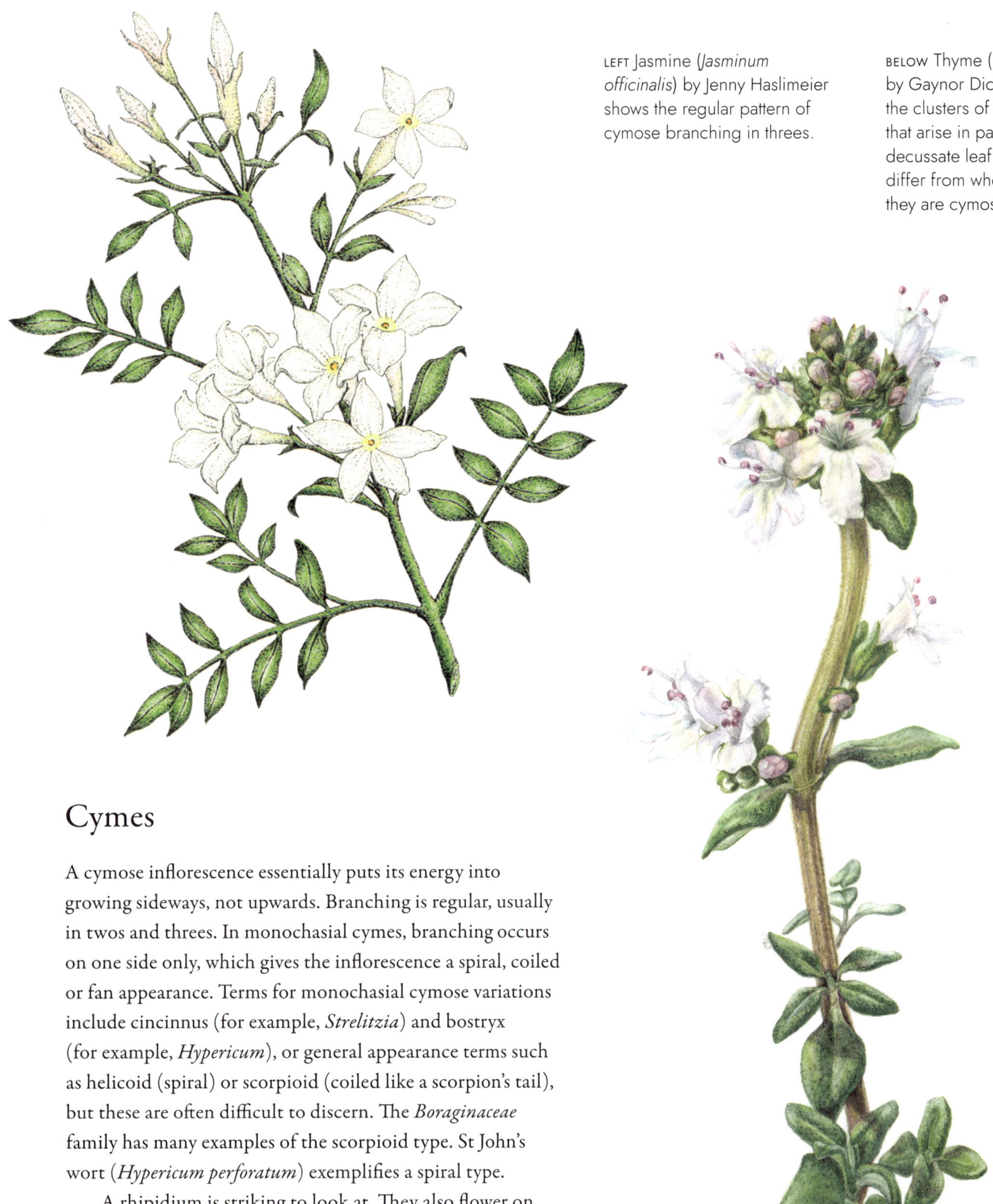

Cymes

A cymose inflorescence essentially puts its energy into growing sideways, not upwards. Branching is regular, usually in twos and threes. In monochasial cymes, branching occurs on one side only, which gives the inflorescence a spiral, coiled or fan appearance. Terms for monochasial cymose variations include cincinnus (for example, *Strelitzia*) and bostryx (for example, *Hypericum*), or general appearance terms such as helicoid (spiral) or scorpioid (coiled like a scorpion's tail), but these are often difficult to discern. The *Boraginaceae* family has many examples of the scorpioid type. St John's wort (*Hypericum perforatum*) exemplifies a spiral type.

A rhipidium is striking to look at. They also flower on one side only, so look for the fading flower on the other side of the stem beneath the others. The lateral branch produces successive flowers in a zigzag pattern on one plane. The bearded iris is a good example of a rhipidium.

Dichasial branching occurs on both sides of the main stem. Jasmine or campions in the *Caryophyllaceae* family show the simplest form. But compound (or polychasial) versions also offer the most beautiful symmetry for an artist – for example, clouds of tiny cymes on baby's breath (*Gypsophila paniculata*) deservedly make it a florist's, as well as an artist's, favourite.

- Look out for both monochasial and dichasial, or even polychasial, forms on one cymose inflorescence.
- In flat-topped cymose inflorescences, the centre flowers open first, followed by the outer ones; for example, the elderflower (*Sambucus nigra*).

When the flowers are organized evenly along the axis like whorls, the cymes are called verticillasters, which is typical of the mint family. The garden plant, *Salvia verticillata* is named for this feature. They are referred to as false whorls because, although they look like whorls, they are really clusters of cymes.

THE SCORPION'S STING

Have a look at a flowering forget-me-not, Russian comfrey or lungwort, and you will see how the flower head curls into itself like a scorpion's tail, with the newest buds tucked into the centre. Look at how each pedicel is inserted on only one side of the axis. However, if you missed an old flower position below the scorpion's tail, on the other side of the stem, you could miss why this inflorescence is a cyme and not a raceme. The old flower is the first, followed by the lateral growth on the other side forming the scorpion's tail.

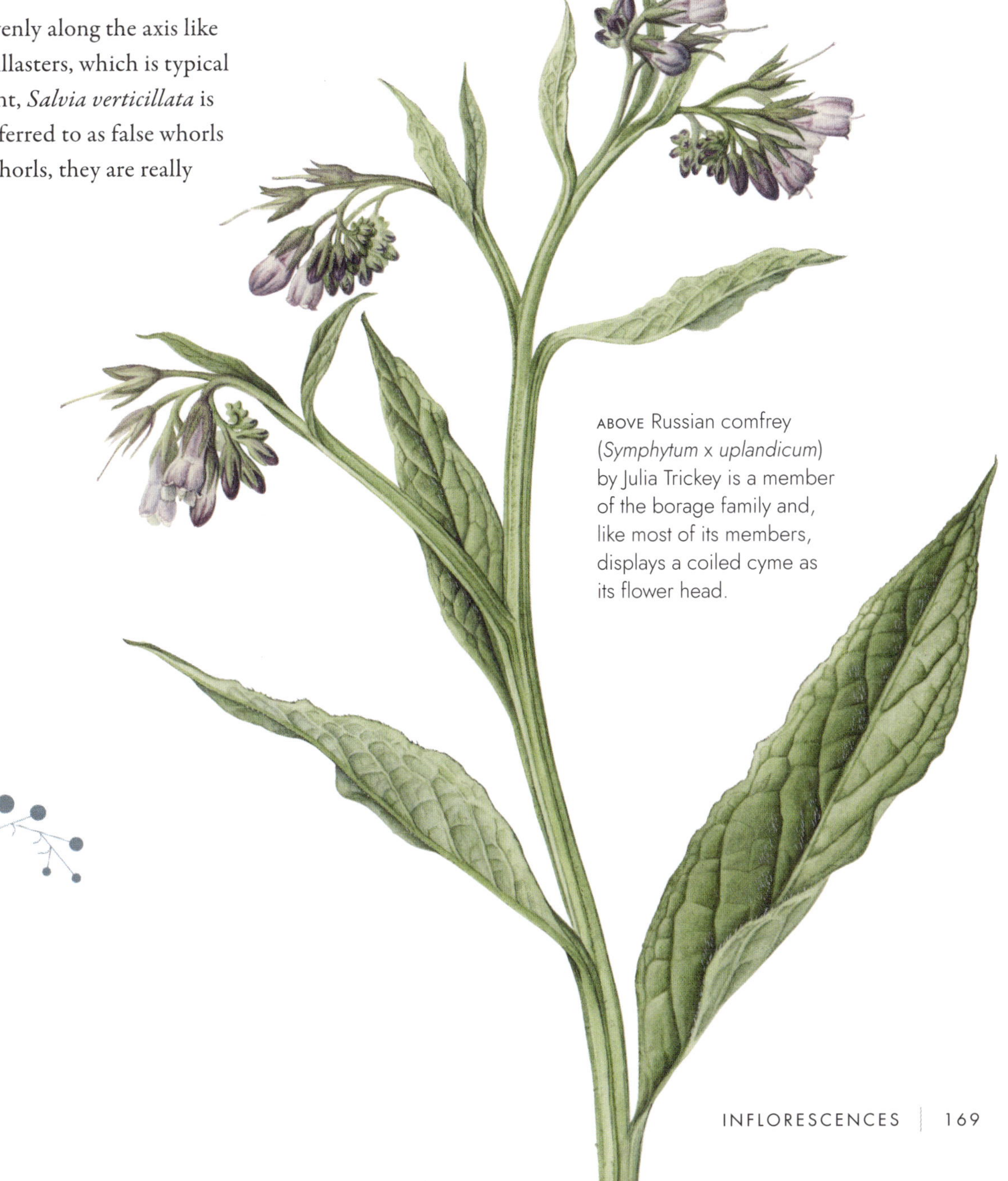

ABOVE Russian comfrey (*Symphytum* x *uplandicum*) by Julia Trickey is a member of the borage family and, like most of its members, displays a coiled cyme as its flower head.

CYMES

Simple cyme
paired

Simple cyme
in threes

Compound
cyme

Coiled
cyme

Painting a cuckoo pint

Cuckoo pint is a woodland-flowering plant species in the family *Araceae*. The flowers cannot be seen without dissection, as they are enclosed within an inflated bract called a spathe. What is visible is the upper portion of the spadix protected by a large spathe (a protective bract). This wildflower has been chosen for its iconic and sculptural shape, plus its beautiful leaves. This tutorial focuses on adding detail to surfaces, such as the small details on the spathe, and creating a dark but shiny leaf.

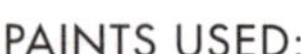

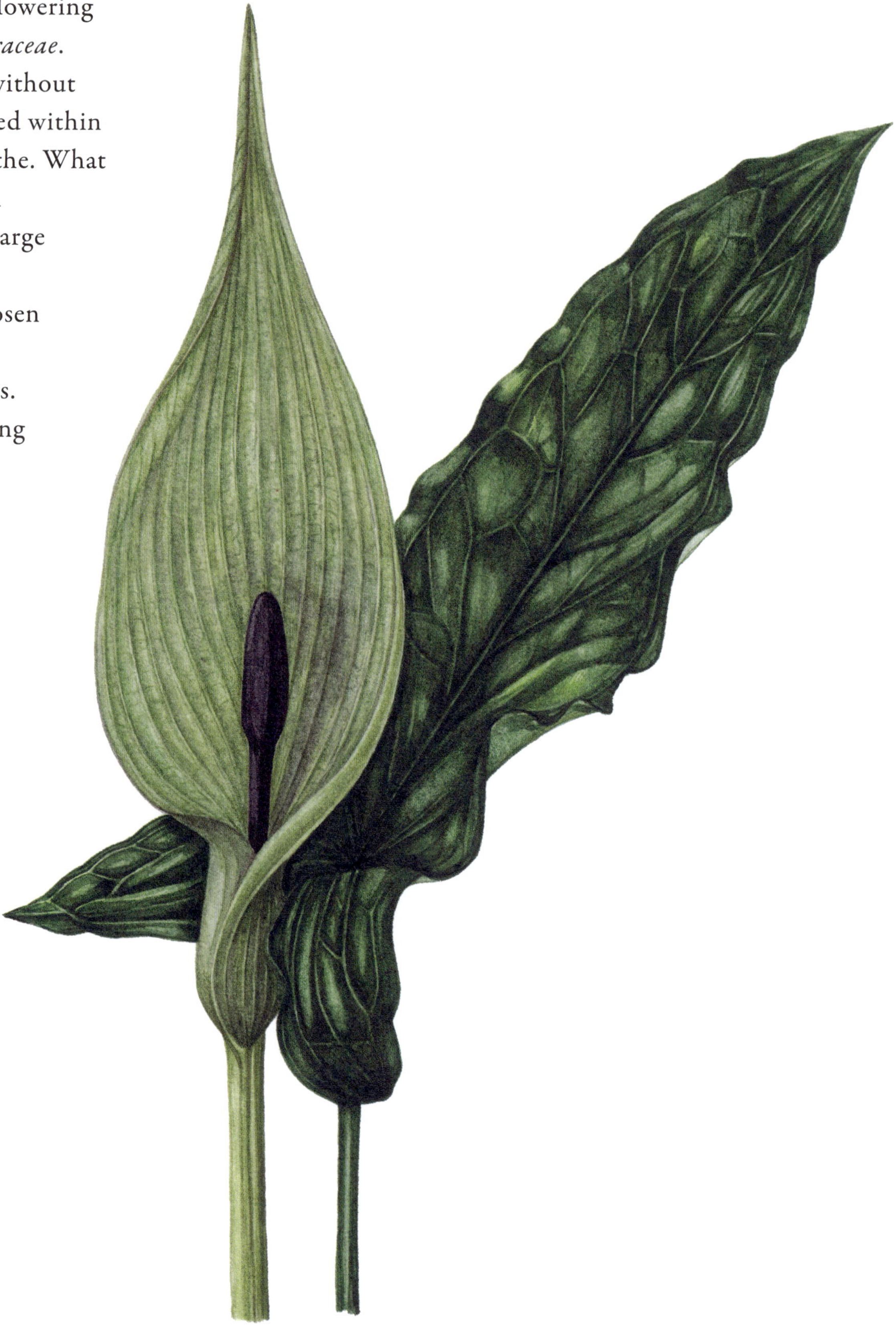

PAINTS USED:

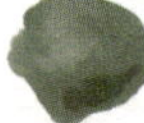 Sap green

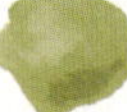 Olive green

 Lemon yellow

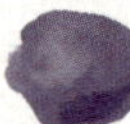 Winsor violet (dioxazine)

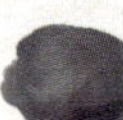 Payne's grey

 Scarlet lake

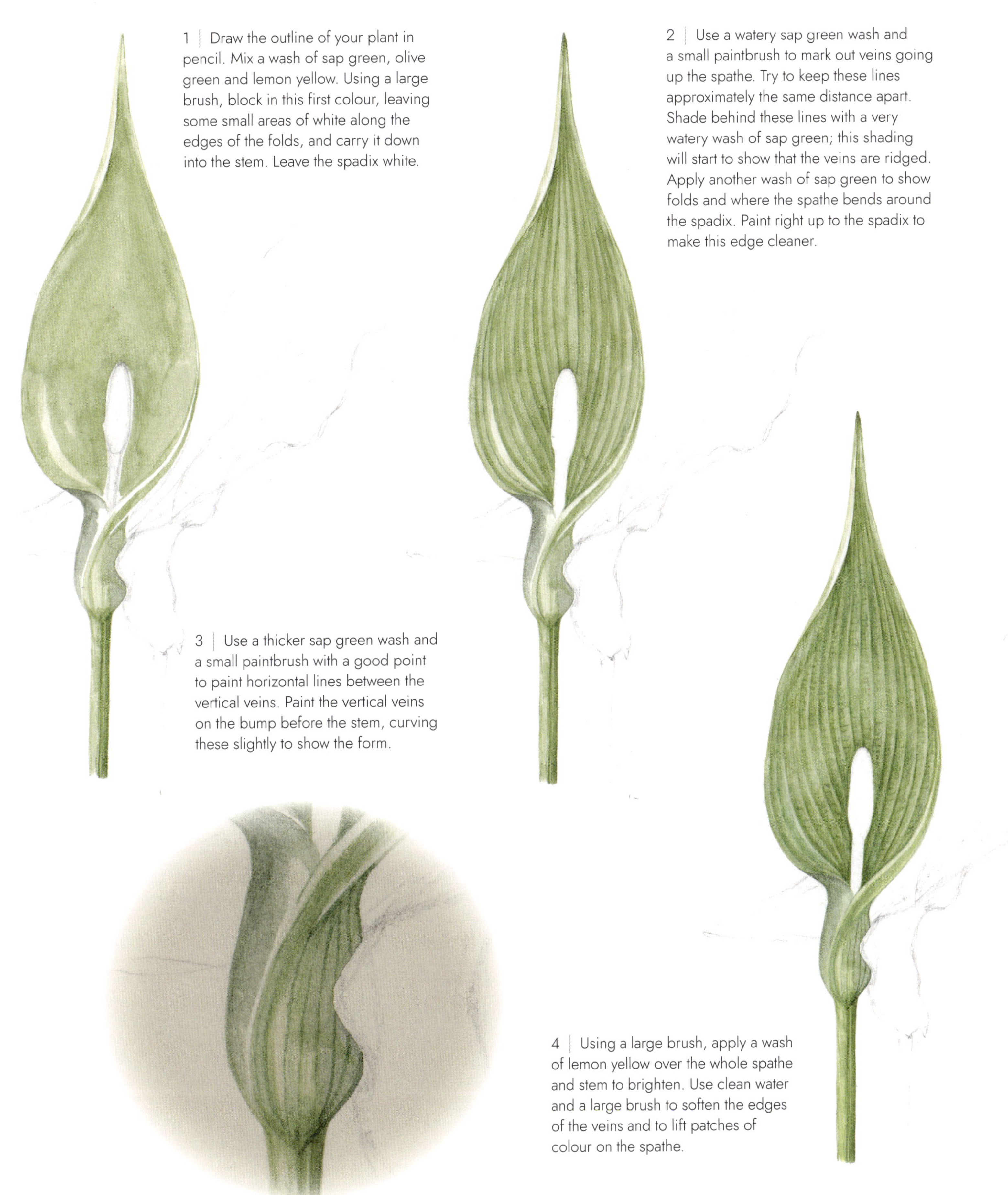

1 | Draw the outline of your plant in pencil. Mix a wash of sap green, olive green and lemon yellow. Using a large brush, block in this first colour, leaving some small areas of white along the edges of the folds, and carry it down into the stem. Leave the spadix white.

2 | Use a watery sap green wash and a small paintbrush to mark out veins going up the spathe. Try to keep these lines approximately the same distance apart. Shade behind these lines with a very watery wash of sap green; this shading will start to show that the veins are ridged. Apply another wash of sap green to show folds and where the spathe bends around the spadix. Paint right up to the spadix to make this edge cleaner.

3 | Use a thicker sap green wash and a small paintbrush with a good point to paint horizontal lines between the vertical veins. Paint the vertical veins on the bump before the stem, curving these slightly to show the form.

4 | Using a large brush, apply a wash of lemon yellow over the whole spathe and stem to brighten. Use clean water and a large brush to soften the edges of the veins and to lift patches of colour on the spathe.

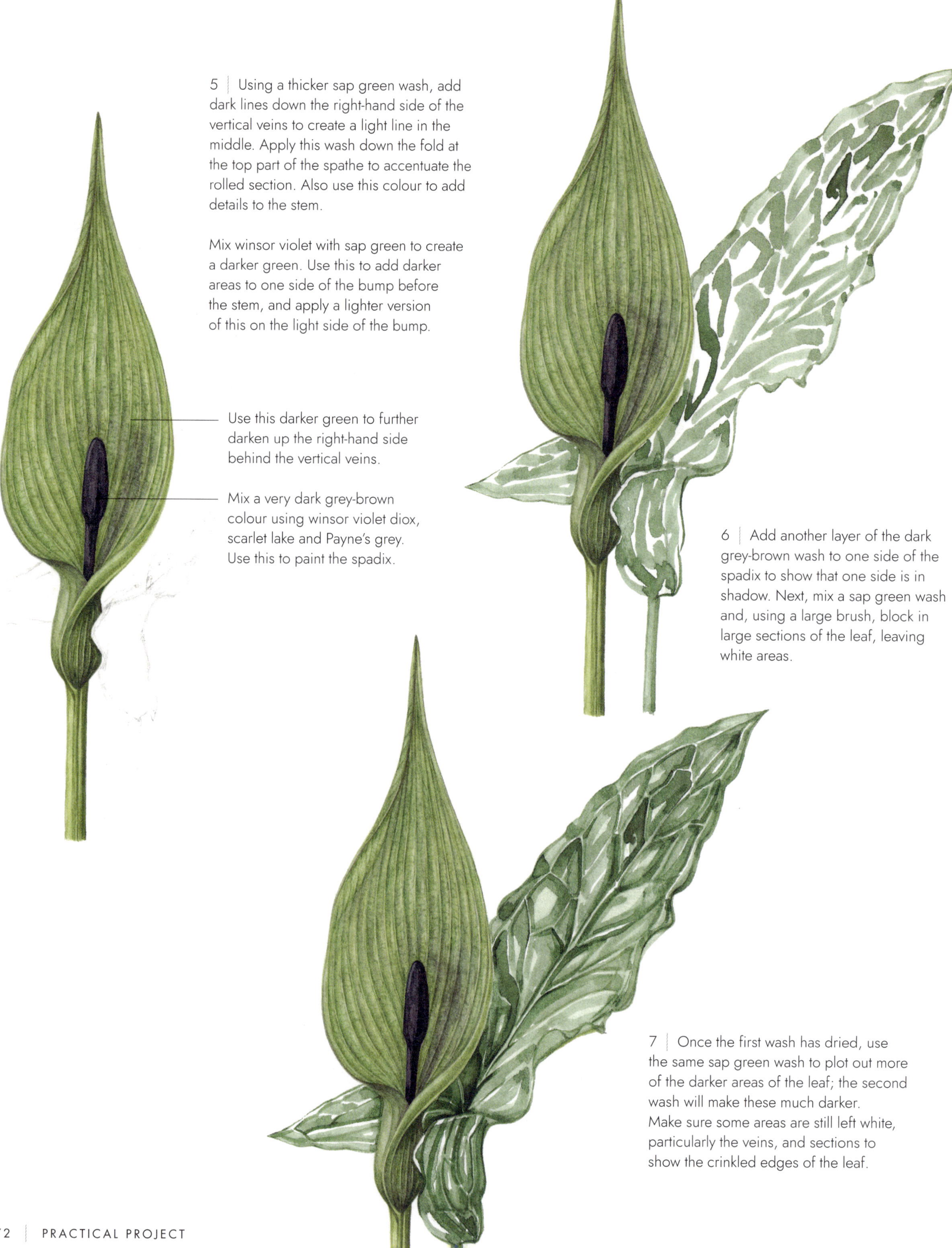

5 | Using a thicker sap green wash, add dark lines down the right-hand side of the vertical veins to create a light line in the middle. Apply this wash down the fold at the top part of the spathe to accentuate the rolled section. Also use this colour to add details to the stem.

Mix winsor violet with sap green to create a darker green. Use this to add darker areas to one side of the bump before the stem, and apply a lighter version of this on the light side of the bump.

Use this darker green to further darken up the right-hand side behind the vertical veins.

Mix a very dark grey-brown colour using winsor violet diox, scarlet lake and Payne's grey. Use this to paint the spadix.

6 | Add another layer of the dark grey-brown wash to one side of the spadix to show that one side is in shadow. Next, mix a sap green wash and, using a large brush, block in large sections of the leaf, leaving white areas.

7 | Once the first wash has dried, use the same sap green wash to plot out more of the darker areas of the leaf; the second wash will make these much darker. Make sure some areas are still left white, particularly the veins, and sections to show the crinkled edges of the leaf.

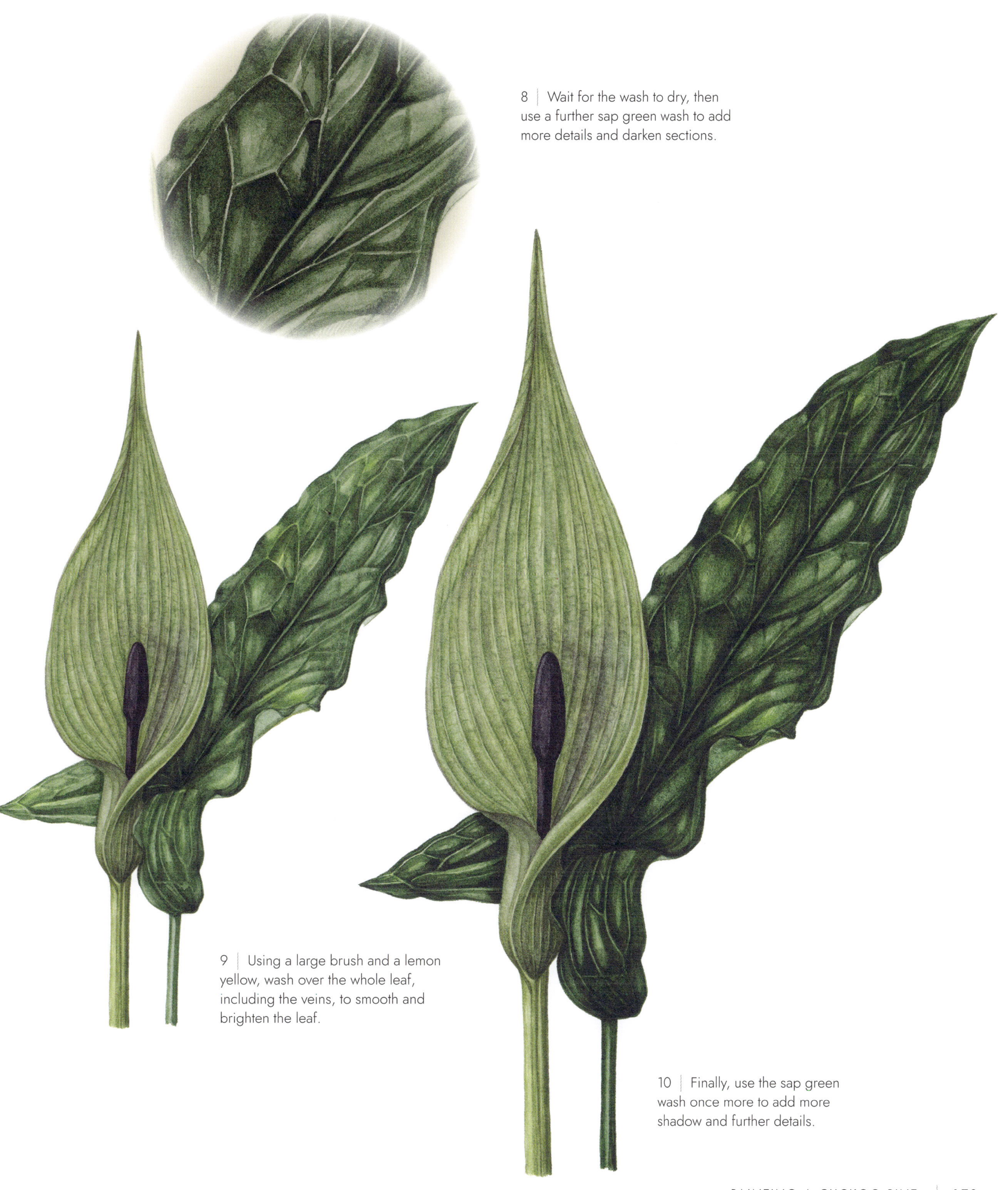

8 | Wait for the wash to dry, then use a further sap green wash to add more details and darken sections.

9 | Using a large brush and a lemon yellow, wash over the whole leaf, including the veins, to smooth and brighten the leaf.

10 | Finally, use the sap green wash once more to add more shadow and further details.

Orchid (x *Papilionanda* Tan Chay Yan)
by Waiwai Hove

Waiwai Hove's style of painting is a perfect fit for the smooth fleshy outlines of an orchid. She paints clearly with a smoothness that does justice to these beautiful plants.

Everyone loves orchids, no more so than in Singapore, where the Singapore Botanic Gardens commissioned Waiwai to paint a series of orchid cultivars, including this one.

This orchid, x *Papilionanda* Tan Chay Yan, is a well-known hybrid between *Vanda dearei* and x *Papilionanda* Josephine van Brero, raised by Tan Hoon Siang in the late 1940s. It was named for his father, a member of a prominent family of pioneer rubber planters in Malaya and a founding father of Singapore. The orchid won an RHS first class certificate at the 1954 Chelsea Flower Show. It is appropriate, therefore, that this orchid should be recorded by an artist of Malaysian origin, who has lived in Singapore and continues to enjoy a cultivar in her current home in Western Europe.

Botanical skill

Waiwai takes our eye straight into the depths of the inflorescence, as though we were bending into the flower for a closer look. We are mesmerized by the lovely golden pink forms, by the firm flowers with their red lips and divided epichilia stretched out horizontally, forming a landing pad for pollinators. There is a suggestion of tessellation in the venation on the petals.

Behind the aesthetic attractions of a lovely painting of a lovely orchid is an artist of real botanical skill, for she has managed to place the flowers in many different angles. They show the twisted inferior ovary beneath the flower, the column (a perfect descriptor in this case), and there is a hint, behind the pinks of the flowers, that the leaves are attached to the stem in an equitant arrangement, indicating the orchid has a monopodial growth form.

ABOUT THE ARTIST

Waiwai Hove grew up in Kuala Lumpur, Malaysia, where her mother nurtured in her a love for plants, and she discovered she had an aptitude for drawing them. While living and working in Hong Kong (China), she began to develop her skills using reference books in her free time. On moving to Singapore, Waiwai completed an online diploma course with distinction and received the Award for Excellence from the UK-based Society of Botanical Artists. She now lives in Denmark but continues to paint commissions for clients in Singapore. She is currently working on illustrated covers for 14 volumes of the flora of Singapore.

RIGHT A cluster of orchid flowers against latticed leaves makes a striking composition. Size: 31 × 41cm (12^1/$_4$ × 16in), watercolour on Lanaquarelle HP 300gsm paper. Commissioned by Singapore Botanic Gardens, National Parks Board.

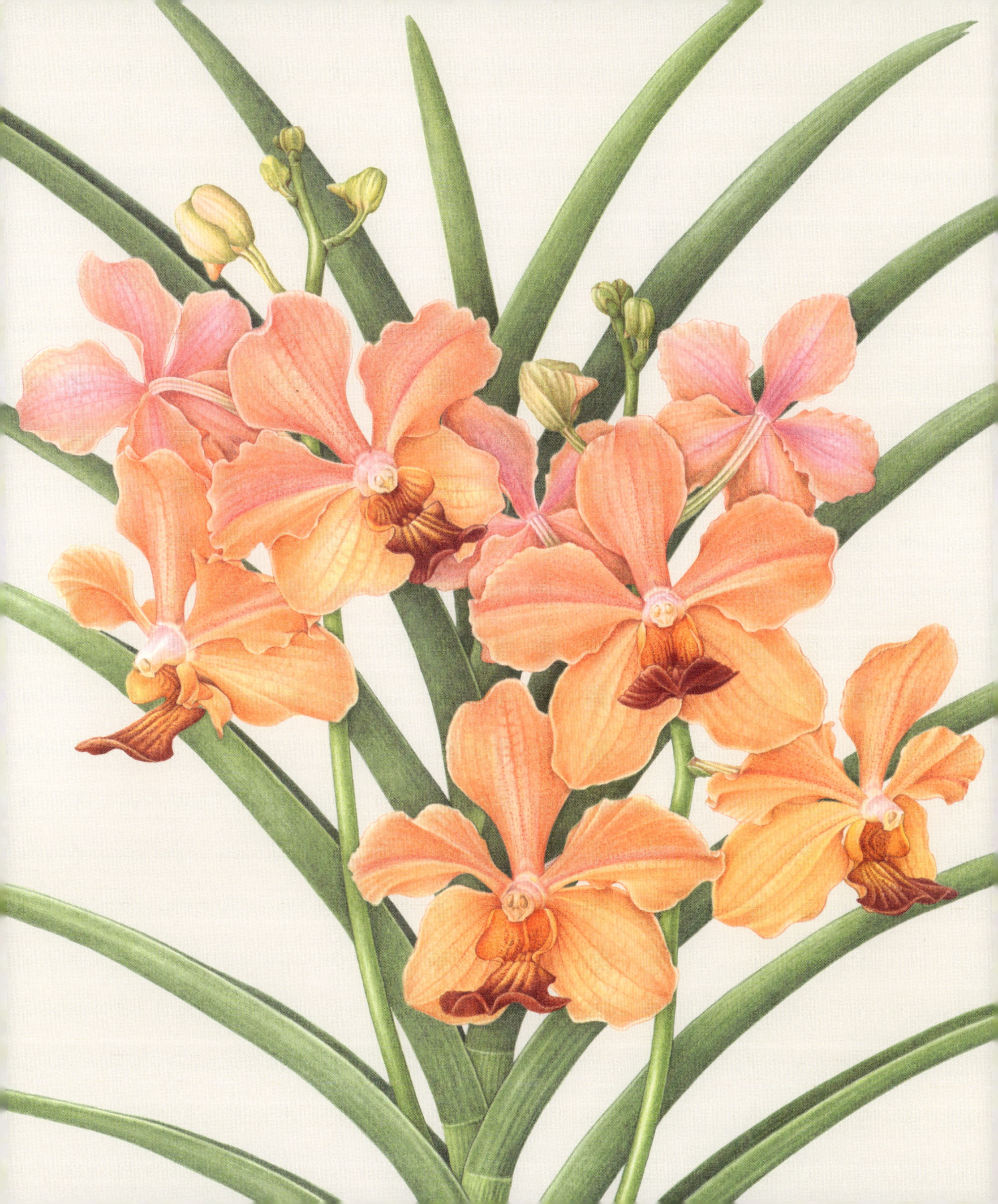

Chapter six

Fruits and seeds

Flowering plants, the angiosperms, reproduce by means of their seeds. They are safely enclosed in an ovary for fertilization and maturation. The structure is generally called a 'fruit' once the ripening process begins.

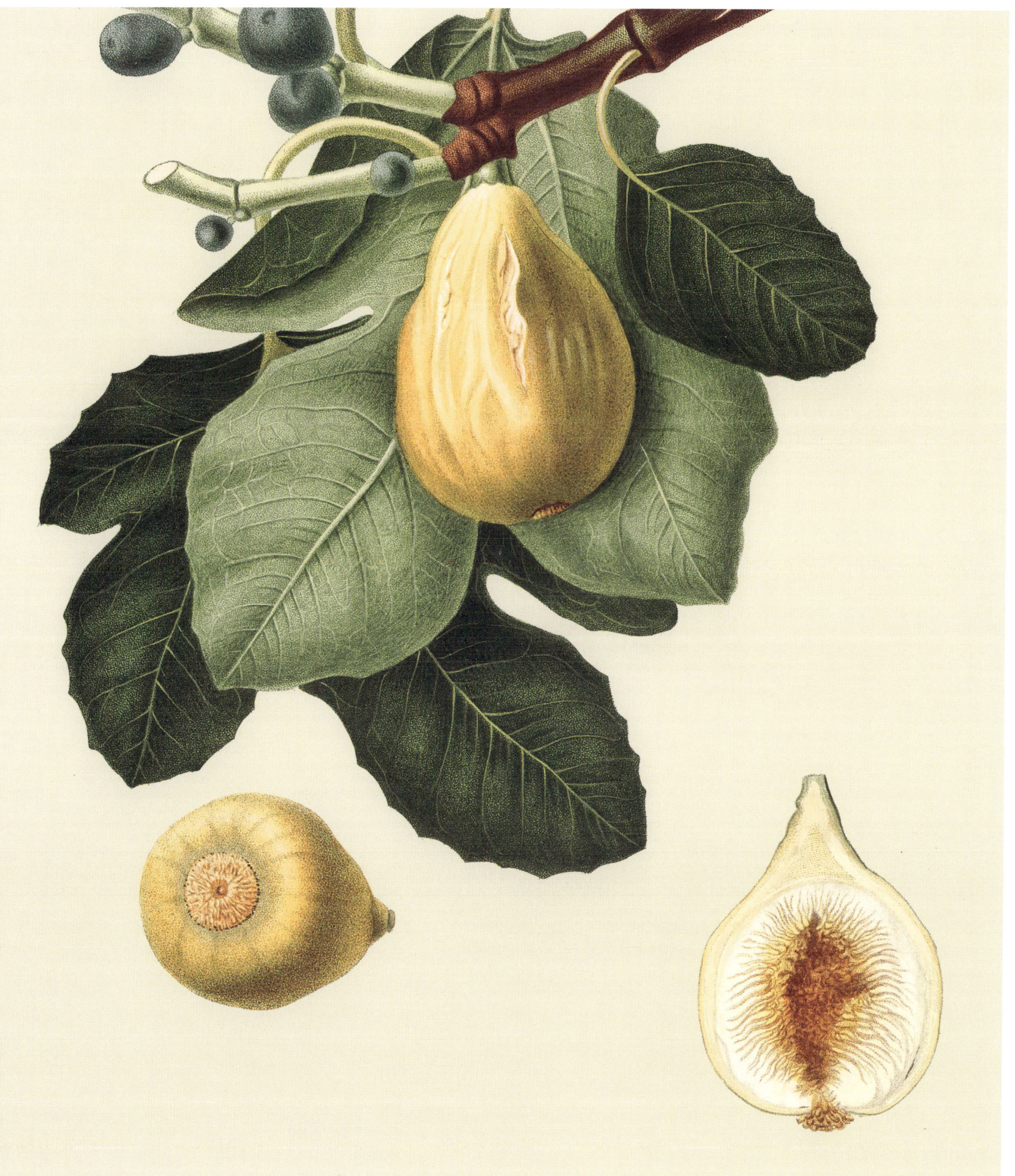

What is a fruit?

Fruits are the ripe ovaries of a plant, which contain and protect the seeds. Fruits of all kinds make beautiful subjects for painting. Occasionally, the terminology defies our common perceptions. For example, why is a coconut not a nut? Why are strawberries not 'true fruits'? It matters only if you want to know what you are really looking at. Understanding the structure of the ovary is a prerequisite. Anne Bebbington and Lizbeth Leech provide the best botanical explanations for artists (see Bibliography).

Many herbaria have carpological collections. These are the dried fruits and seeds used for study. But collections of your own will make excellent painting subjects. Different kinds of cone, together with their seeds, provide flowing Fibonacci patterns. Various maple samaras, with their nutlets, can flutter pleasingly on the paper. Meanwhile, the legume pods of the New Zealand 'kowhai' (*Sophora* spp.) are distinctively constricted at intervals along their length.

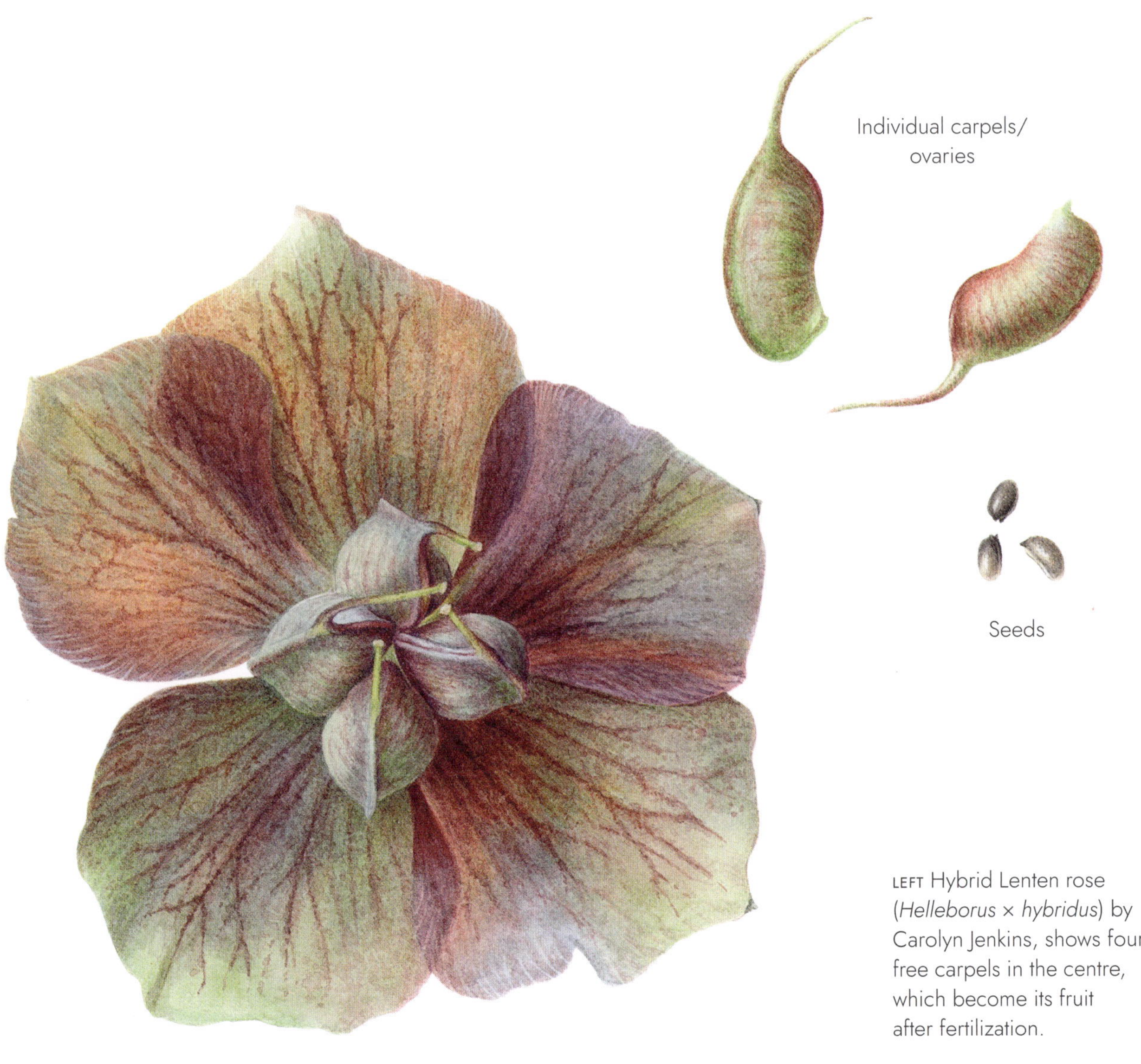

LEFT Hybrid Lenten rose (*Helleborus* × *hybridus*) by Carolyn Jenkins, shows four free carpels in the centre, which become its fruit after fertilization.

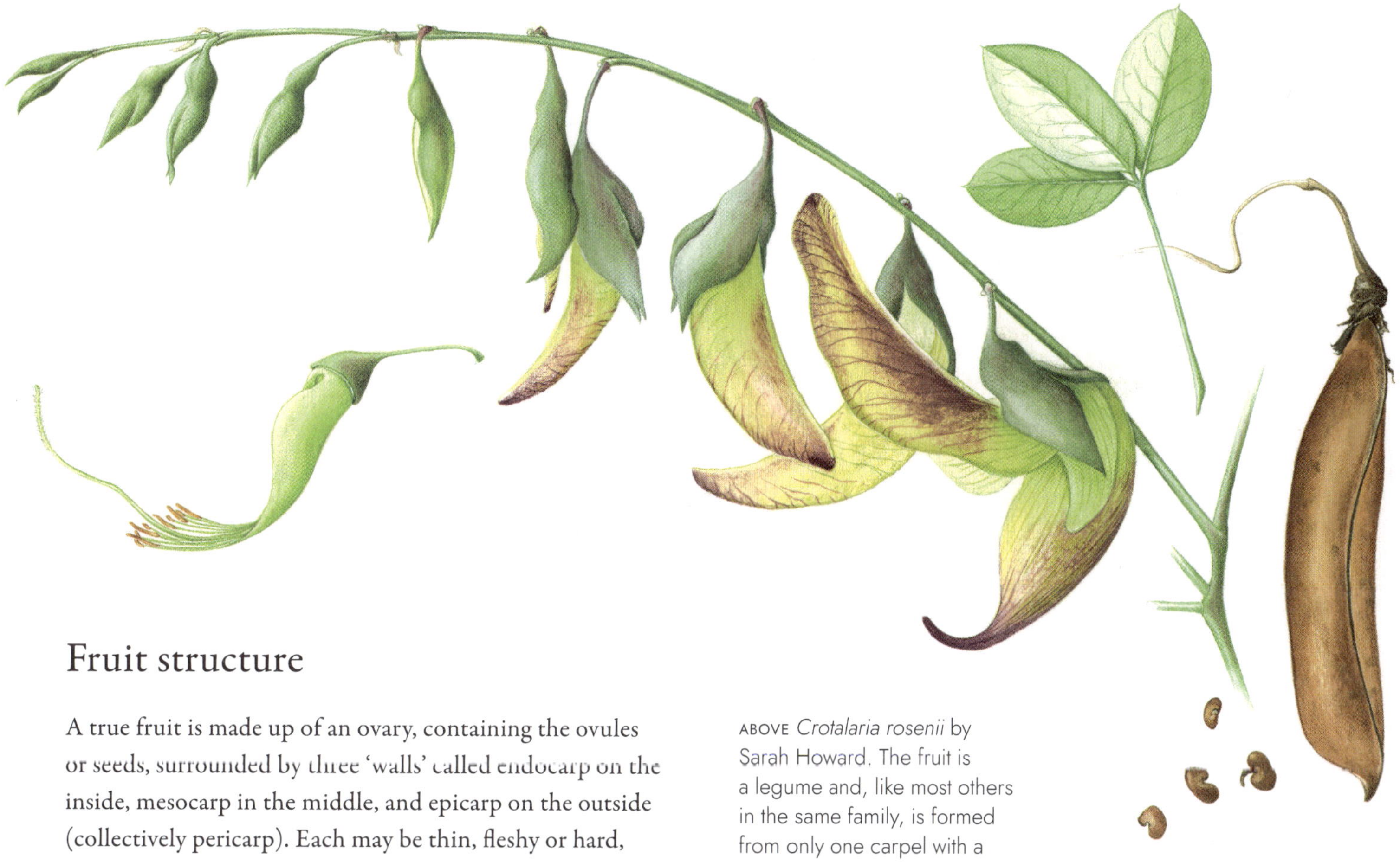

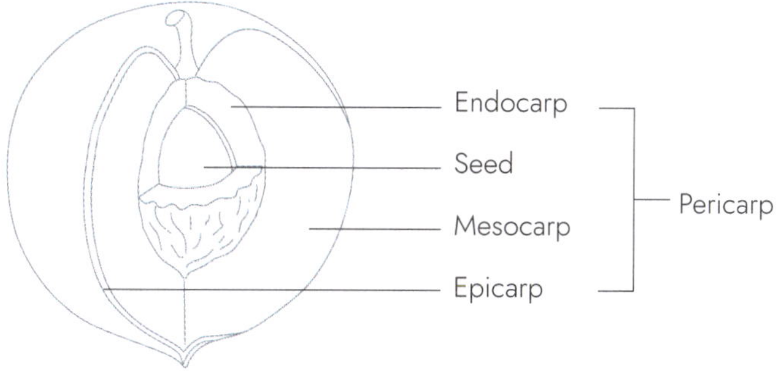

Fruit structure

A true fruit is made up of an ovary, containing the ovules or seeds, surrounded by three 'walls' called endocarp on the inside, mesocarp in the middle, and epicarp on the outside (collectively pericarp). Each may be thin, fleshy or hard, depending on the fruit. It also helps to understand that flowers may have more than one ovary or carpel, and that each may be formed of more than one part. Most flowers have more than one carpel. They are syncarpous if all the carpels are fused, as they are in lilies, tomatoes and bananas; or apocarpous if the ovaries are free, for example, in hellebores. They are monocarpous when there is only one carpel, as in most legumes.

The number of style arms is one way of deciding the number of carpels. If there is only one, the fruit is 'monocarpous', like a fleshy peach, coffee 'cherry' or a pea pod. If there are several style arms, it may be syncarpous. Examine the three arms of a daffodil, or the five that appear from inside a chickweed.

Counting the styles and making transverse dissections of ripening fruit help to discern what's what, but it is easy to be confused by a plant's ability to hide its true nature. Several flowers may even merge to form a multiple fruit, for example, pineapples, mulberries and figs.

FRUIT PARTS

Endocarp

Seed

Mesocarp

Epicarp

Pericarp

Dry fruits

A common method of distinguishing between different fruits is whether they are dry or fleshy. Dry fruits can be divided first by how they split open (dehiscence):

- Legumes in the *Fabaceae* family open on both margins (for example, broom, edible bean pods).

- Follicles open on one side only.

- Capsules are dry, dehiscent fruit with two or more carpels, opening on one side by valves, slits or pores (for example, iris, poppy, horse chestnut and peony).

- Siliqua(ae) is an elongated pod of two fused carpels, divided inside by a partition on which the seeds are attached, for example, willowherbs (*Epilobium* spp.).

Dry fruits that do not split open (indehiscent) include the following types:

- Achenes (for example, sunflower, clematis) are small fruit containing one seed. The seed is free of the endocarp wall, except where it's attached to the placenta. Cypsela is a special term for *Asteraceae* fruit.

- Nuts (such as acorn, hazel and beech) derive from flowers containing one seed. The pericarps have very hard casings.

- Samara are like an achene except that part of the pericarp extends to form a wing (for example, maples, sycamores).

There is also a raft of specialist terms referring to dry fruit which, when ripe, splits into parts called mericarps. They include important spice plants in the *Umbelliferae* family.

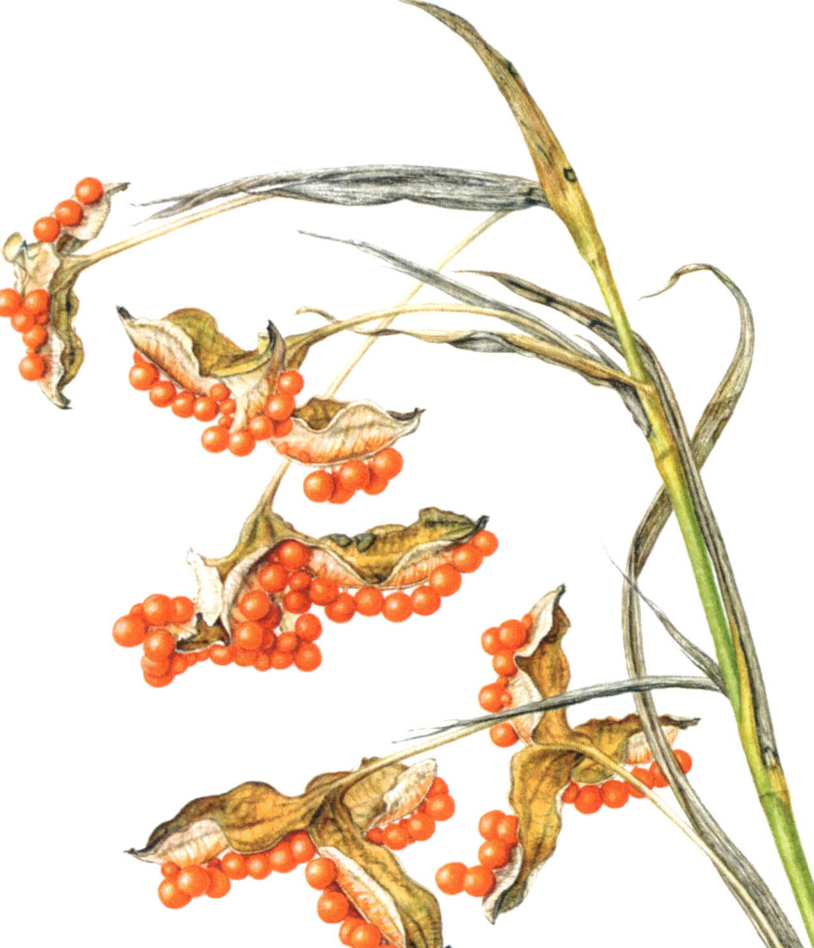

ABOVE Hornbeam (*Carpinus betulus*) detail by Agita Keiri. The samara of hornbeam consist of two nuts attached to wings to 'helicopter' them away from the parent tree.

LEFT Stinking iris (*Iris foetidissima*) by Linda Pitkin. Its charm is in the capsules that open in autumn to reveal bright red seeds.

Fleshy fruits

The culinary use of fruits predates botanical studies, so common names, rather than botanically correct names, make the subject confusing. Botanically, fleshy fruit can be divided by pericarp characters: those with a hard endocarp, surrounded by fleshy or thin exocarps and mesocarps (drupes); and those with a soft endocarp (berries).

Drupes are fruit with a hard endocarp embedded in a fleshy exo- and mesocarp. The stone containing the seed is inside, for example, peaches or plums, sloes, dates, coffee, walnuts and Brazil nuts. A drupe's sweet flesh is attractive to animals, who eat the fruit and help to distribute the seeds inside through their digestive tracts. Drupes can be combined into an aggregate fruit (see below).

A coconut is a drupe, rather than a nut. The outer layers of the green covering and the coir, form the exo- and mesocarp. The inner endocarp is the hard shell with a soft germination hole at one end. The white 'meat' inside is thickened endocarp in which is embedded a small seed.

Berries, strictly, are formed from a single ovary with many seeds inside a thin-skinned exocarp. The inner two layers are fleshy and undifferentiated, for example, tomatoes, kiwi fruit and honeysuckle. A variant is the hesperidium, common to citrus fruit, which all have a thick rind for the exocarp and a fleshy interior divided into segments.

An avocado is a berry, not a drupe, despite the hard seed inside. That is because the brown skin or endocarp that surrounds the white seed is soft, rather than hard.

Even a pomegranate is a berry, with many seeds in its one ovary.

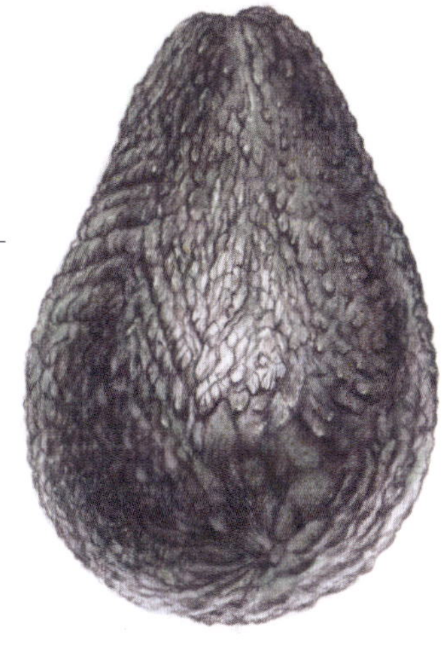
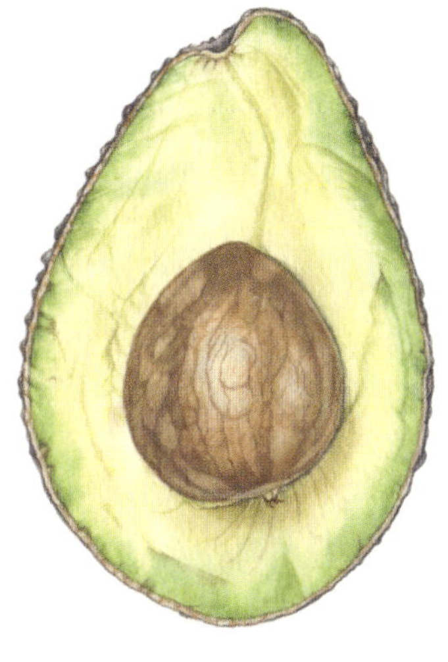

ABOVE Avocado (*Persea americana*) by Toni Dade. The fruit is botanically a berry because the brown skin — the endocarp surrounding the white seed inside — is soft rather than hard.

BELOW Lemon 'Mano di Buddha' (*Citrus medica* 'Fingered') by Simonetta Occhipinti. Like other citrus fruit, it has a thick rind (the exocarp) with a fleshy, segmented interior and is called a hesperidium.

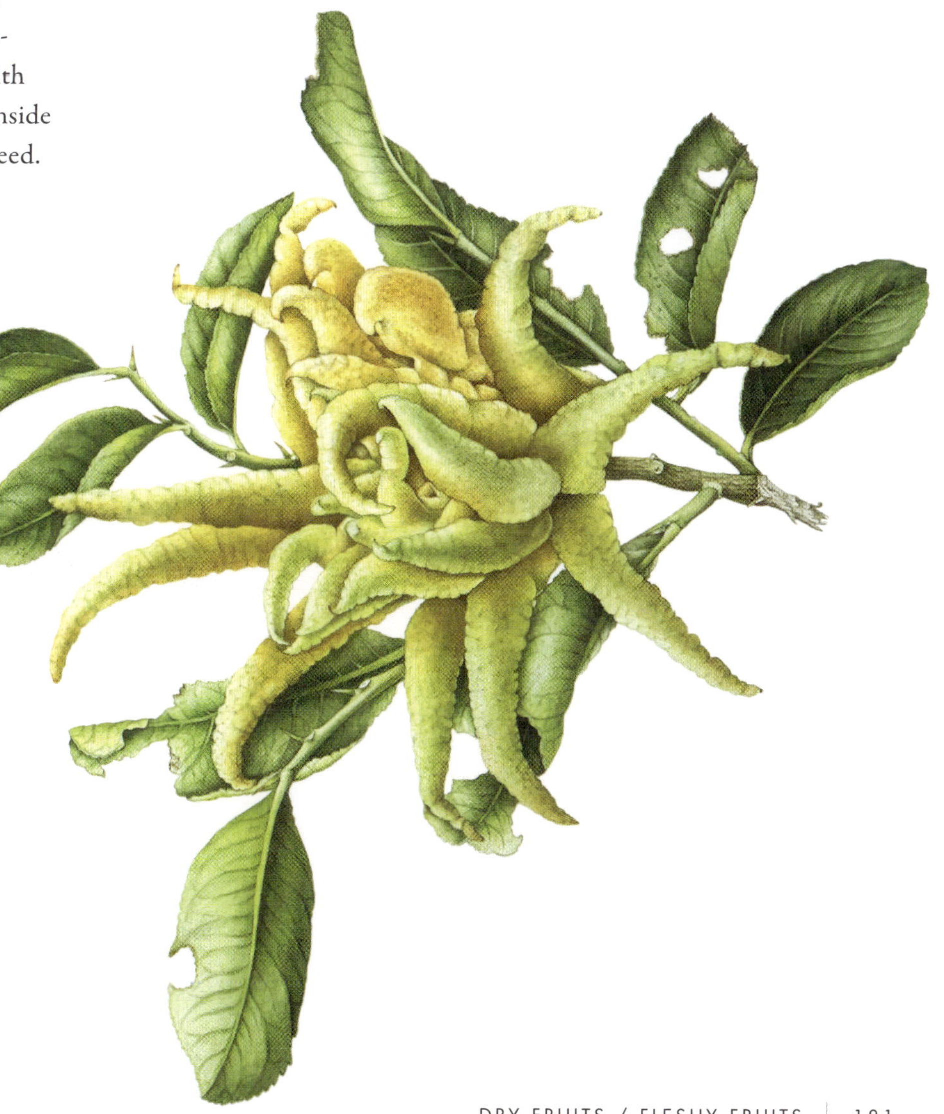

False, aggregate and multiple fruits

In practice, an artist is unlikely to need to know why a fruit is false, aggregate or multiple, but it can be helpful to understand their structure. An explanation often depends on having a powerful microscope to see the differences in tissue structure. False (accessory) fruits are formed from parts in addition to the ovary. Aggregate fruits are, as their name suggests, aggregated carpels. Multiple fruits comprise several fruits fused into a single one. Aggregate fruits include true and false fruits. An example of an aggregate true fruit is the raspberry (*Rubus* spp). It has one flower, but several carpels, so each swelling represents a drupe from a single carpel, with its pip or seed inside.

WHAT TO LOOK FOR

- Look for an apple's five fused carpels in a transverse section. The pentamerous character of *Malus* will be repeated in its five sepals and five style arms.

- Identify the faint mark of the pericarp that surrounds the carpels. It should be evident in both a longitudinal section and a transverse section.

- Note the skin colour and any bleeding into the flesh on a half apple. The skin, in a pome's case, is the epidermis, or outer skin of the hypanthium, not part of the pericarp.

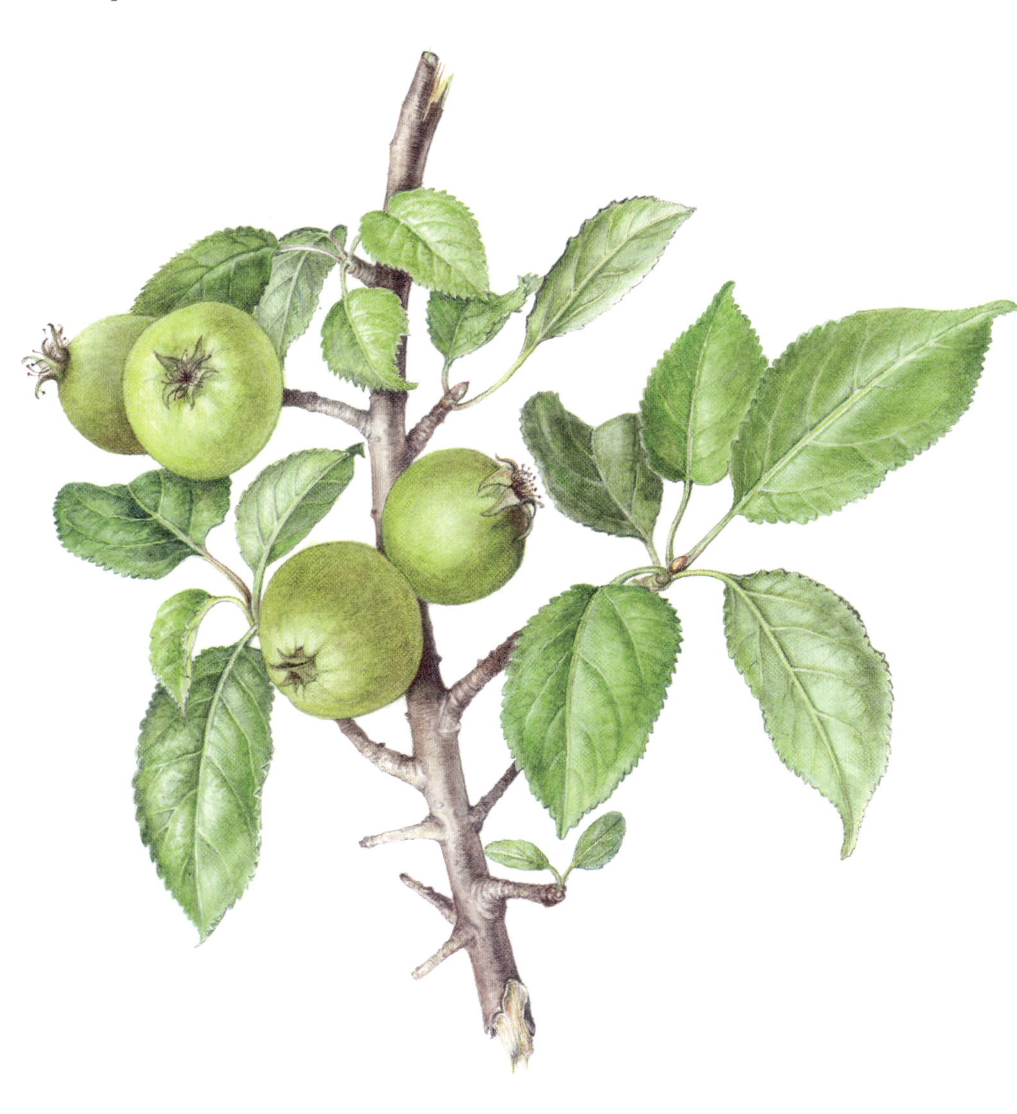

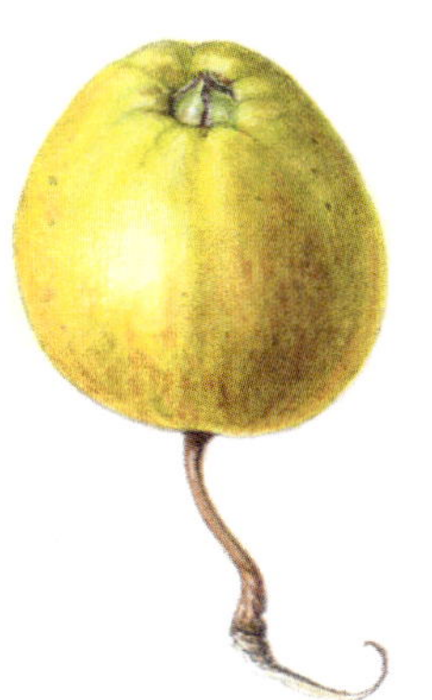

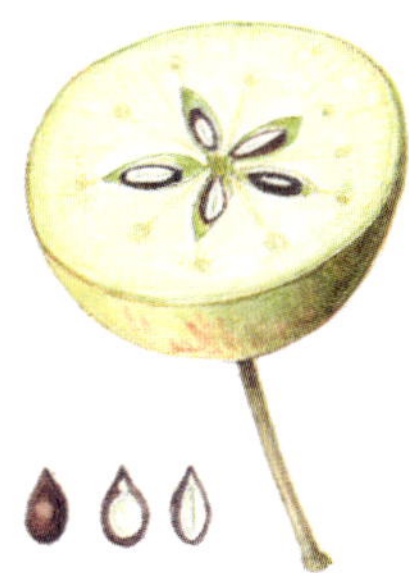

A cut through a pome

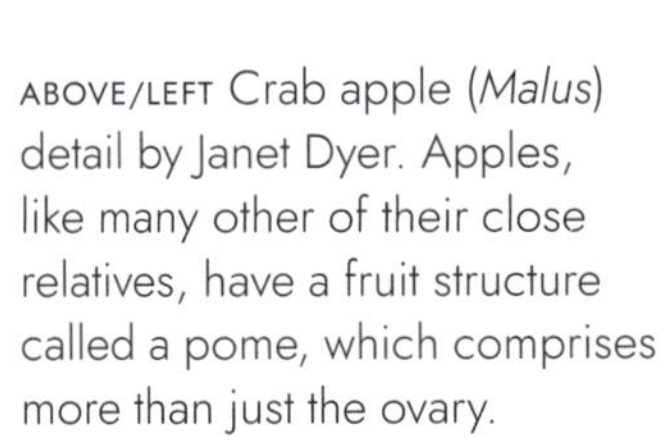

ABOVE/LEFT Crab apple (*Malus*) detail by Janet Dyer. Apples, like many other of their close relatives, have a fruit structure called a pome, which comprises more than just the ovary.

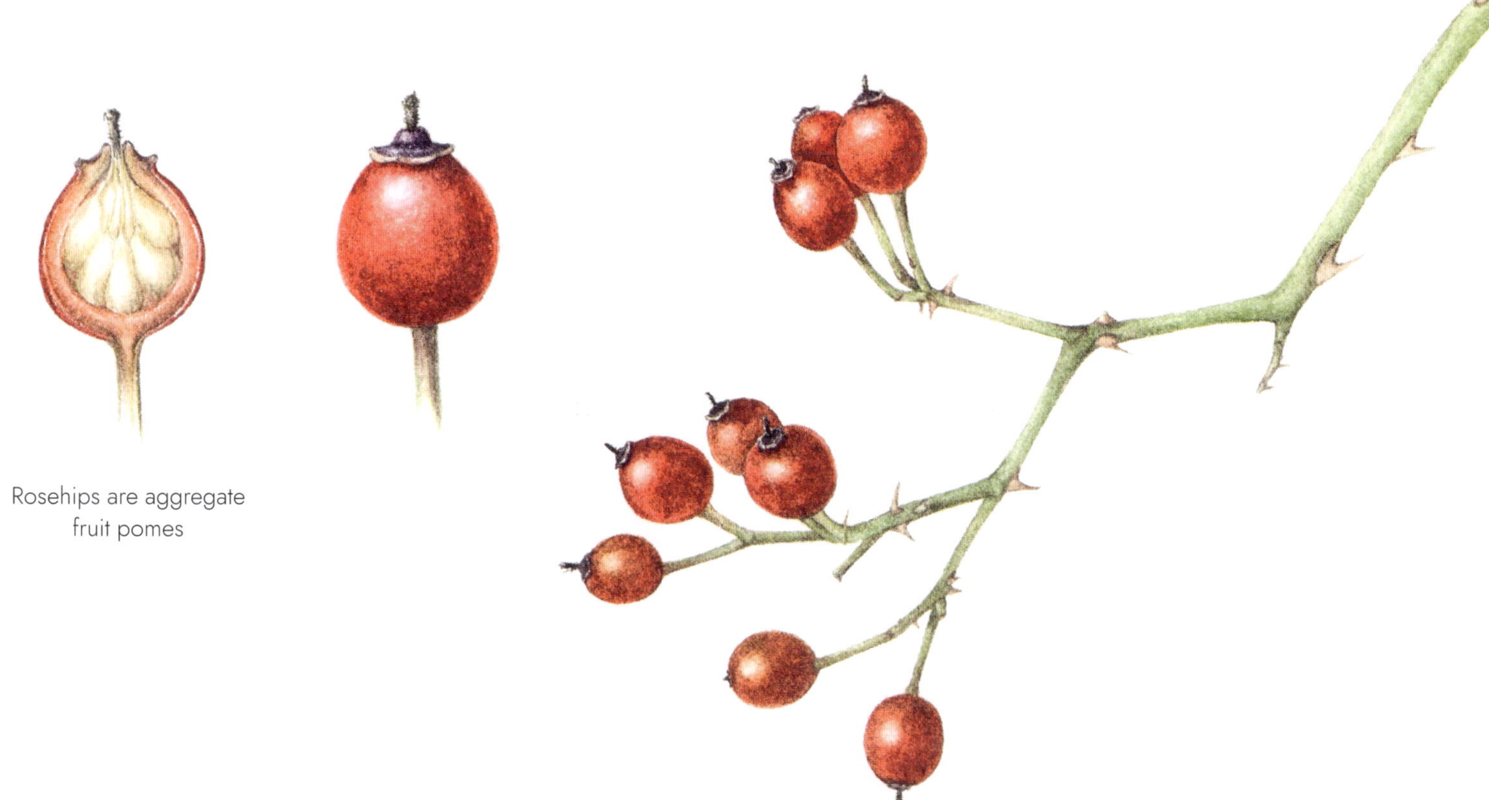

Rosehips are aggregate
fruit pomes

The best-known accessory fruit is the pome, formed from the swollen hypanthium that surrounds the ovary in apples, pears, quinces, rowans and whitebeam, for example. In an apple, the core is the ovary and the fleshy hypanthium is the edible part.

Rosehips are also pomes, the outer red layer being the hypanthium. Inside, the 'seeds' are botanically achenes, each with a style arm. The many style arms in the flowers, often mixed up with the numerous stamens, are difficult to count.

The calyx, pedicel and receptacle can also be part of the makeup of false fruits.

A strawberry is an aggregation of many drupes, which are the dark specks on the outside; but it is also false because the fleshy bit inside is a swollen receptacle.

Multiple fruits such as pineapples, mulberries and figs are made up of a cluster of flowers that aggregate and form one edible fruit. The individual parts of a pineapple are botanically berries formed from several flowers. The fruit of mulberries and figs are multiple fruits, but also 'accessory'. The fleshy components of a mulberry are swollen calyces, and the edible part of a fig is a swollen inflorescence stem with the crunchy flowers inside.

GROWING IN THE DARK

When you eat a fig, you are eating the swollen top of an inflorescence stalk, which has curled inwards to form a hollow with a well-guarded opening. The structure is called a syconium (*sykon* means fig in Greek). The flowers never see the light of day, for they form inside the hollow lining. A small wasp is adapted to pollinate the female flowers in return for laying her eggs inside.

Painting a raspberry

The raspberry has soft, bright green leaves in varying shades, depending on their age. The stems are also a bright green with a darker side to show the rounded structure. The raspberries themselves need to be painted carefully, one section at a time, to show the shape and shiny texture of the berries. It is imperative that each section of the berry has a lighter and darker side to accentuate this.

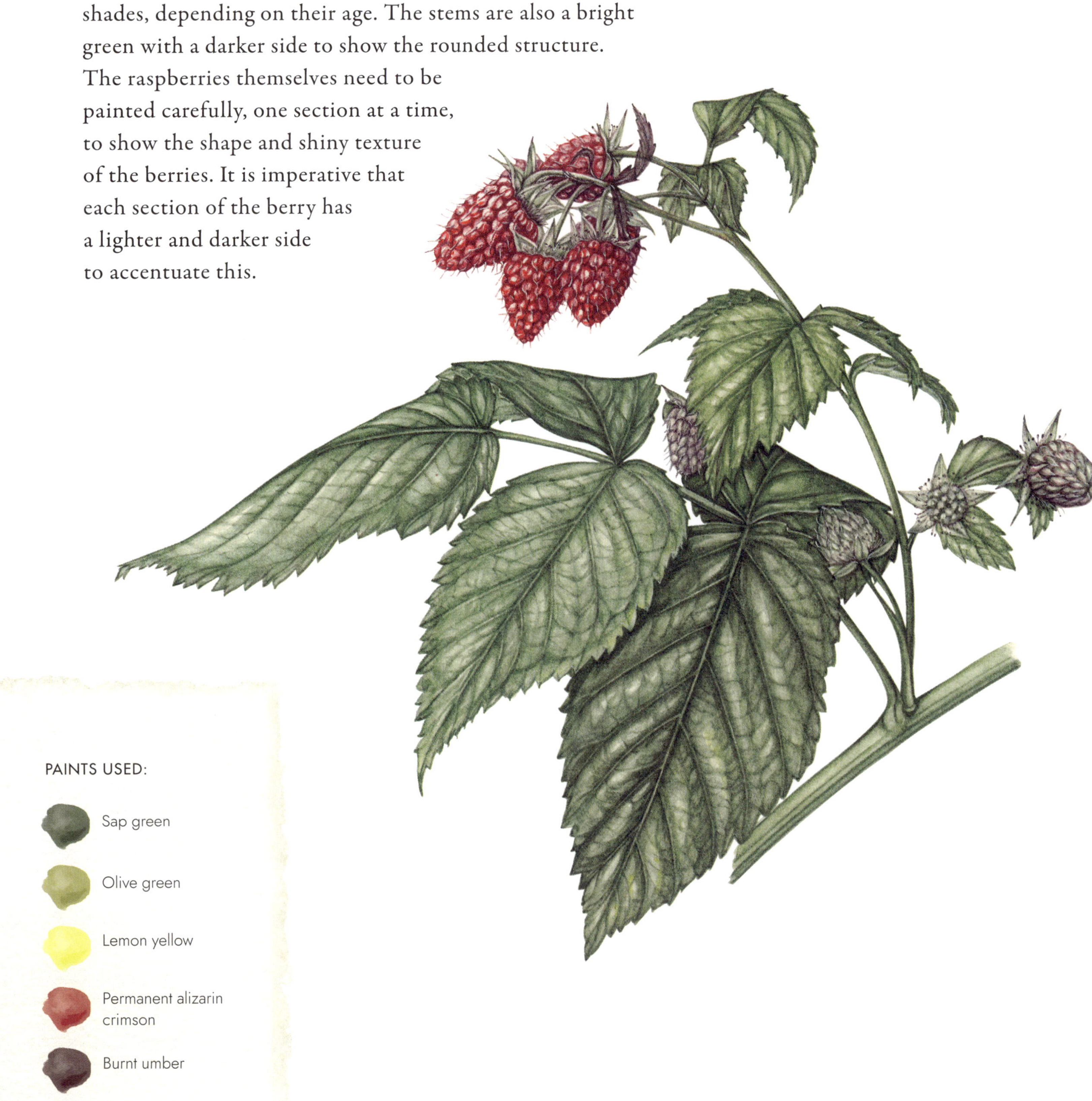

PAINTS USED:

Sap green

Olive green

Lemon yellow

Permanent alizarin crimson

Burnt umber

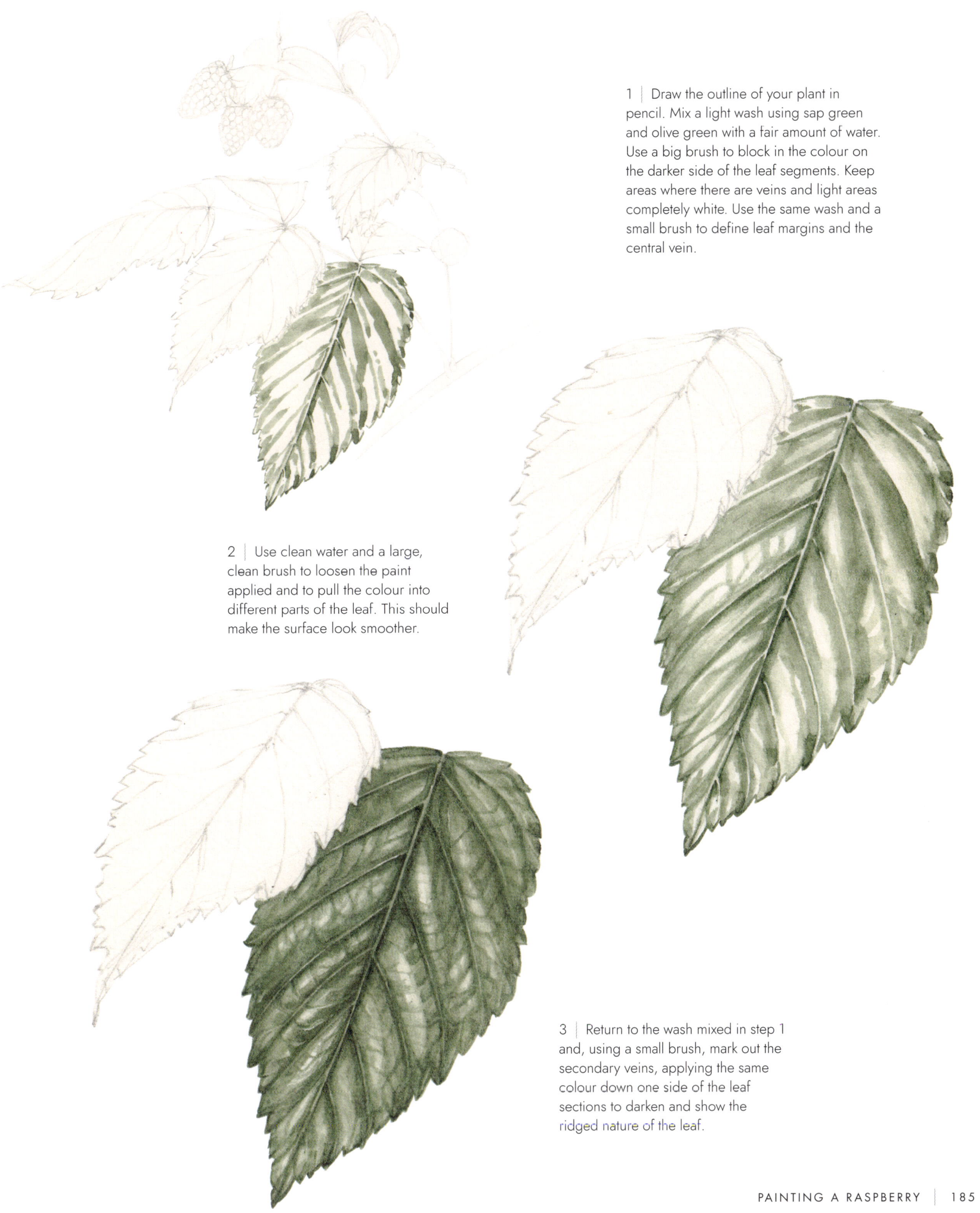

1 | Draw the outline of your plant in pencil. Mix a light wash using sap green and olive green with a fair amount of water. Use a big brush to block in the colour on the darker side of the leaf segments. Keep areas where there are veins and light areas completely white. Use the same wash and a small brush to define leaf margins and the central vein.

2 | Use clean water and a large, clean brush to loosen the paint applied and to pull the colour into different parts of the leaf. This should make the surface look smoother.

3 | Return to the wash mixed in step 1 and, using a small brush, mark out the secondary veins, applying the same colour down one side of the leaf sections to darken and show the ridged nature of the leaf.

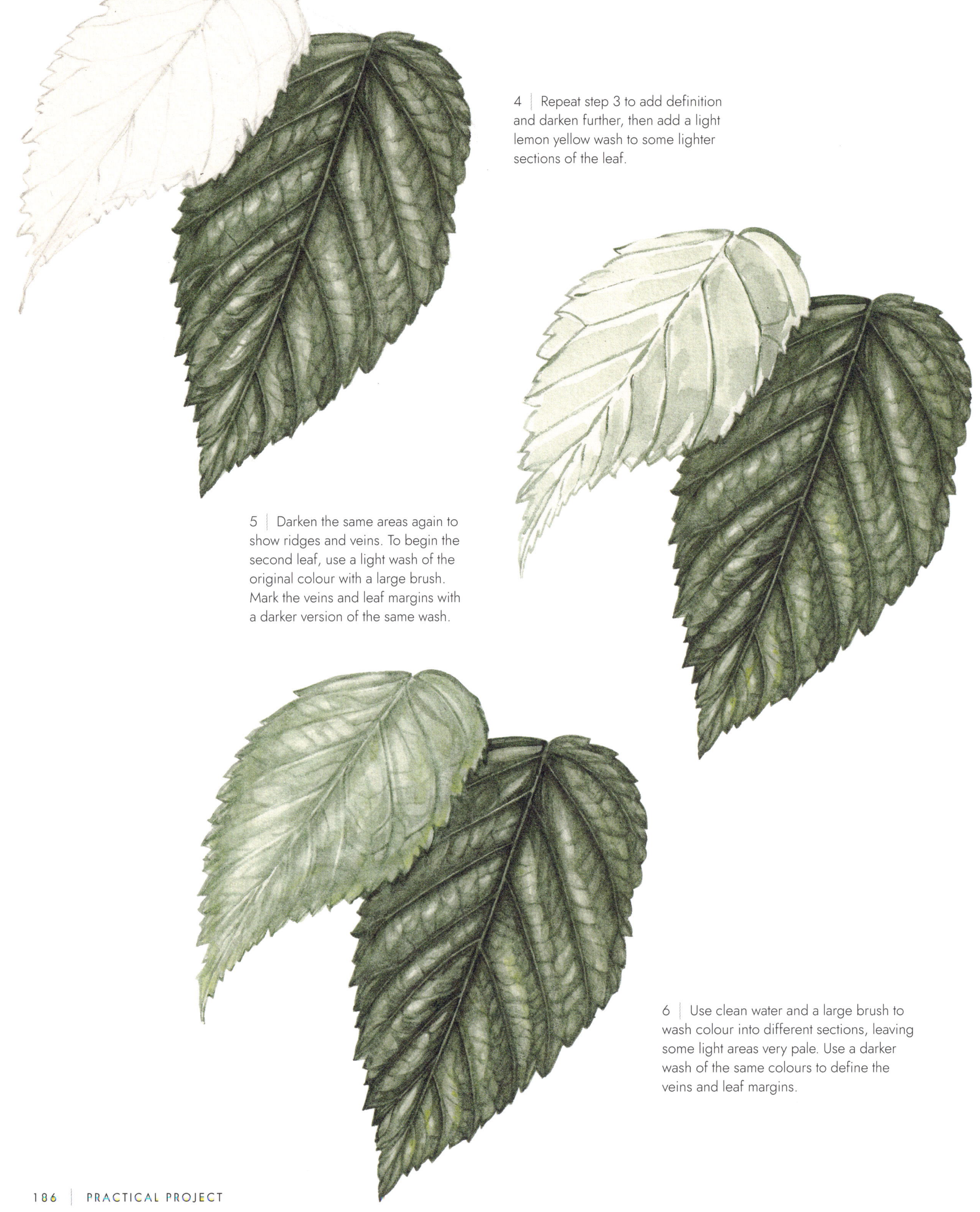

4 | Repeat step 3 to add definition and darken further, then add a light lemon yellow wash to some lighter sections of the leaf.

5 | Darken the same areas again to show ridges and veins. To begin the second leaf, use a light wash of the original colour with a large brush. Mark the veins and leaf margins with a darker version of the same wash.

6 | Use clean water and a large brush to wash colour into different sections, leaving some light areas very pale. Use a darker wash of the same colours to define the veins and leaf margins.

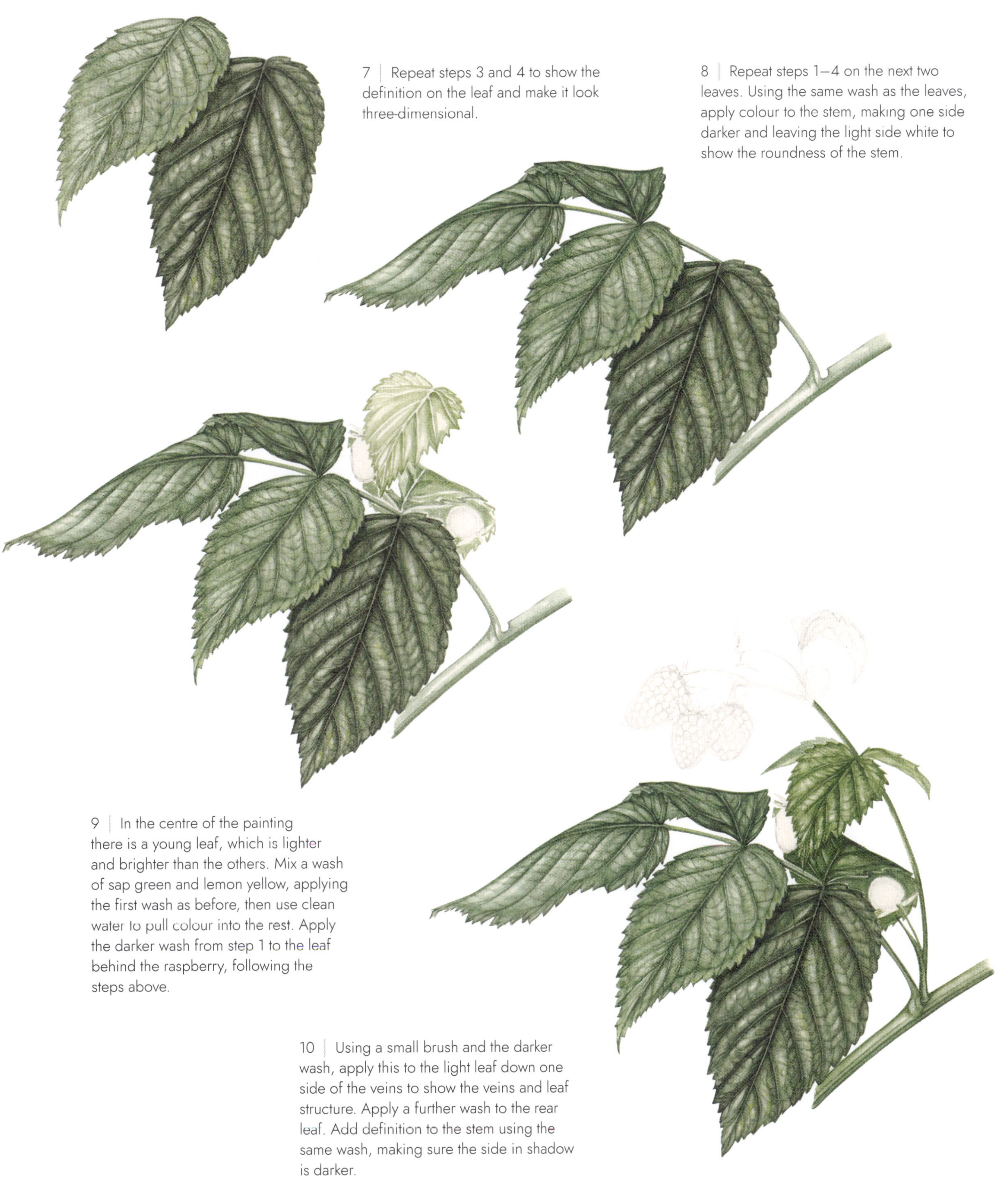

7 | Repeat steps 3 and 4 to show the definition on the leaf and make it look three-dimensional.

8 | Repeat steps 1–4 on the next two leaves. Using the same wash as the leaves, apply colour to the stem, making one side darker and leaving the light side white to show the roundness of the stem.

9 | In the centre of the painting there is a young leaf, which is lighter and brighter than the others. Mix a wash of sap green and lemon yellow, applying the first wash as before, then use clean water to pull colour into the rest. Apply the darker wash from step 1 to the leaf behind the raspberry, following the steps above.

10 | Using a small brush and the darker wash, apply this to the light leaf down one side of the veins to show the veins and leaf structure. Apply a further wash to the rear leaf. Add definition to the stem using the same wash, making sure the side in shadow is darker.

11 | Paint all further small young leaves following steps 9 and 10. Use the light green wash to paint an initial wash on the stems leading to the fruit and the sepals behind the fruit.

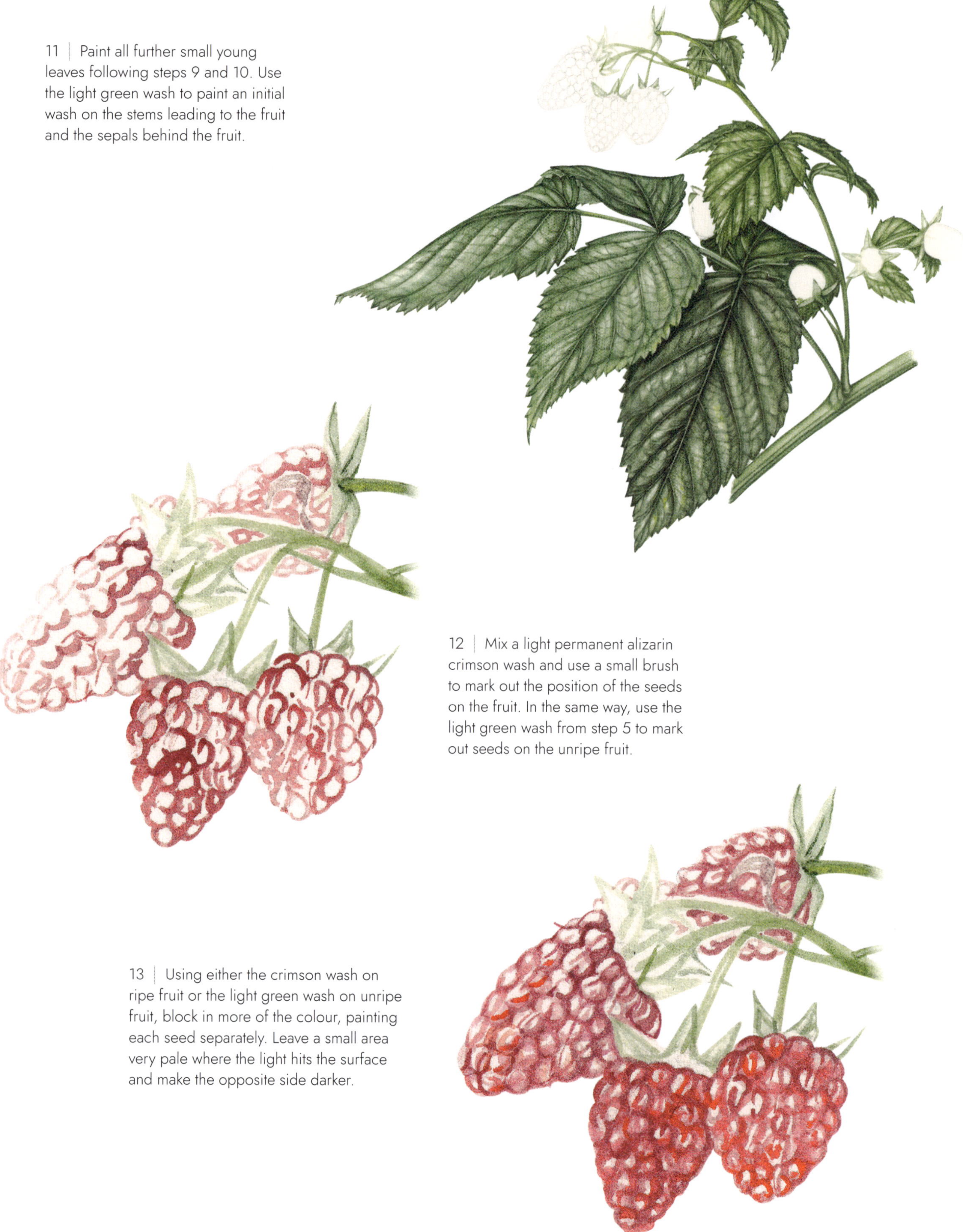

12 | Mix a light permanent alizarin crimson wash and use a small brush to mark out the position of the seeds on the fruit. In the same way, use the light green wash from step 5 to mark out seeds on the unripe fruit.

13 | Using either the crimson wash on ripe fruit or the light green wash on unripe fruit, block in more of the colour, painting each seed separately. Leave a small area very pale where the light hits the surface and make the opposite side darker.

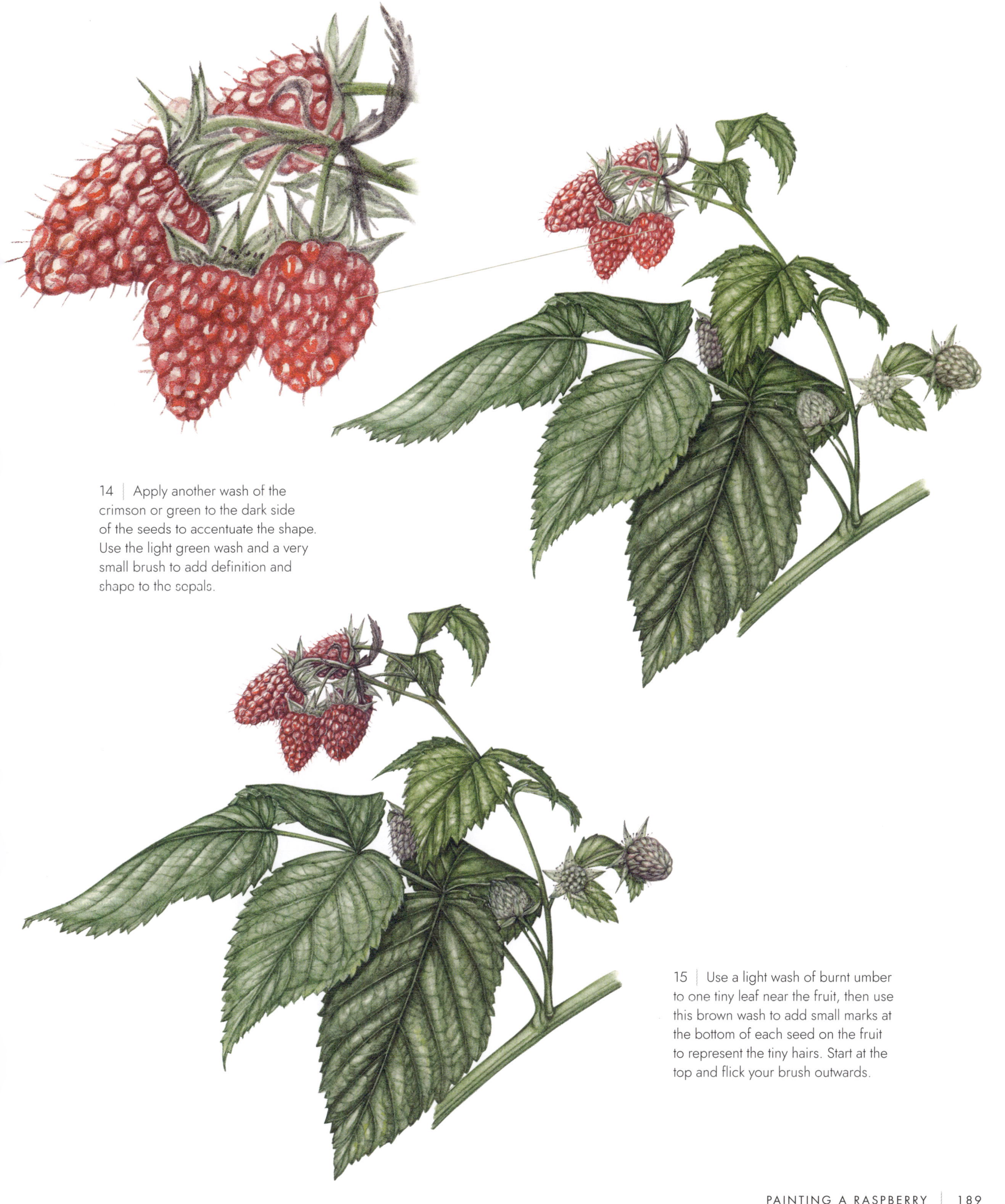

14 | Apply another wash of the crimson or green to the dark side of the seeds to accentuate the shape. Use the light green wash and a very small brush to add definition and shape to the sepals.

15 | Use a light wash of burnt umber to one tiny leaf near the fruit, then use this brown wash to add small marks at the bottom of each seed on the fruit to represent the tiny hairs. Start at the top and flick your brush outwards.

What is a seed?

A seed is a mature fertilized ovule containing an embryo for reproduction. A distinguishing mark of seeds is the scar where it was attached to a placenta (rather than a scar from a fruit's attachment to a stalk).

If you're asked to paint a fruit or a seed, be sure to know the difference. You may choose an edible fruit, where the seeds are usually discarded. Conversely, you may choose a hazel or a walnut, where the edible part is the seed and the rest of the fruit — the hard shell — is discarded. You may choose a tomato, normally eaten as a vegetable, but which is anatomically a fruit.

Seeds form a large part of the human diet: grains, spices, legumes and pulses, and those used to make beverages, such as coffee and cocoa, are a few examples. It can be interesting to make drawings and paintings of different shapes, sizes, colours and surface patterns for any one of these, to demonstrate their diversity and the basis from which 'heritage' foods are bred.

SEED STRUCTURE

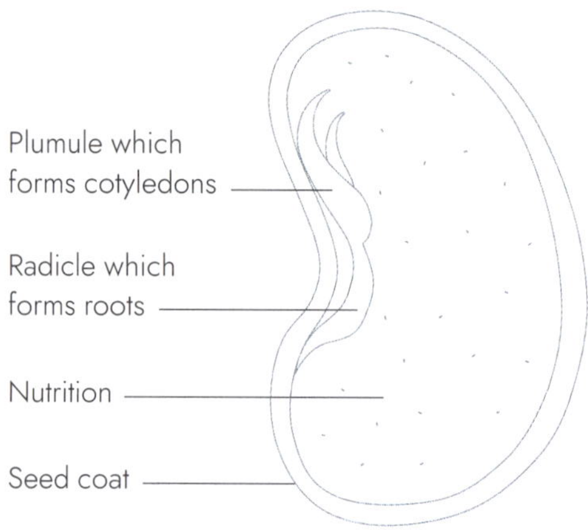

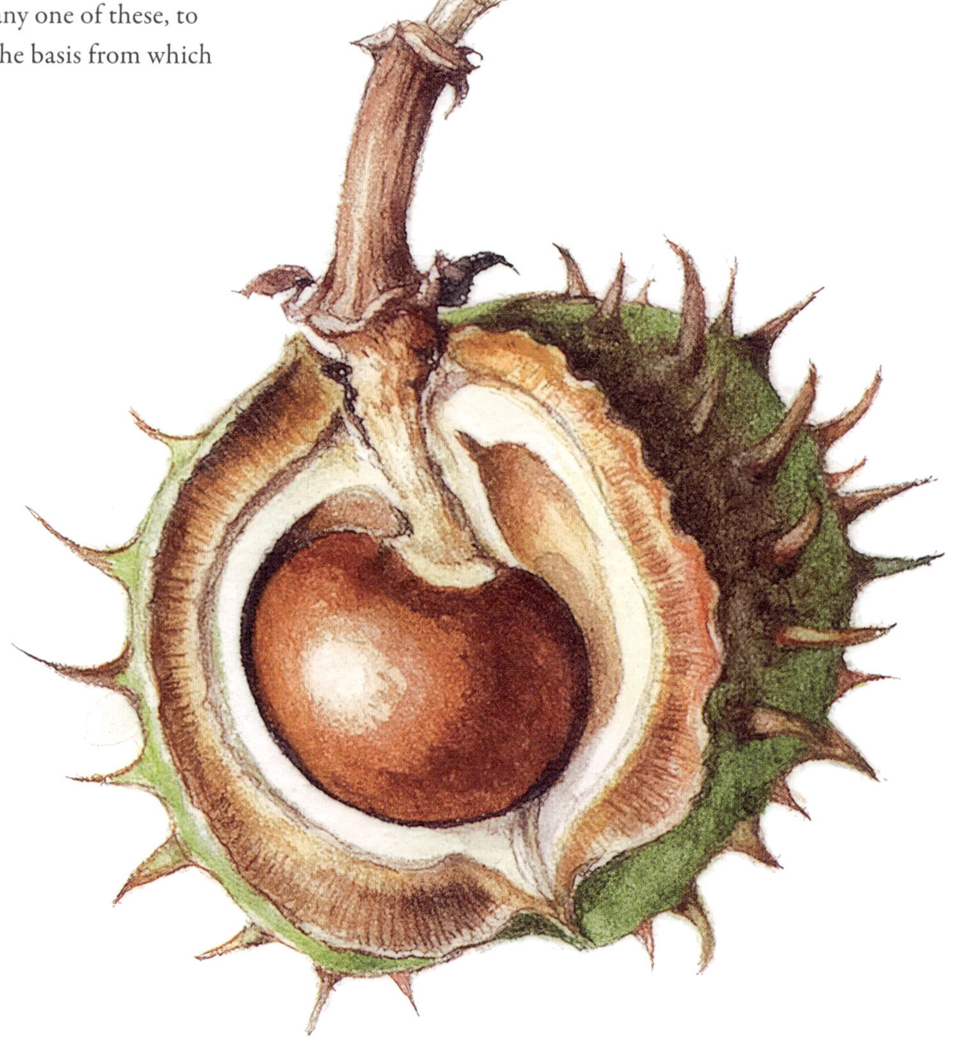

RIGHT Horse chestnut (*Aesculus hippocastanum*) by Janet Dyer shows the seed, the brown 'conker', inside its prickly green casing, the fruit.

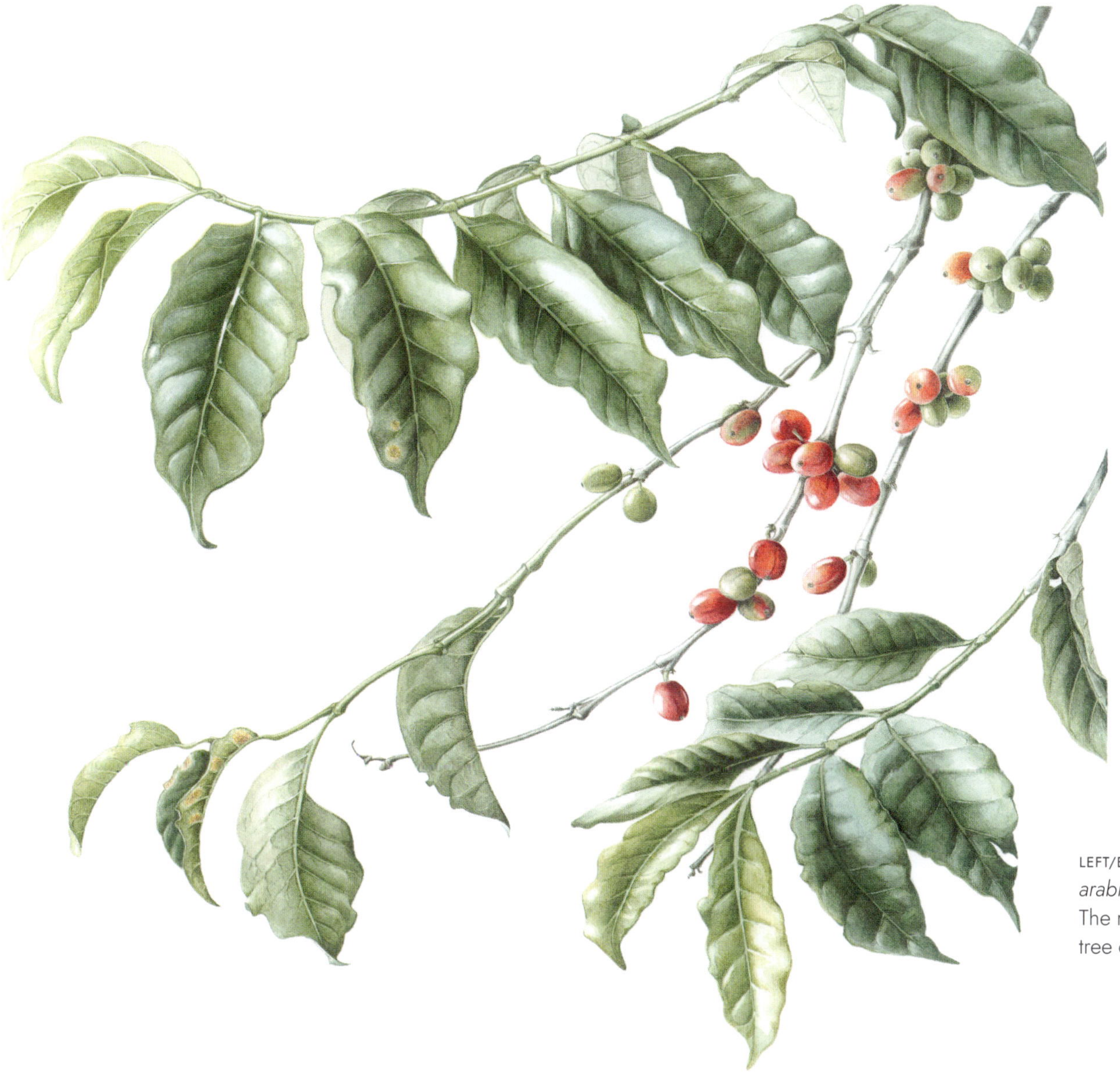

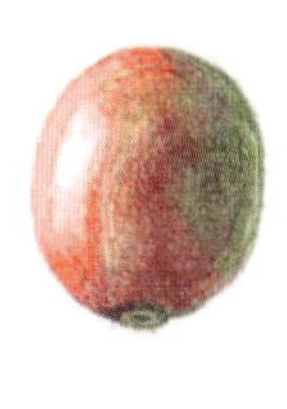

FROM CHERRY TO BEAN

A coffee bean is not like an edible bean. It is the hard seed contained within a fleshy fruit (a drupe). The red outer skin of the 'cherry' is the exocarp and the mucilage inside is the mesocarp. The mucilage is removed on site by washing, or by drying in the sun, to reveal a hard husky 'parchment' which, unusually for drupes, protects two seeds inside. Before shipment, the parchment is removed and the beans (the seeds) are bagged for onward travel and roasting.

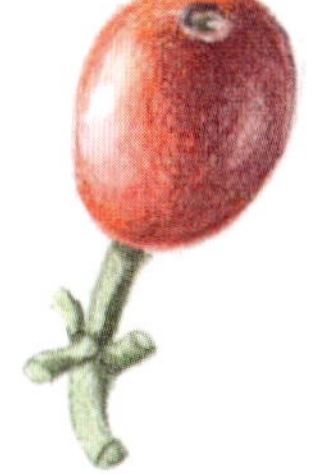
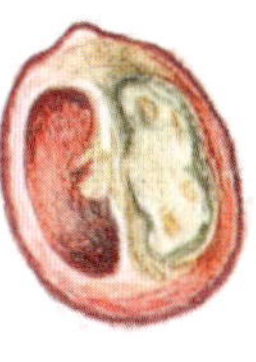
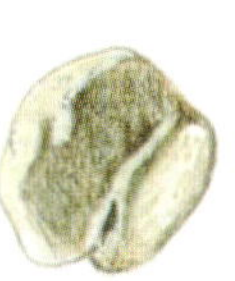

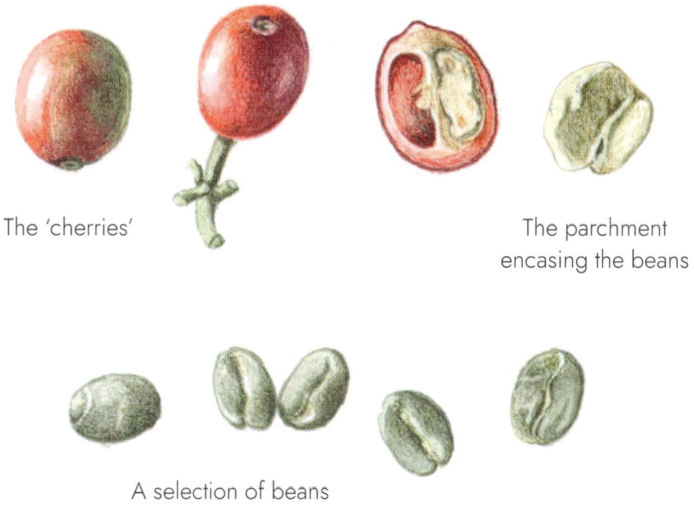

The 'cherries'

The parchment encasing the beans

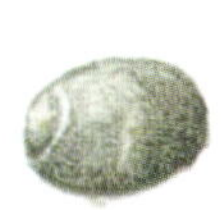
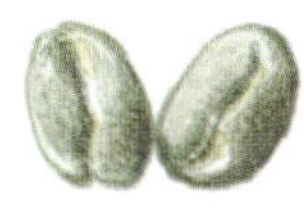
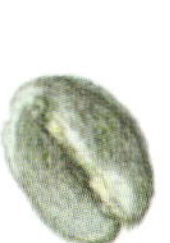
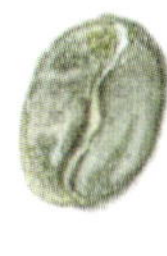

A selection of beans

Seed dispersal

Seeds are protected inside the fruit and must be released
to start new growth. Their different shapes and appendages
relate to different dispersal methods, which give plants the
best chance of survival if their seeds are widely dispersed. At
least one might survive even if other seeds cannot germinate,
or the parent plant is lost. An understanding of how seeds
are dispersed helps the artist identify and appreciate their
important features.

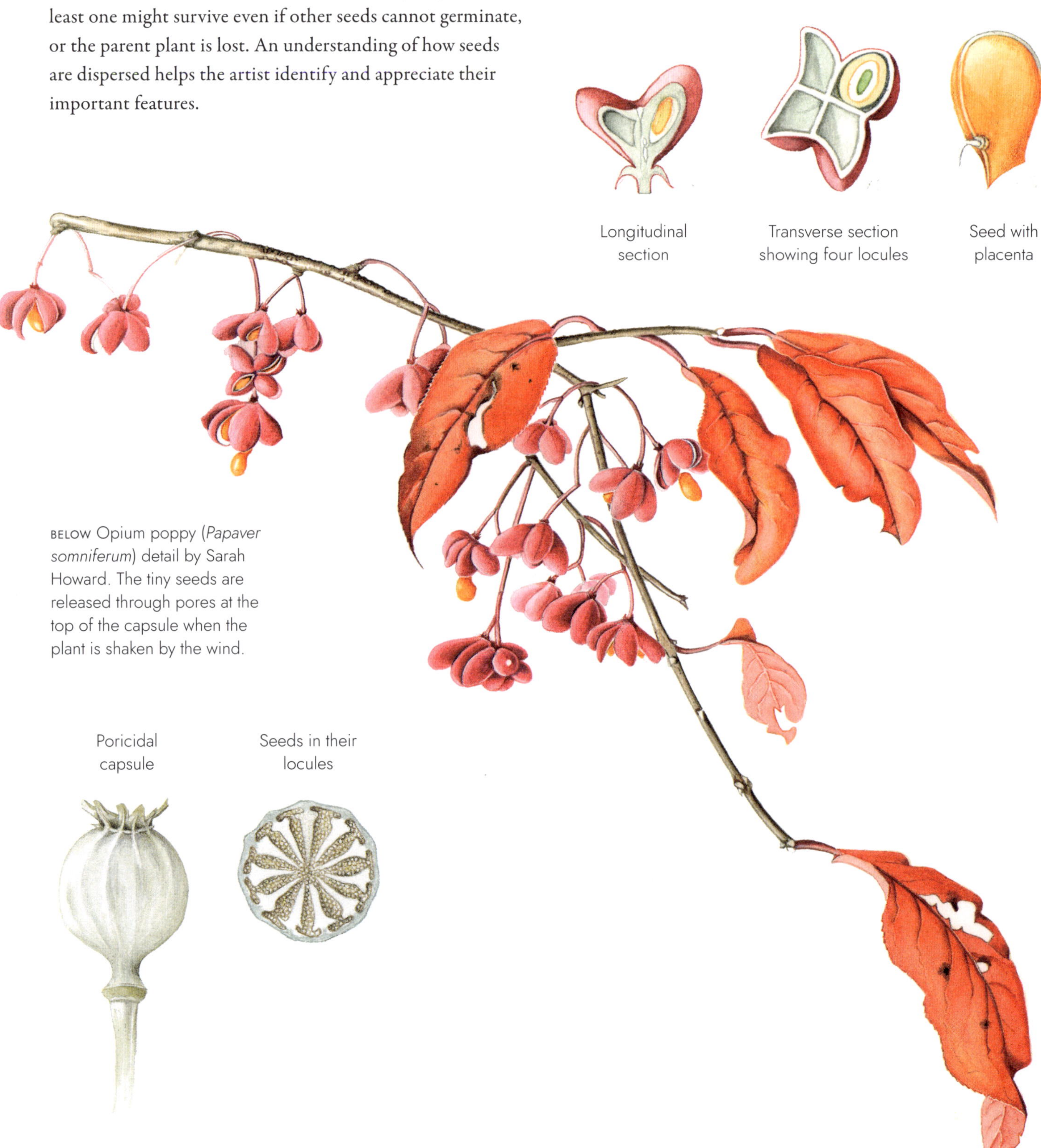

BELOW, DETAILS AND MAIN IMAGE
Spindle (*Euonymus europaeus*)
by Sarah Howard. Birds eat the
seeds and disperse them in their
droppings.

BELOW Opium poppy (*Papaver
somniferum*) detail by Sarah
Howard. The tiny seeds are
released through pores at the
top of the capsule when the
plant is shaken by the wind.

WHAT TO LOOK FOR

- Light, tiny seeds, such as those of the vanilla orchid, poppies and grasses, are dispersed by the **wind**. Poppy capsules have pores at the top of the fruit from which the tiny seeds are easily shaken out by movement.

- Seeds that carry wings or other appendages to help them fly or helicopter to a different site, for example, cypsela or samara, are **aerodynamically** dispersed.

- Seeds with hooks or barbs, such as cleavers, or the awns on a grass, can hitch rides on passing **animals**.

- **Edible** seeds are often brightly coloured, such as the orange *Euonymus* seeds dispersed by birds.

- Many plants have ingenious mechanisms for **ballistic** dispersal of seeds. Storksbill (*Geranium*) seeds, for example, are set at the base of the style in a capsule attached to the style tip. When ripe, the seed covers split apart towards the tip, catapulting them out at top speed. The touch-me-not balsam gets its name from the mechanism which, when touched, 'fires' its seeds.

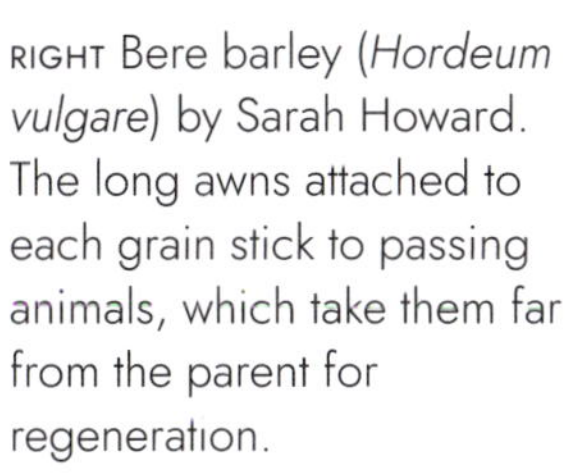

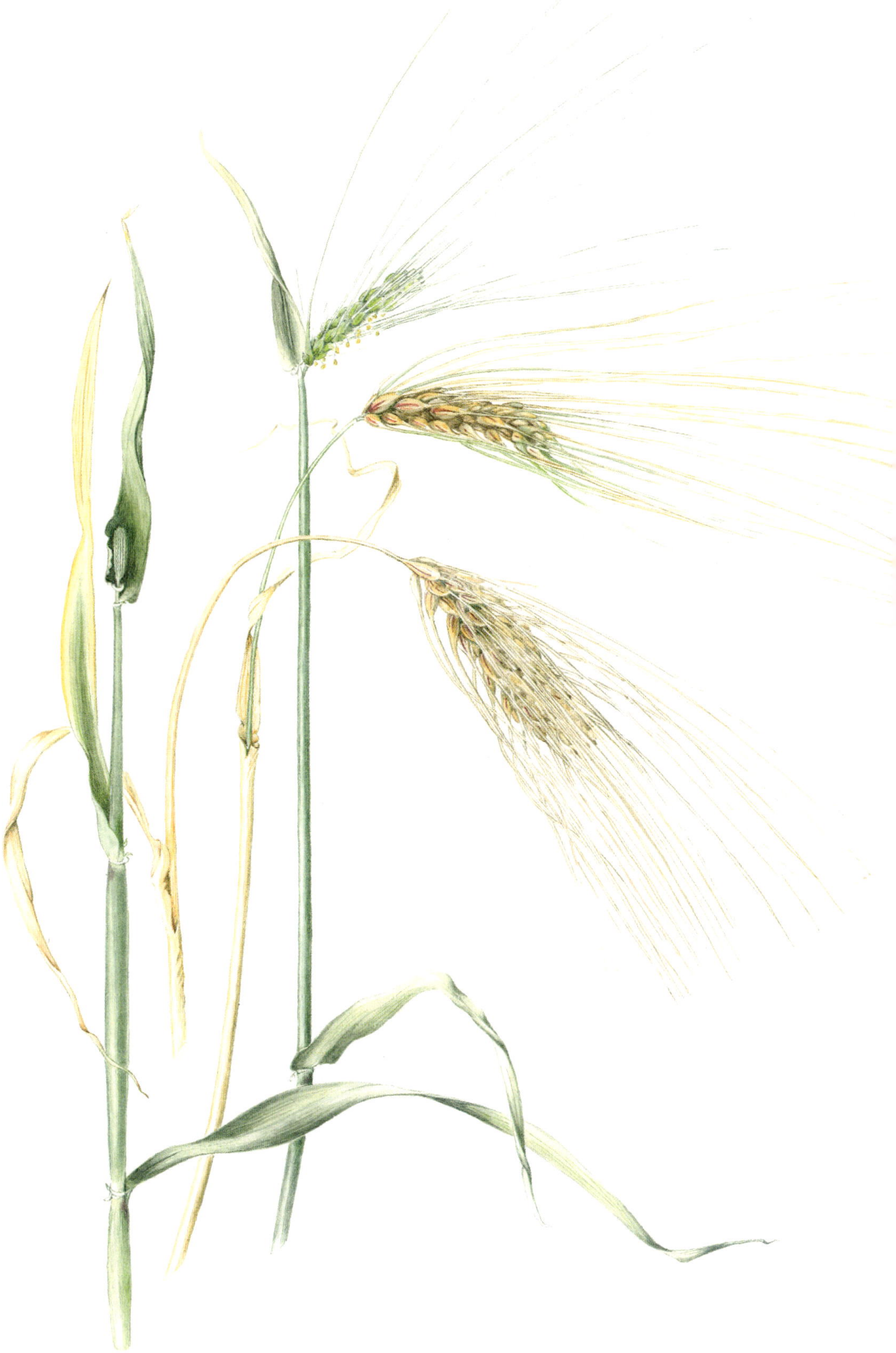

RIGHT Bere barley (*Hordeum vulgare*) by Sarah Howard. The long awns attached to each grain stick to passing animals, which take them far from the parent for regeneration.

Painting peony fruits

When painting hard and shiny objects, the form and the sheen
are depicted using light areas. It is imperative to leave these
areas white. Gradually build the darker areas and those in
shadow in washes, making the areas where the seeds are tucked
under the capsules the darkest. To maximize the shine on the
seeds, look carefully for areas of light, and leave these completely
white. I don't use masking fluid, but many artists find it useful.

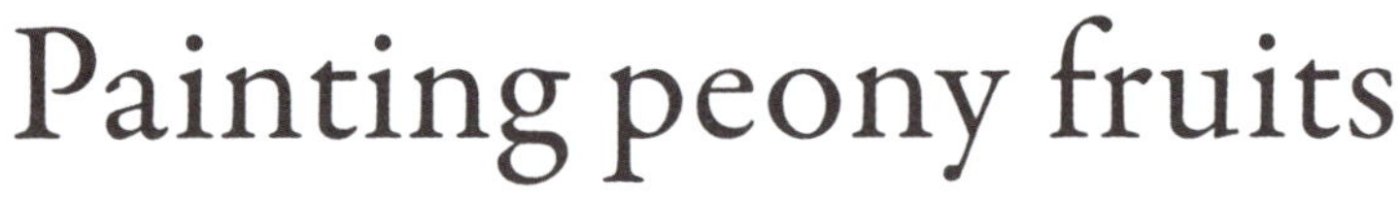

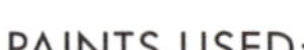

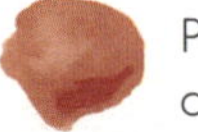
Permanent alizarin crimson

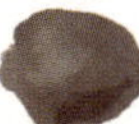
Burnt umber

Lemon yellow

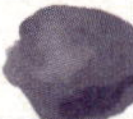
Winsor violet (dioxazine)

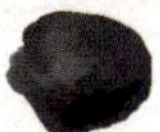
Ivory black

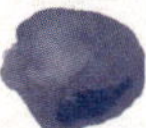
French ultramarine

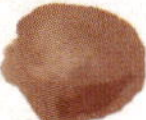
Indian red

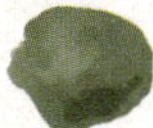
Sap green

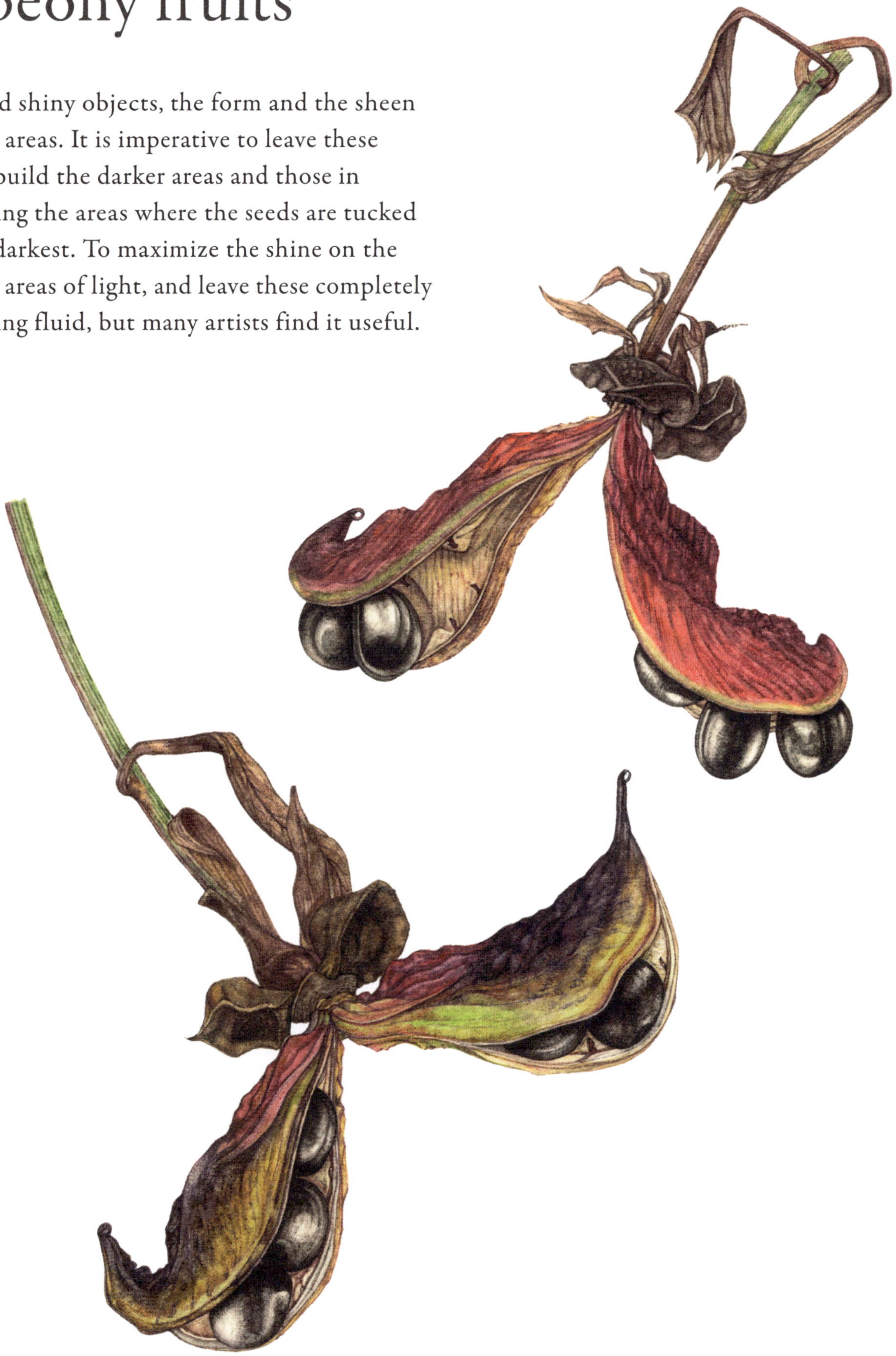

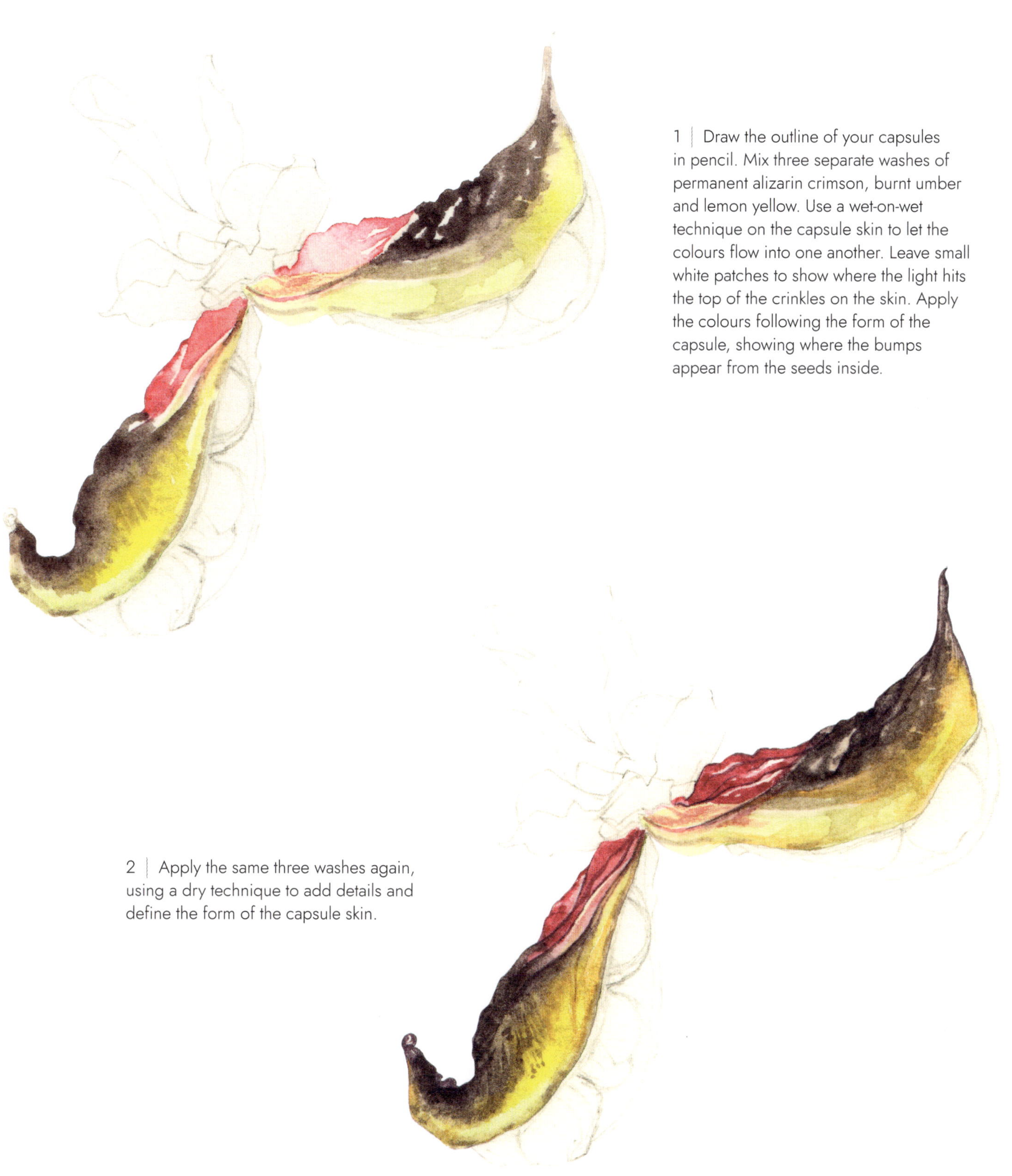

1 | Draw the outline of your capsules in pencil. Mix three separate washes of permanent alizarin crimson, burnt umber and lemon yellow. Use a wet-on-wet technique on the capsule skin to let the colours flow into one another. Leave small white patches to show where the light hits the top of the crinkles on the skin. Apply the colours following the form of the capsule, showing where the bumps appear from the seeds inside.

2 | Apply the same three washes again, using a dry technique to add details and define the form of the capsule skin.

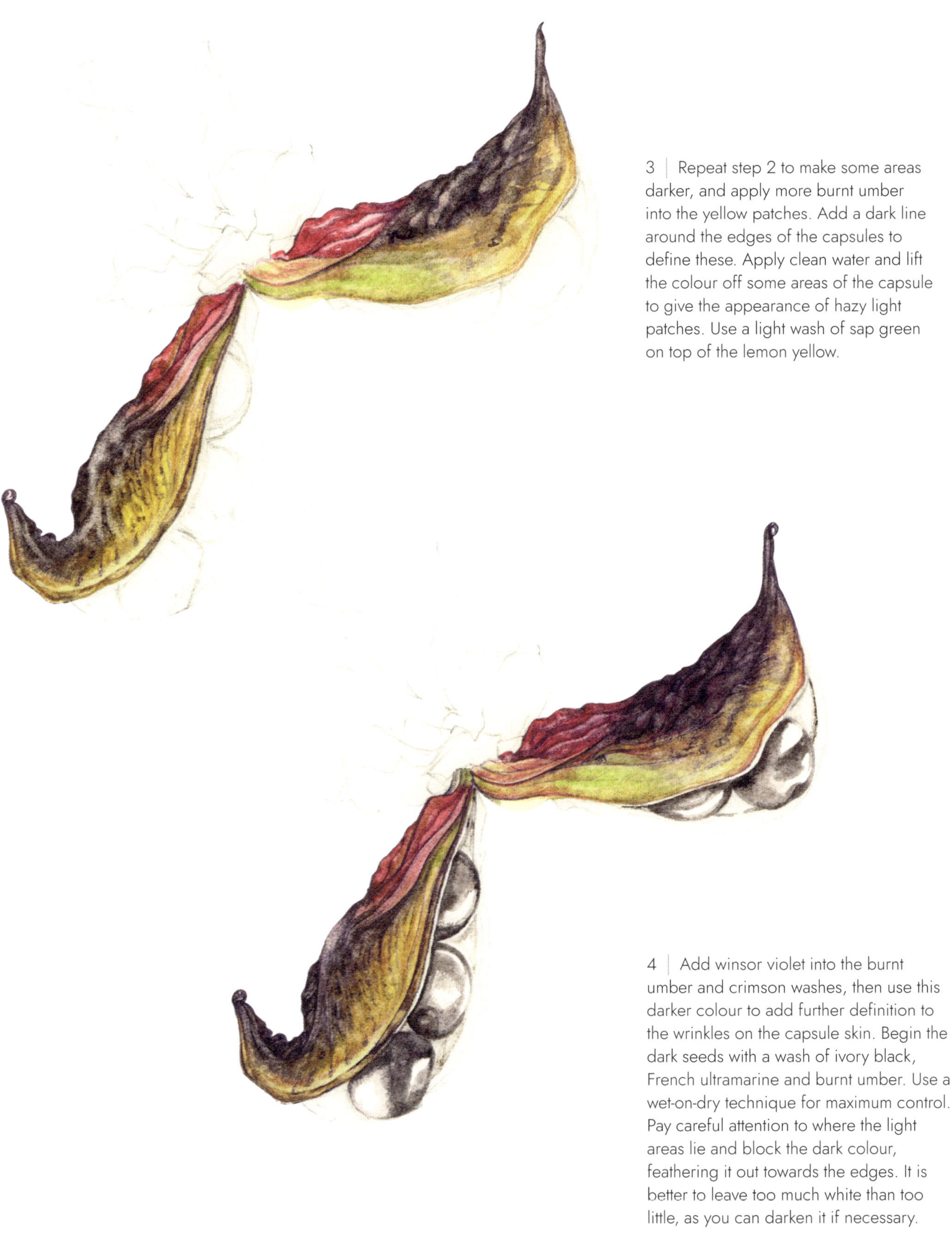

3 | Repeat step 2 to make some areas darker, and apply more burnt umber into the yellow patches. Add a dark line around the edges of the capsules to define these. Apply clean water and lift the colour off some areas of the capsule to give the appearance of hazy light patches. Use a light wash of sap green on top of the lemon yellow.

4 | Add winsor violet into the burnt umber and crimson washes, then use this darker colour to add further definition to the wrinkles on the capsule skin. Begin the dark seeds with a wash of ivory black, French ultramarine and burnt umber. Use a wet-on-dry technique for maximum control. Pay careful attention to where the light areas lie and block the dark colour, feathering it out towards the edges. It is better to leave too much white than too little, as you can darken it if necessary.

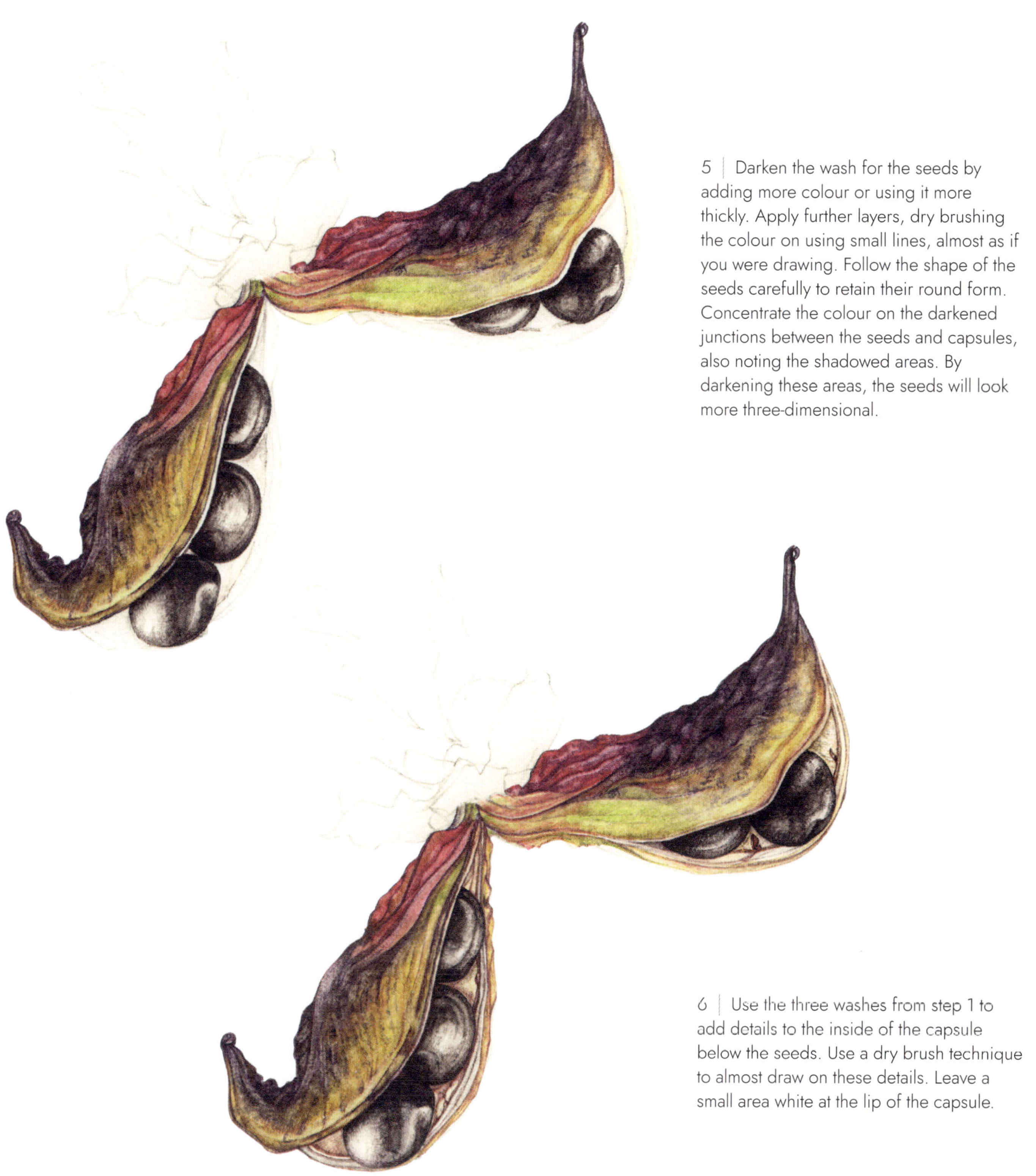

5 | Darken the wash for the seeds by adding more colour or using it more thickly. Apply further layers, dry brushing the colour on using small lines, almost as if you were drawing. Follow the shape of the seeds carefully to retain their round form. Concentrate the colour on the darkened junctions between the seeds and capsules, also noting the shadowed areas. By darkening these areas, the seeds will look more three-dimensional.

6 | Use the three washes from step 1 to add details to the inside of the capsule below the seeds. Use a dry brush technique to almost draw on these details. Leave a small area white at the lip of the capsule.

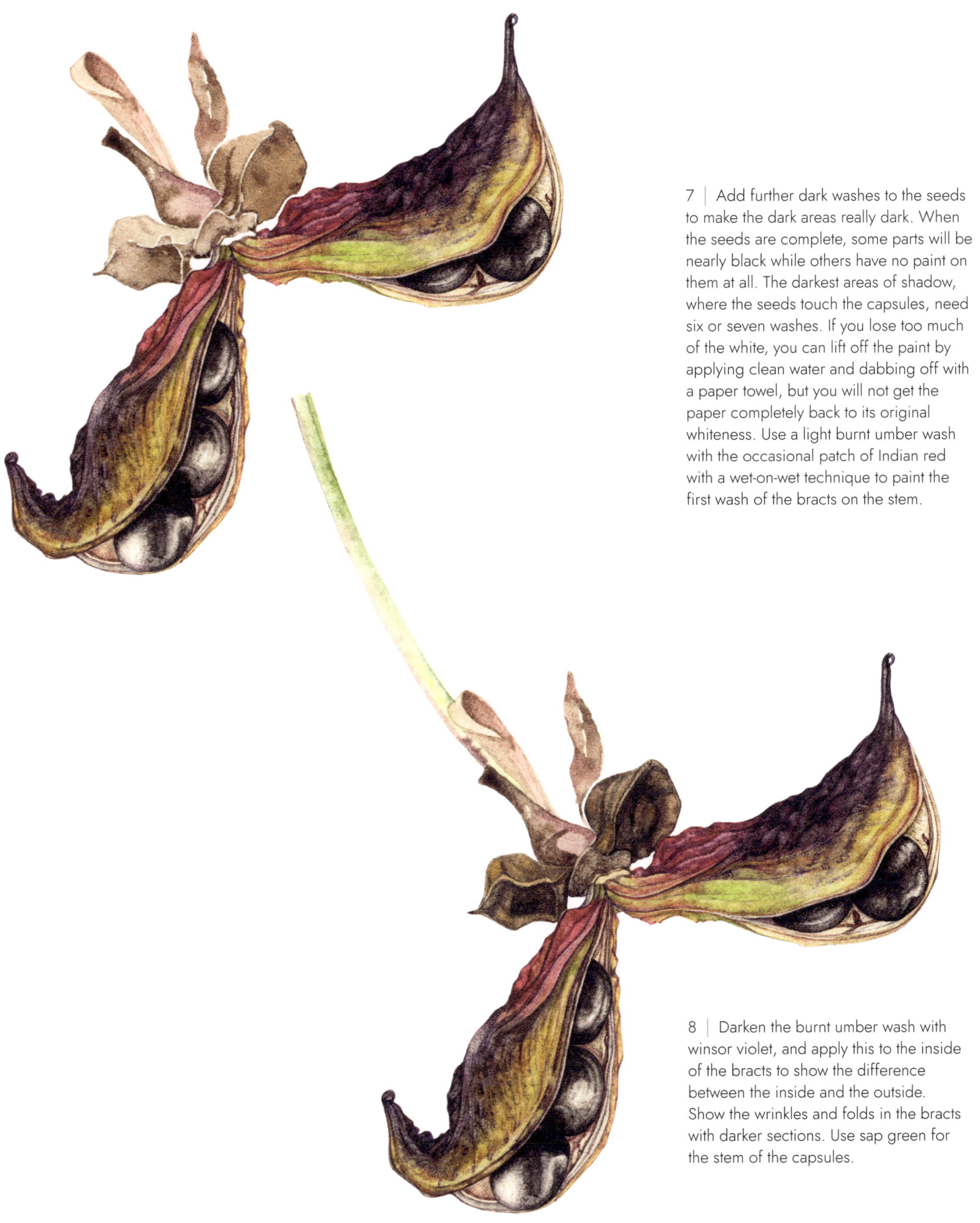

7 | Add further dark washes to the seeds to make the dark areas really dark. When the seeds are complete, some parts will be nearly black while others have no paint on them at all. The darkest areas of shadow, where the seeds touch the capsules, need six or seven washes. If you lose too much of the white, you can lift off the paint by applying clean water and dabbing off with a paper towel, but you will not get the paper completely back to its original whiteness. Use a light burnt umber wash with the occasional patch of Indian red with a wet-on-wet technique to paint the first wash of the bracts on the stem.

8 | Darken the burnt umber wash with winsor violet, and apply this to the inside of the bracts to show the difference between the inside and the outside. Show the wrinkles and folds in the bracts with darker sections. Use sap green for the stem of the capsules.

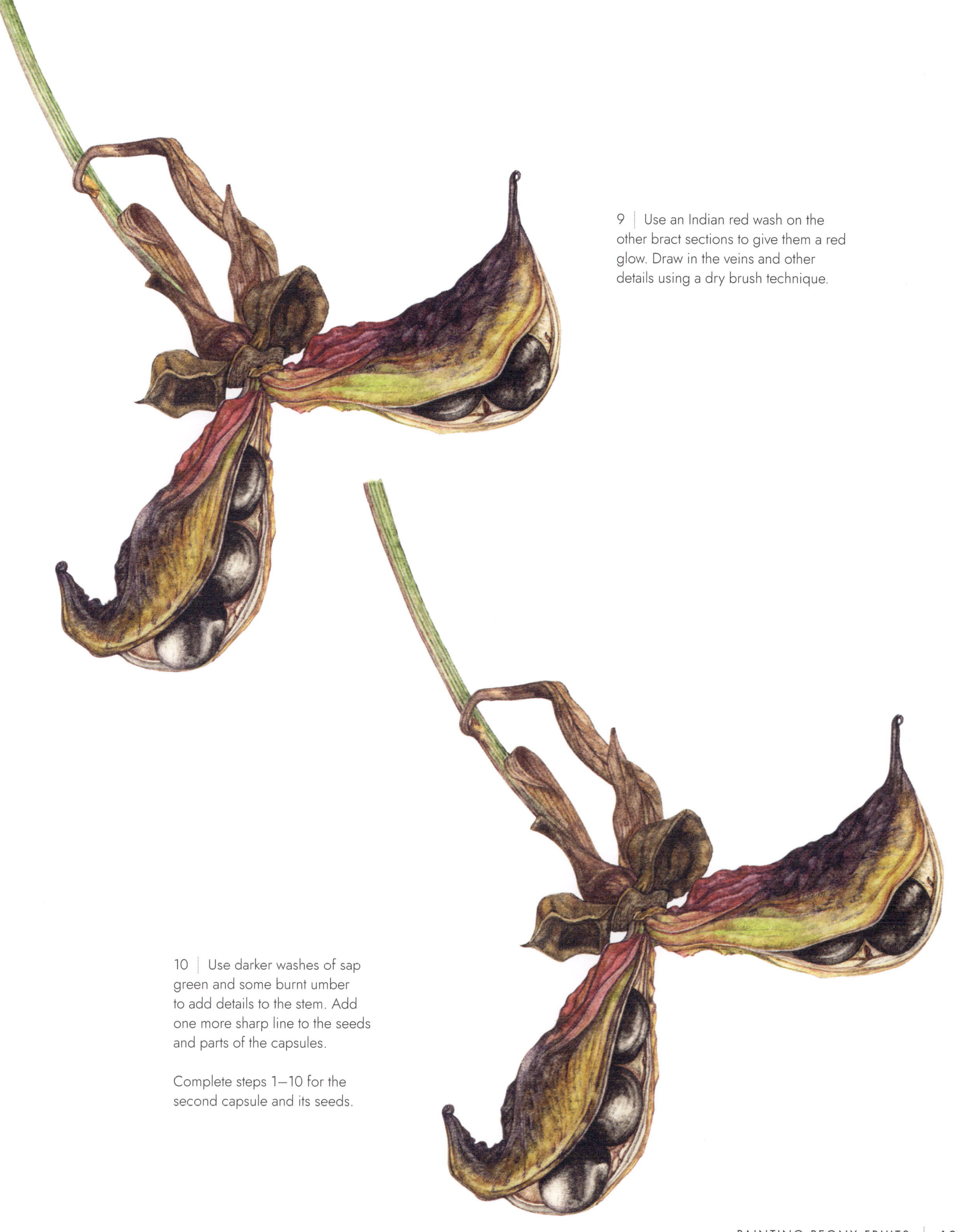

9 | Use an Indian red wash on the other bract sections to give them a red glow. Draw in the veins and other details using a dry brush technique.

10 | Use darker washes of sap green and some burnt umber to add details to the stem. Add one more sharp line to the seeds and parts of the capsules.

Complete steps 1–10 for the second capsule and its seeds.

Common screwpine (*Pandanus utilis*)

by Mariko Ikeda

Mariko Ikeda, a specialist in screwpines (*Pandanus*), has won many awards and has work represented in collections from the Shirley Sherwood Collection at Kew Gardens and the Royal Botanic Garden Edinburgh in the UK, to the Hunt Institute in Pittsburgh, USA.

Pandanus utilis demonstrates Mariko's prowess and the care with which she approaches her work. Smooth vellum provides an ideal substrate for tiny strokes of paint smoothing out every bump and blemish on a surface. Every bract, with its toothed margins, guards a white male inflorescence spike with sword-sharp precision. Every phalange (or key) from a female tree looks as uncompromisingly angular and hard as it is supposed to. Screwpines are monocots, not pines and not gymnosperms. They are native to coastal regions of the Indian Ocean, and provide local people with leaves for thatching.

Depicting the fruit

In cultivation, the interest is in the scented male flowers and aggregate cone-like fruit. Mariko provides an arresting depiction of a fruit where some keys have been removed. Revealed are strong yellow bases to each key, and how they are inserted with honeycomb efficiency in Fibonacci-related spirals.

The keys are clue to the aggregate nature of screwpine fruit. The female tree produces female 'flowers' which have no perianth. Instead, what is seen are dense, multi-locular ovaries or carpels, topped with sessile stigmas, which ripen into drupes, each containing small seeds — much like an enlarged raspberry.

Painting on vellum requires patience. Mariko stretches her unusually large vellum on a frame after soaking it. She will use up to seven new brushes for each painting. Painting layer upon layer with a dry brush, and only after the previous layer has thoroughly dried, is how to get results on vellum. But the translucent effect of the paint and the beautiful off-white colour of vellum is reward enough.

RIGHT Screwpines provide strong shapes in their globose fruit that contrast with needle-sharp bracts of the male flowers. Size: 57.3 x 39.4cm (22^1/2 x 15^1/2in), watercolour on vellum.

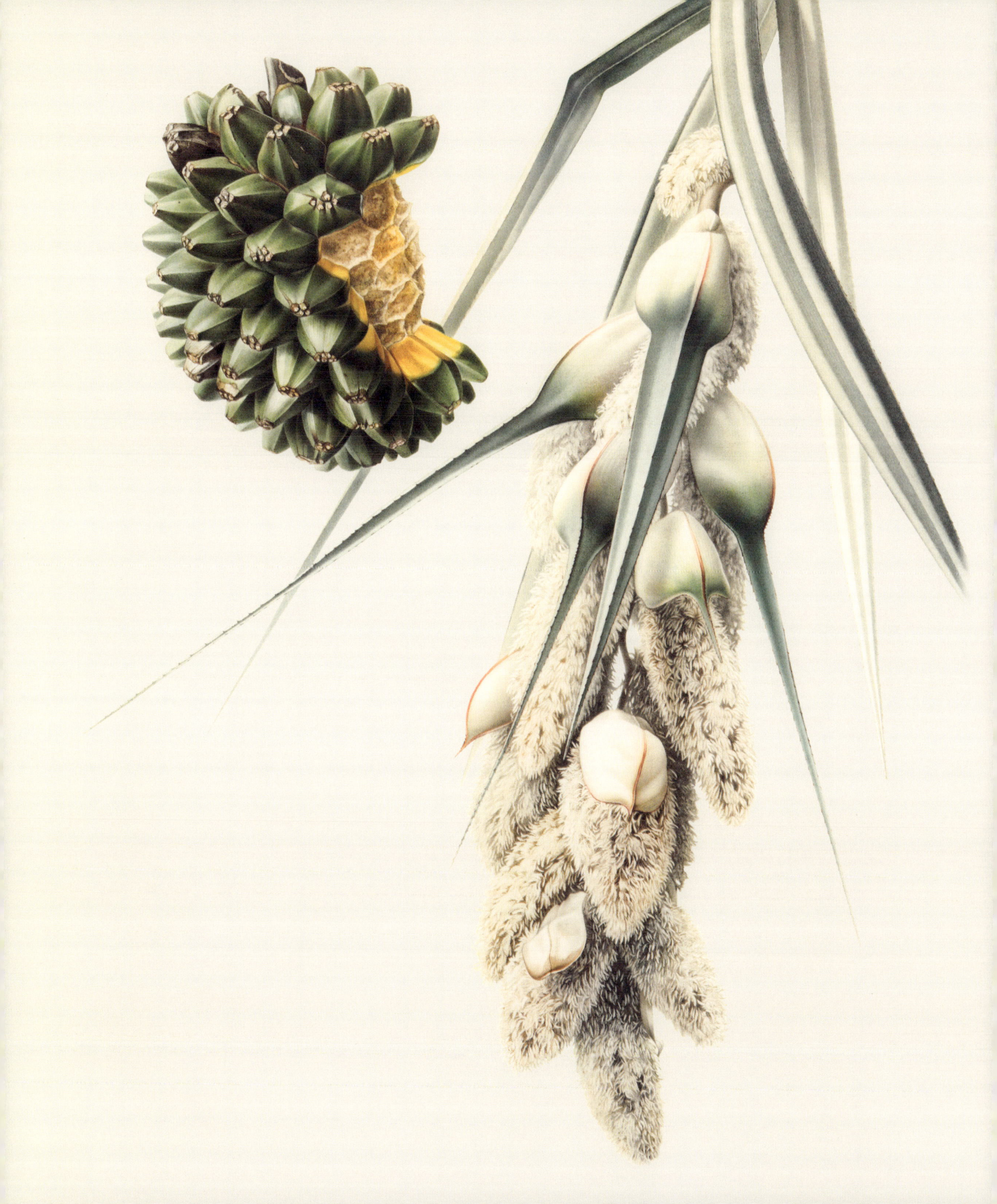

Chapter seven

Plants that reproduce asexually

Asexual reproduction refers to regeneration by means other than sexual. There is no exchange of genetic material between separate plants, and offspring will be identical to their parents. It's also called vegetative reproduction, with leaves, stems, bulbs, corms and roots all deployed in the service of regeneration.

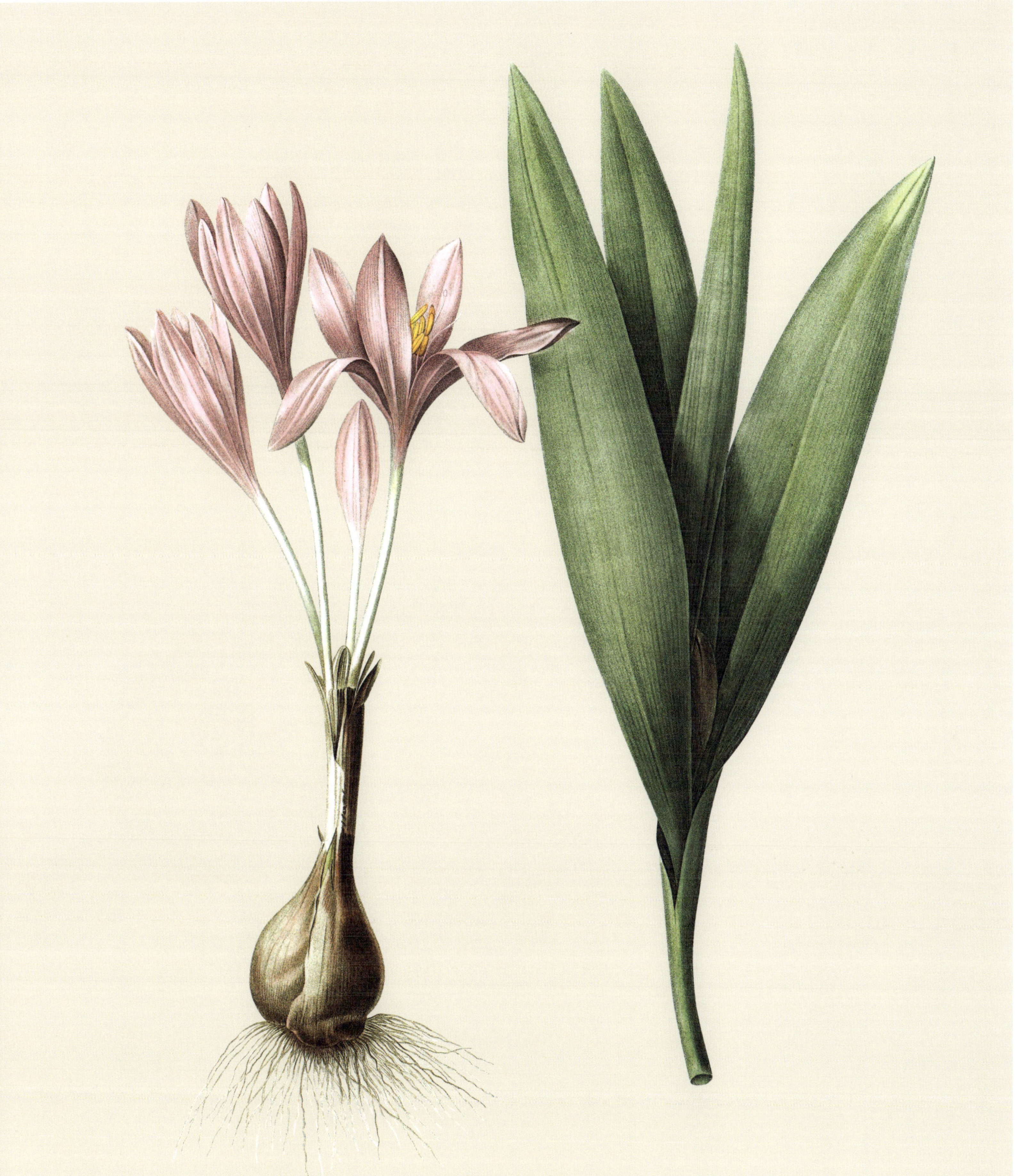

Types of asexual reproduction

Plants often have a second option of regenerating without exchanging pollen, especially when the flowering season is short or the conditions are so good there is no need for variety in the offspring.

Above ground features

Above ground, stolons are lateral growing stems, which put down roots at nodes to form a new plant. Strawberries are well known for their extended stems and the plantlets that grow at their ends. Other examples include spider plants (*Chlorophytum comosum*) with their aerial hanging plantlets.

Ivy stems creep along walls and fences, putting down new roots as they progress. Offsets, plantlets, pups and suckers are words often used for similar features, but they can have more specific meanings.

A bulbil is a small bulb found in leaf axils, or amongst flowers, that can go on to make new plants.

Pseudobulbs are swollen bulb-like thickenings of the stem, unlike true bulbs, which are modified underground leaves, or bulbils, which are modified axillary buds. In orchids, pseudobulbs store water and nutrients to feed the mother plant in dry conditions. They can be removed and replanted to produce new plants.

BELOW Few-flowered garlic (*Allium paradoxum*) detail by Janet Dyer. The many bulbils that form at the base of the pedicel provide an effective way of extending its range.

RIGHT Wild strawberry (*Fragaria vesca*) by Betty Gadsby, produces runners on which new plants form at nodes far from the parent, and these in turn can produce more runners.

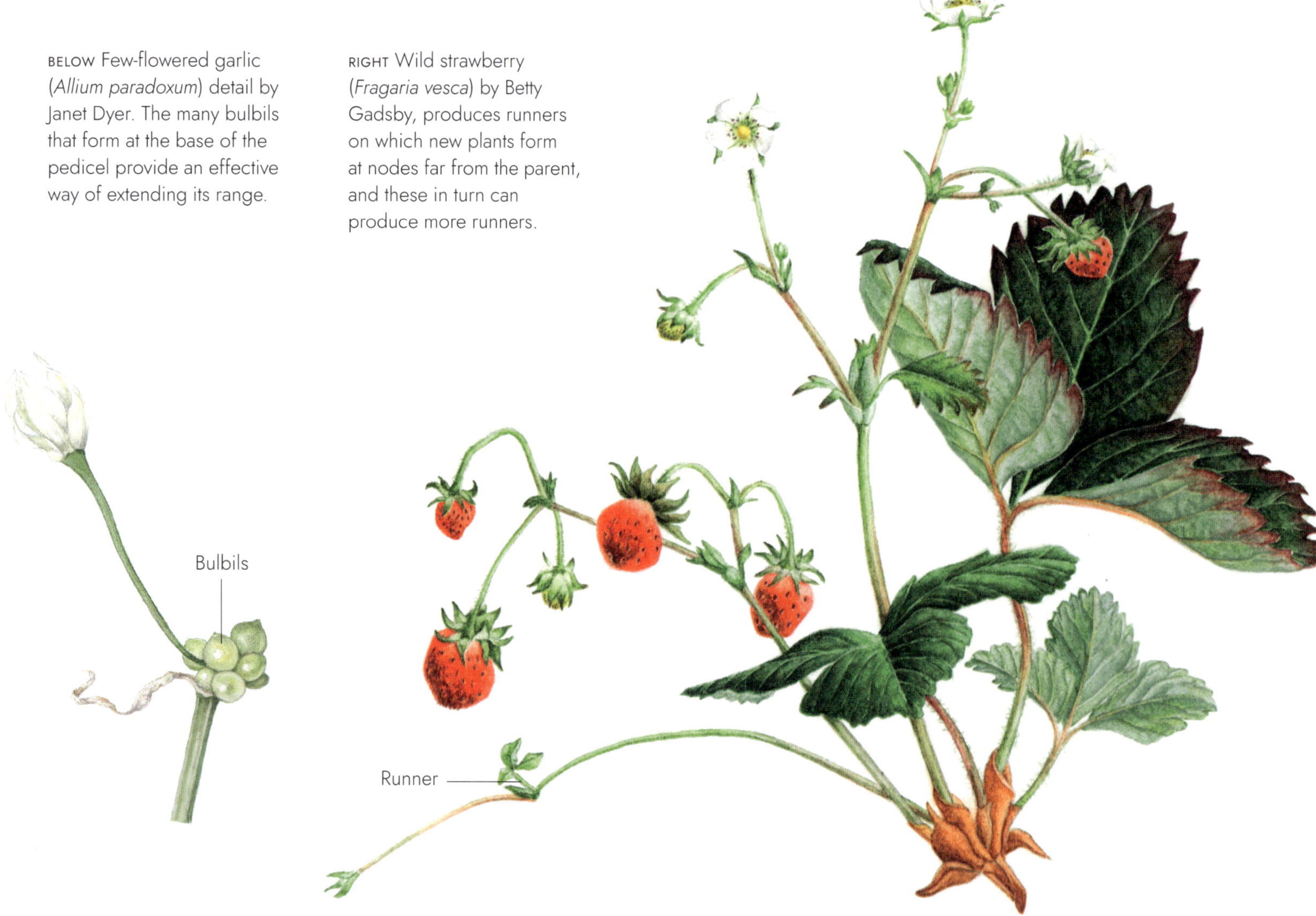

Below ground features

Underground, or in water, the rhizome is the equivalent of the above-ground stolons. Gardeners use the word 'layering' to describe how subsidiary plants can be made from the rooting of lateral stems.

Many species of grasses, including bamboo, and some members of the mint family, spread by rhizomes. Vernal water starwort (*Callitriche palustris*) puts down nodal roots in water. Rhizomes can be a blessing as well as a curse in a garden, for their sneaky ability to spread without being seen.

Thickened, tuberous rhizomes are characteristic of bearded irises, gingers and turmeric. They are often seen partially in the ground as well as in it. They are distinguished from roots by their nodes which, compared to those on above-ground stolons, are short. Turions are swollen buds on underground or aquatic stems, which easily detach in optimal conditions to form a new plant. Many bladderworts (*Utricularia*) have globular swellings at the end of their shoots, which disperse easily in running water.

Marsh willowherb (*Epilobium palustre*) makes turions in the autumn to tide it over the winter. They form at the end of the parent plant's stolons, well away from the parent plant and, during floods, will detach and be carried far away to another site.

Suckers are new plants generated from an extended root close to the parent plant.

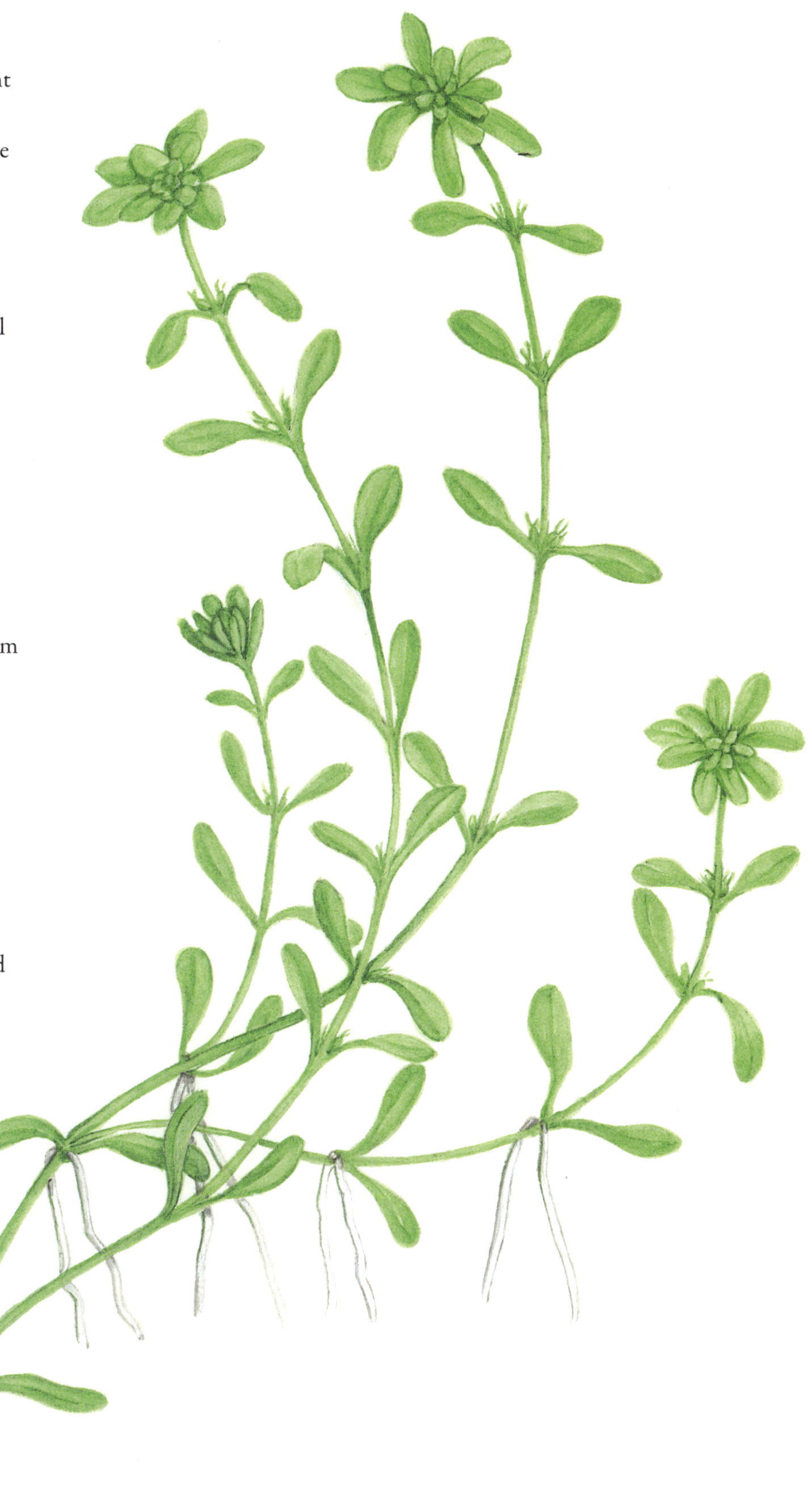

RIGHT Water chickweed (*Callitriche palustris*) by Tina Bone. This aquatic plant forms mats of foliage by sinking its rhizomous roots into soft mud. When painting, keep the roots damp to prevent them drying out and curling unnaturally.

Reproduction by bulbs, corms and tubers

Bulbs

Bulbs are layers of fleshy underground leaves at the base of a stem. Bulblets are small bulbs that develop within the scales, or on the sides, of parent bulbs, to become independent plants. It can sometimes take a few years for them to bulk up and reach flowering size. Snowdrops and bluebells are two examples of bulbs that produce bulblets. Shallots and some onions are often grown on from bulblets.

Corms

Corms are swollen underground stems with roots. Their tunics can be papery or thick and leathery (for example, *Watsonia*), but inside they are solid. Most grow a new corm above the old one for the following year, but they can also grow from cormels. These are small corms at the end of short stolons that grow out of the base of a developing new corm. They appear once leaves are present. Cormels, like bulblets, possibly provide insurance against any mishap to the main corm.

Seen mostly in monocots such as gladiolus and crocus, corms are also found in the North American eudicot *Liatris spicata*, and in the *Asteraceae* family. These plants can also regenerate from cormels if necessary. Corms can be great fun to paint with their whiskery tunics, which are the old leaf sheaths, and their slender roots. *Gladiolus* corms can be an attractive pink.

Tubers

Tubers are swollen stems or roots, positioned horizontally or vertically in the soil. They are primarily food storage organs for the plant, but also serve a reproductive function. Root tubers are found on dahlias, which can be propagated vegetatively. As an edible part, tuberous yams, for example, are full of starches. Rhizome (stem) tubers are found on potatoes, which can be divided.

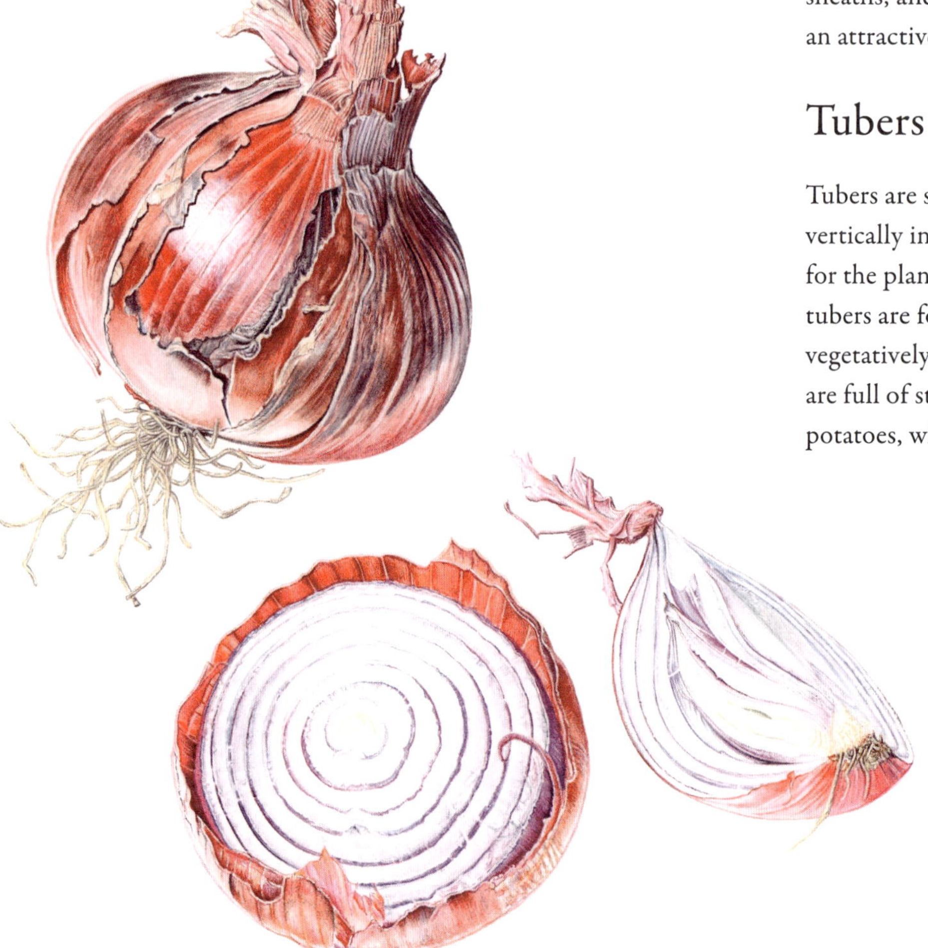

LEFT Red onions (*Allium cepa*) detail by Sarah-Jane Tweed. They are fleshy underground leaves set at the base of a stem, seen best in cross-section.

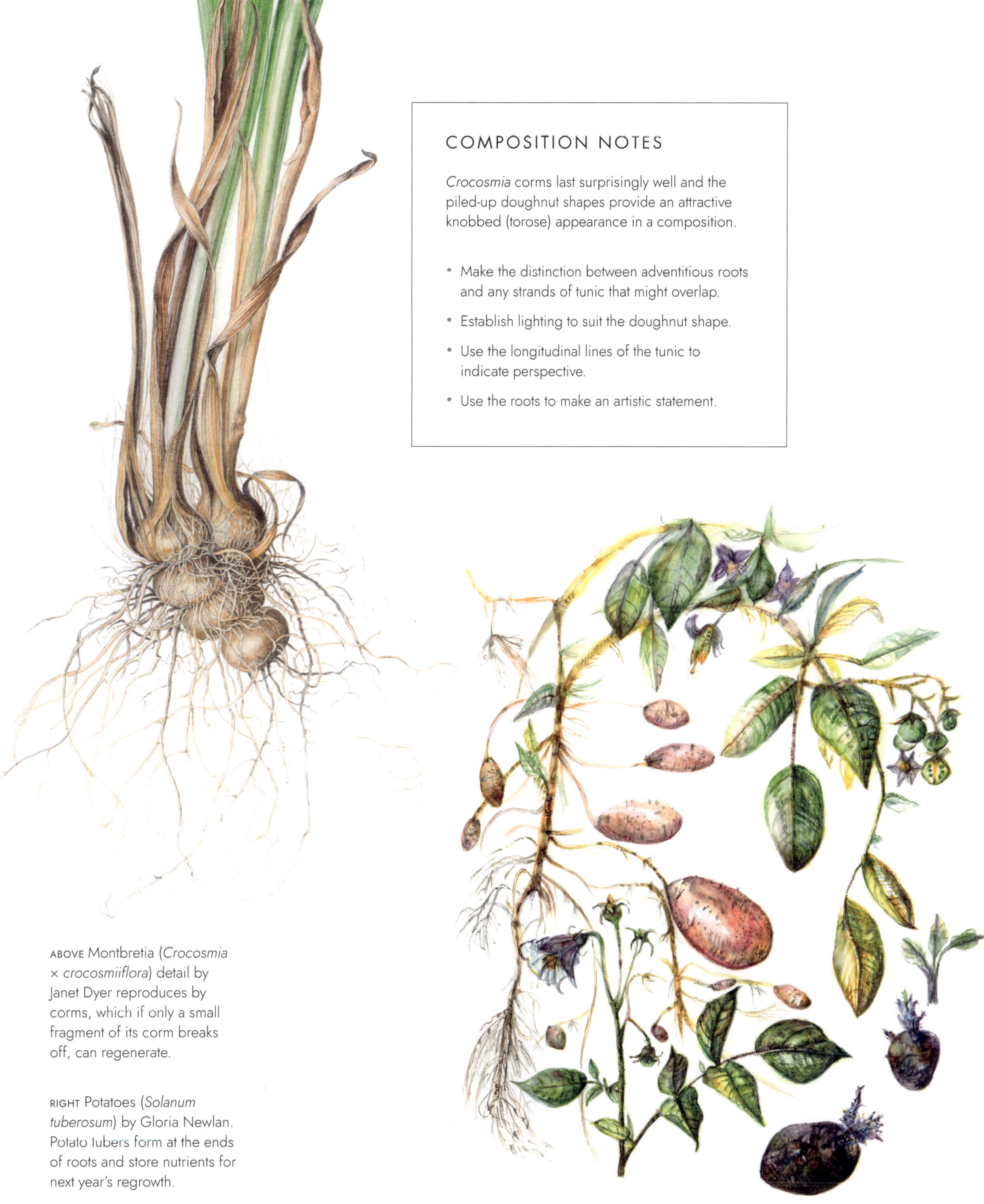

ABOVE Montbretia (*Crocosmia × crocosmiiflora*) detail by Janet Dyer reproduces by corms, which if only a small fragment of its corm breaks off, can regenerate.

RIGHT Potatoes (*Solanum tuberosum*) by Gloria Newlan. Potato tubers form at the ends of roots and store nutrients for next year's regrowth.

Painting an onion

This onion is from my garden, and I really like
the contrast between the dark red, hard, shiny
surface and the soft, light-green leaves. The
darkness of the onion is a chance to mix up
some dark washes and apply them liberally.
In contrast, the other side of the onion needs
to be almost completely white to show the shiny
surface. The roots are painted in a very light
shadow colour down one side to
show the rounded shape.

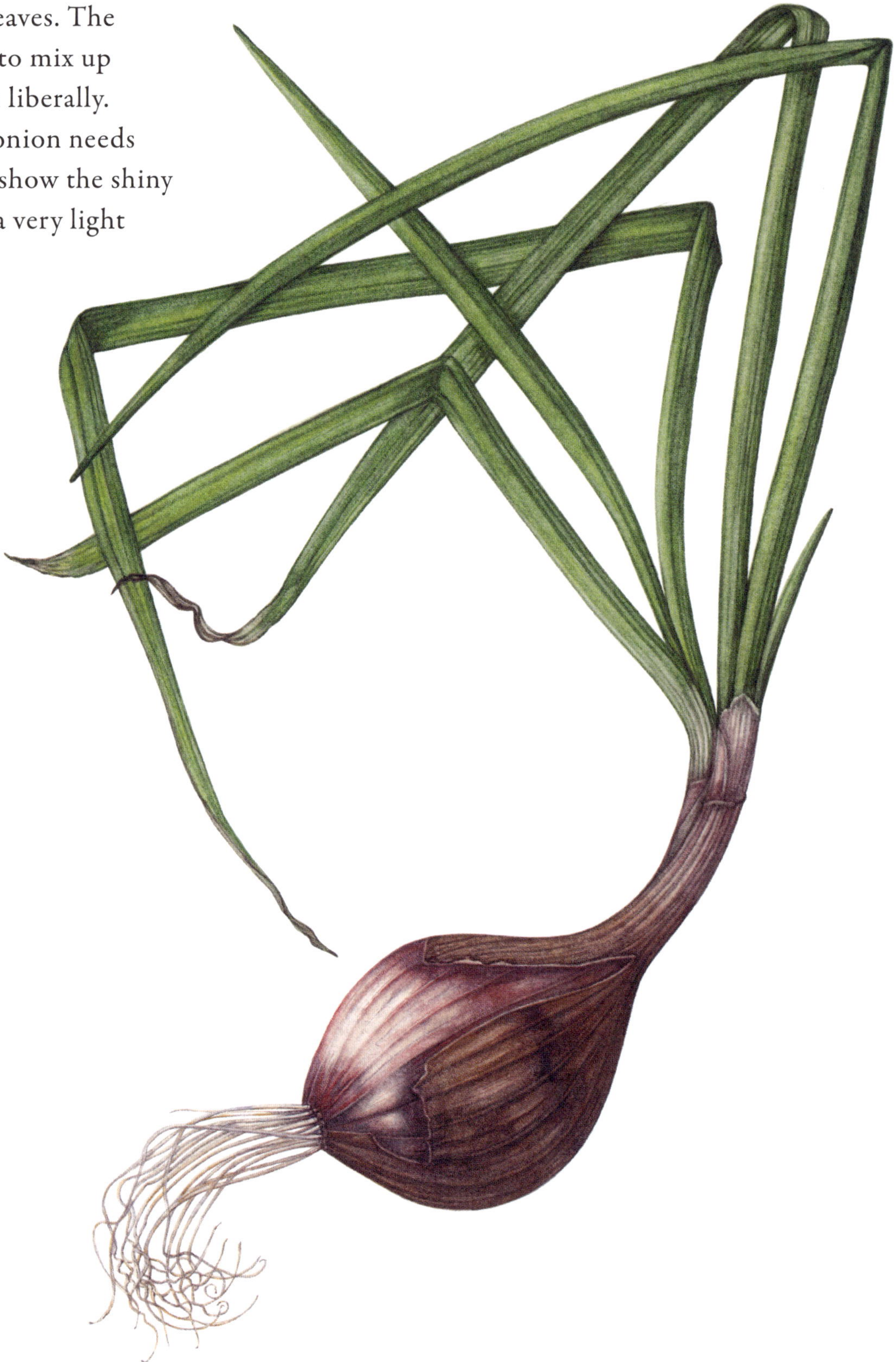

PAINTS USED:

Permanent rose

Permanent alizarin
crimson

Burnt sienna

Burnt umber

Winsor violet
(dioxazine)

Cadmium red

Sap green

Lemon yellow

Yellow ochre

Olive green

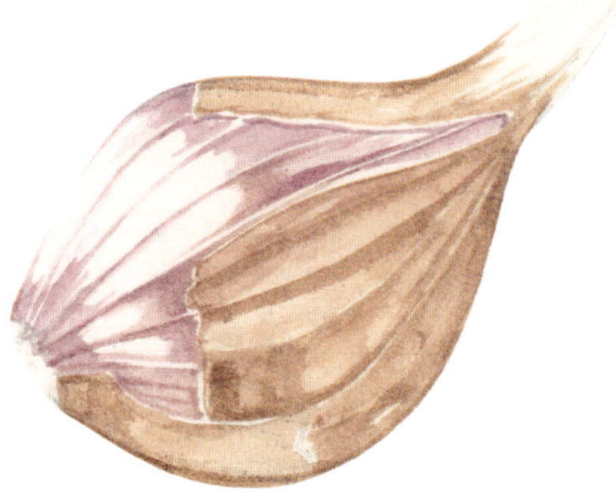

1 | Draw the outline of your plant in pencil. Create a wash from permanent rose and alizarin crimson and, using a large brush, apply to the surface of the onion following the form of the skin, leaving some areas white to show the light areas. Mix burnt sienna and burnt umber in a light wash for the outer skin, then apply this wash in a similar fashion, following the form of the skin.

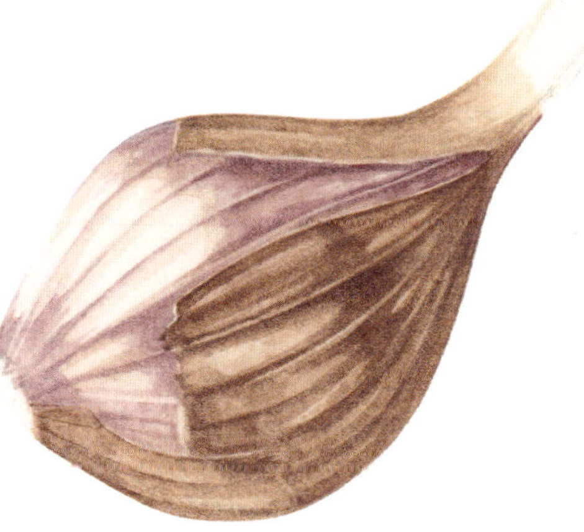

2 | Apply the first two colour washes again, taking into consideration the structure of the onion skin, making some areas darker around the veins. Darken one side of the onion to show the form of the whole object. Keep the top side really light to show the shininess of the skin's surface.

3 | Add winsor violet and more burnt umber to the brown outer skin wash and darken again, leaving lighter areas for the veins. Brighten the red wash with cadmium red and apply this wash to the skin, making sure not to lose some areas of complete white. Use a pale version of these washes and a yellow ochre wash with a very small brush to depict the roots. Paint the underside of the roots a darker colour and take care to show how each separate root sprouts from the base.

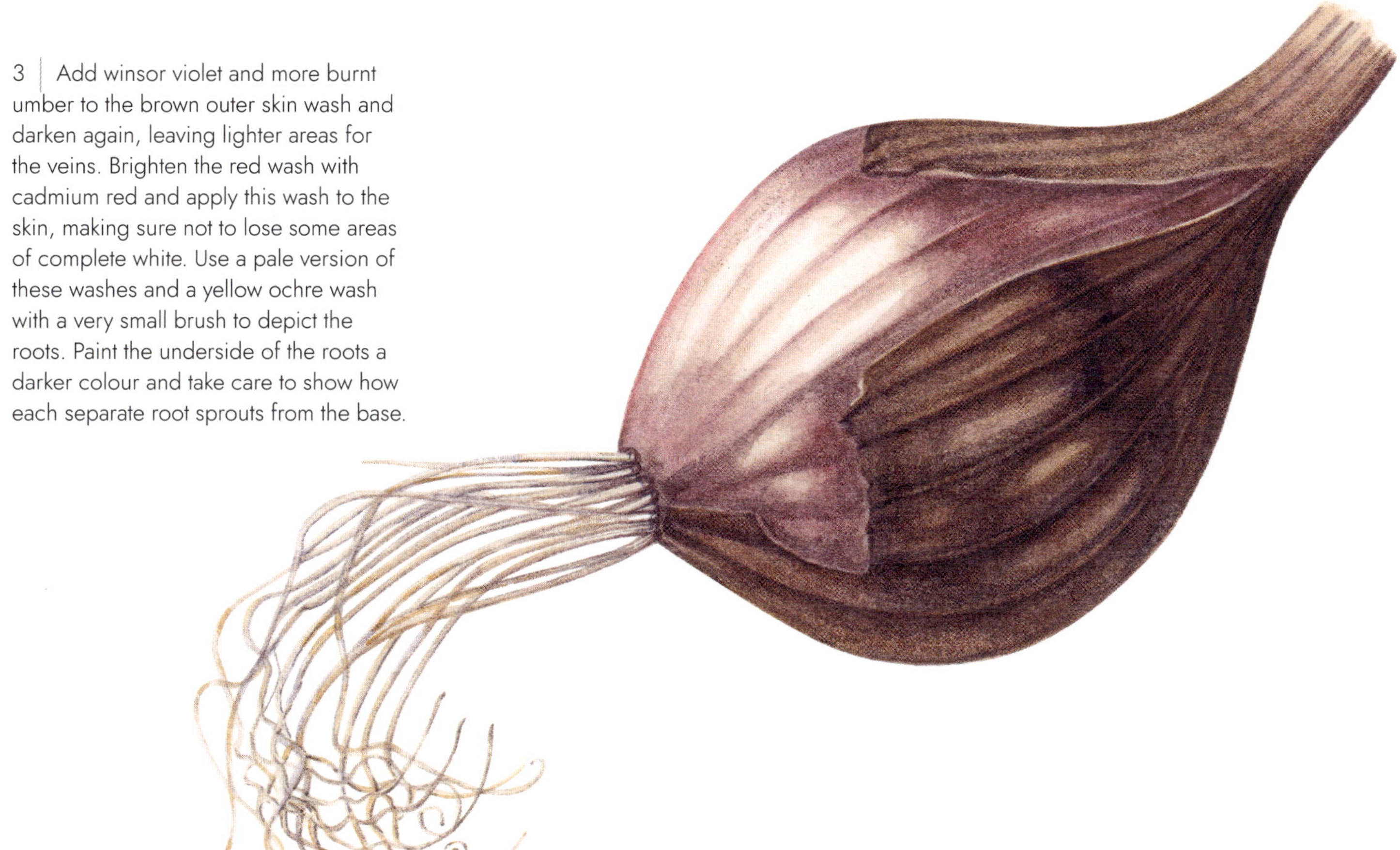

4 | Add a wash of permanent alizarin crimson to the onion skin to really brighten it. Use the purple wash created in step 3 on the sections of onion skin more in shadow. Define the lines of the roots carefully using the purple wash.

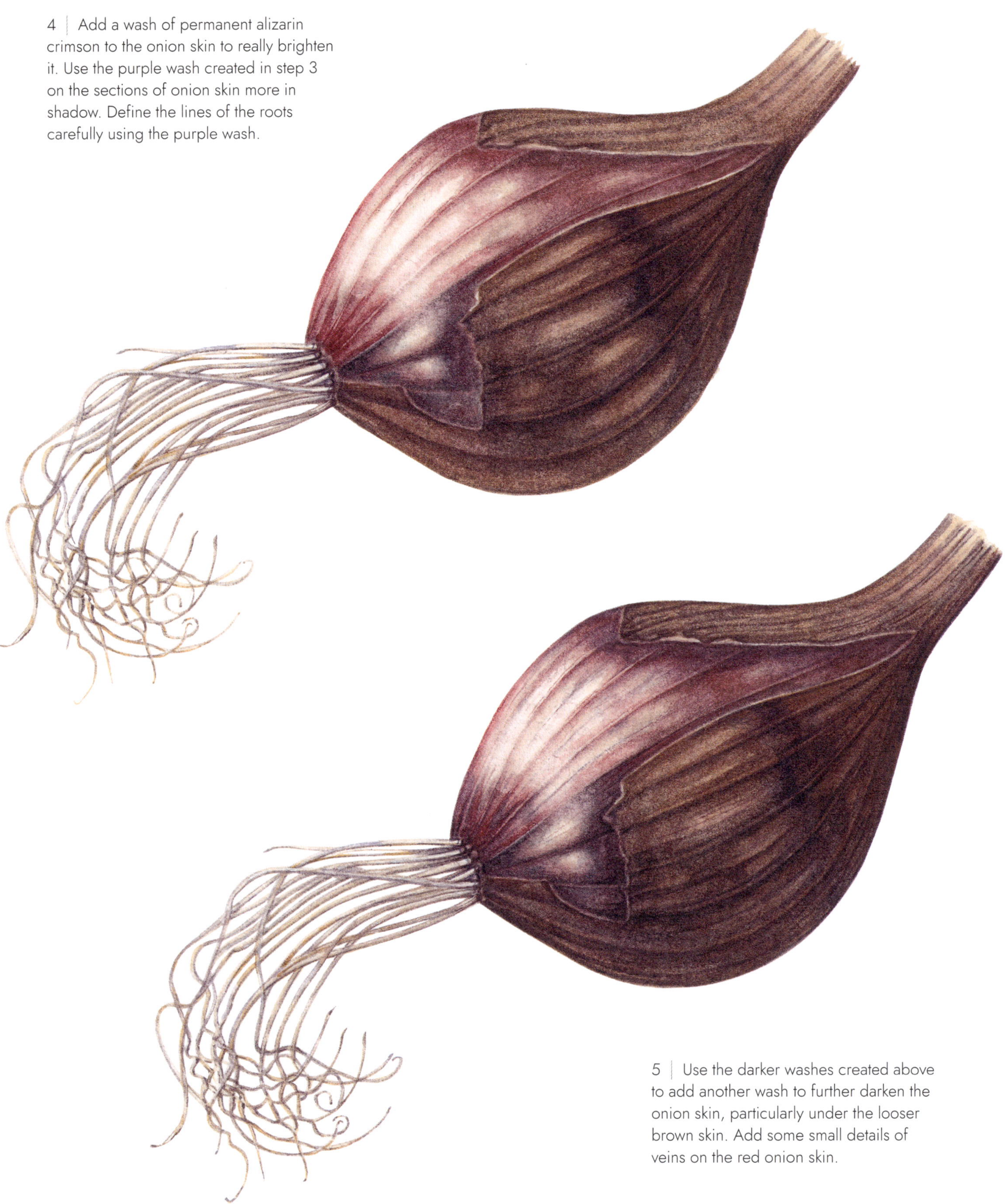

5 | Use the darker washes created above to add another wash to further darken the onion skin, particularly under the looser brown skin. Add some small details of veins on the red onion skin.

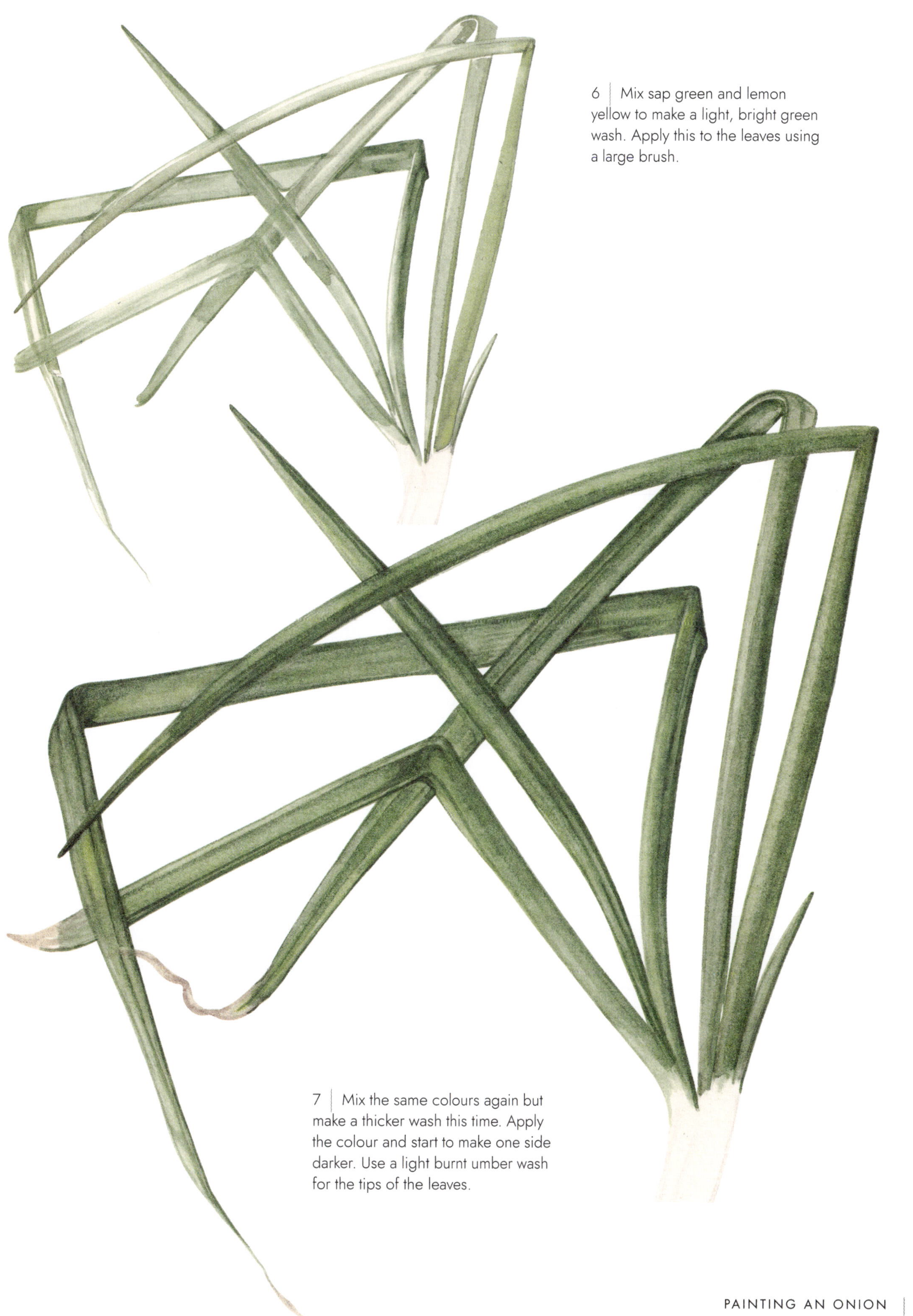

6 | Mix sap green and lemon yellow to make a light, bright green wash. Apply this to the leaves using a large brush.

7 | Mix the same colours again but make a thicker wash this time. Apply the colour and start to make one side darker. Use a light burnt umber wash for the tips of the leaves.

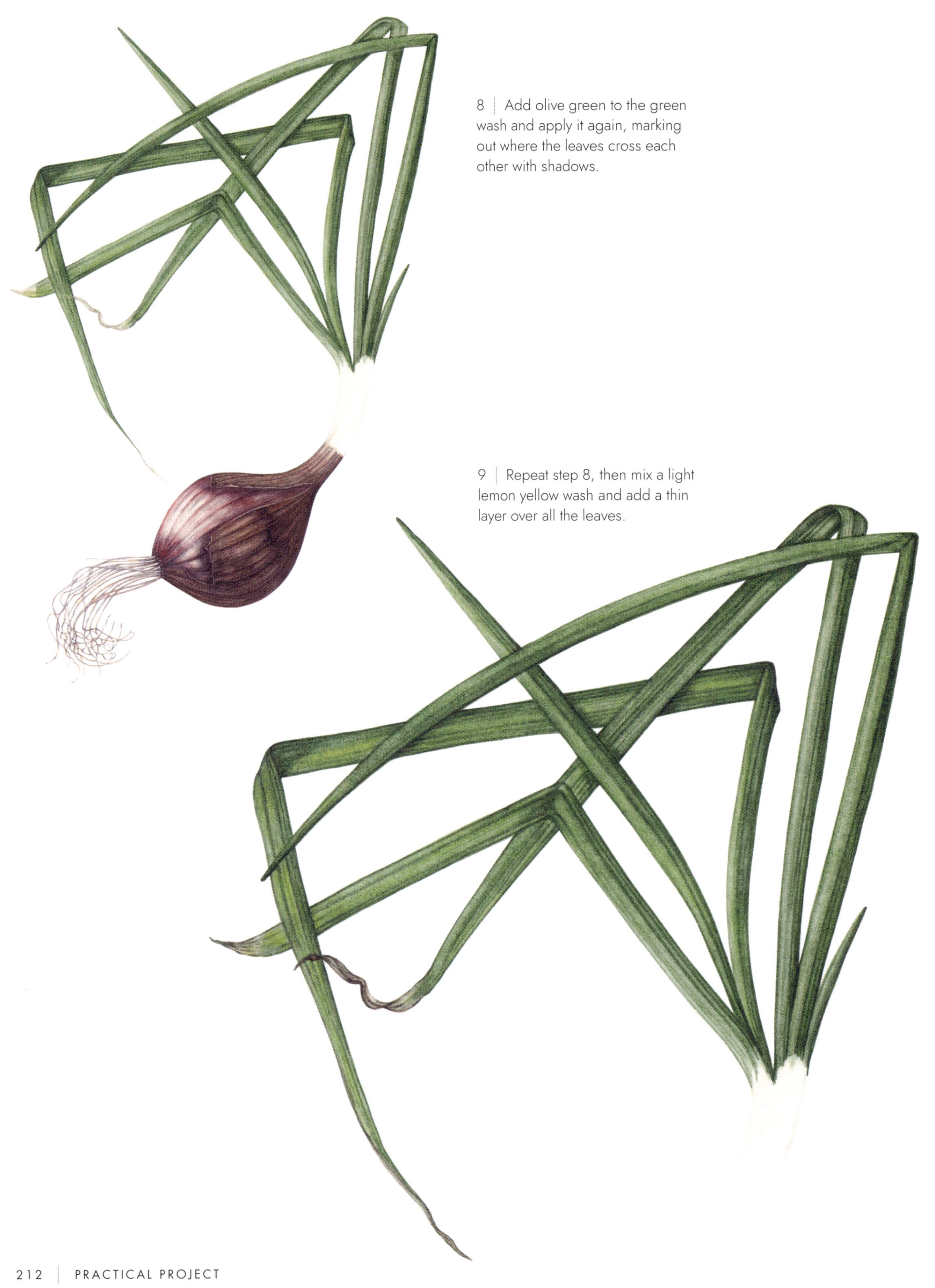

8 | Add olive green to the green wash and apply it again, marking out where the leaves cross each other with shadows.

9 | Repeat step 8, then mix a light lemon yellow wash and add a thin layer over all the leaves.

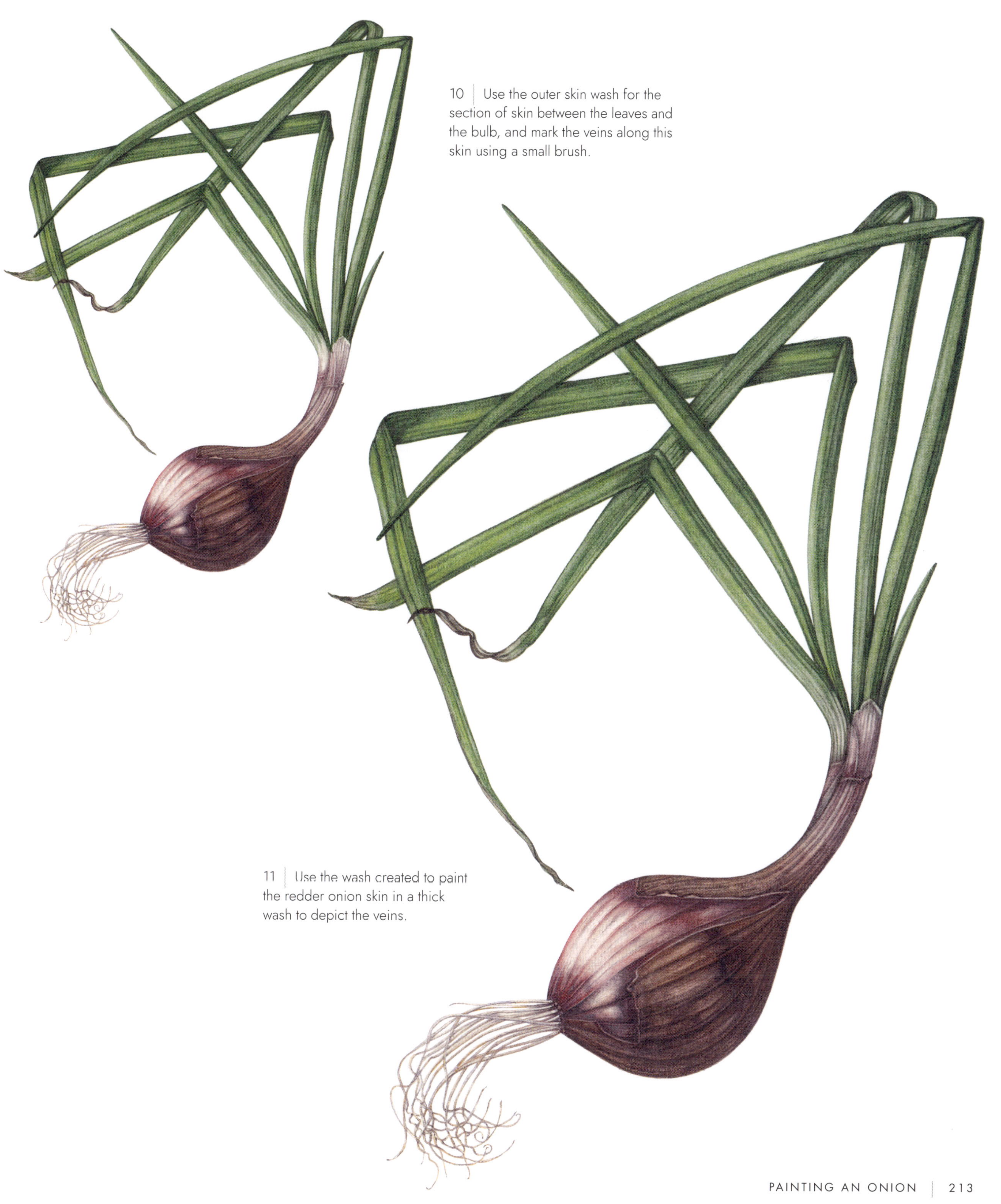

10 | Use the outer skin wash for the section of skin between the leaves and the bulb, and mark the veins along this skin using a small brush.

11 | Use the wash created to paint the redder onion skin in a thick wash to depict the veins.

East Indian galingale (*Kaempferia galanga*)

by Nataya Udompat

The plant

Nataya has given us a beautiful breakdown of an important member of the ginger family, one that is grown in gardens throughout Southeast Asia. The leaves are cooked as a fresh vegetable and the rhizomes are used to flavour fish curry. An oil extracted from the rhizome is used in rituals.

Many gingers are propagated by exploiting the nodes on the underground rhizomes, where new leafy shoots sprout when conditions are right. It makes members of *Zingiberaceae*, such as gingers and turmeric, easy to grow for the kitchen. This painting was identified as *Kaemferia galanga*, a ginger which is distinctive in having no stem and having dark, rounded rhizomes, both of which are clearly shown in the painting.

The success of the painting lies in Nataya's analytical approach. We have a view of the plant's habit as though looking down on it. Each flower part is separately depicted, revealing the ginger family's characteristic ephemeral zygomorphic flowers. They are trimerous and complex. There are two whorls to the perianth, consisting of a tubular calyx and a tubular corolla with lobes, one lobe bigger than the other two. Two of the three stamens look like petals, are sterile and form a lip. One stamen is fertile. The inferior ovary has two nectaries on top of a funnel-shaped stigma.

Nataya has composed the galingale's parts into several rectangles, all of which conform to the golden ratio or golden section. The heaviest, comprising green and buff colours in a rectangle at the bottom right, is where we see the utilitarian parts necessary for cooking and growing. The delicate flowers, together with the rhizomes, leaves and shoots, make up the largest rectangle. The insect bites and holes in the parallel veined leaves provide a good clue that the plant was painted from life and that it might well have come from her kitchen garden.

ABOUT THE ARTIST

Nataya Udompat is a self-taught botanical artist from Thailand. After retirement she began painting on porcelain. She then took classes in 2019 and 2020 with the pioneer Thai botanical artist, Sansanee Deekrajang, who helped her understand the underlying botany of her work. She successfully submitted this painting for the Royal Botanic Garden Edinburgh florilegium collection 2020 exhibition.

RIGHT Adding various parts and aspects of a plant squares off the composition and provides the parts necessary to identify it.
Size: 42 x 29.7cm (16$^{1}/_{2}$ x11$^{3}/_{4}$in), watercolour on paper

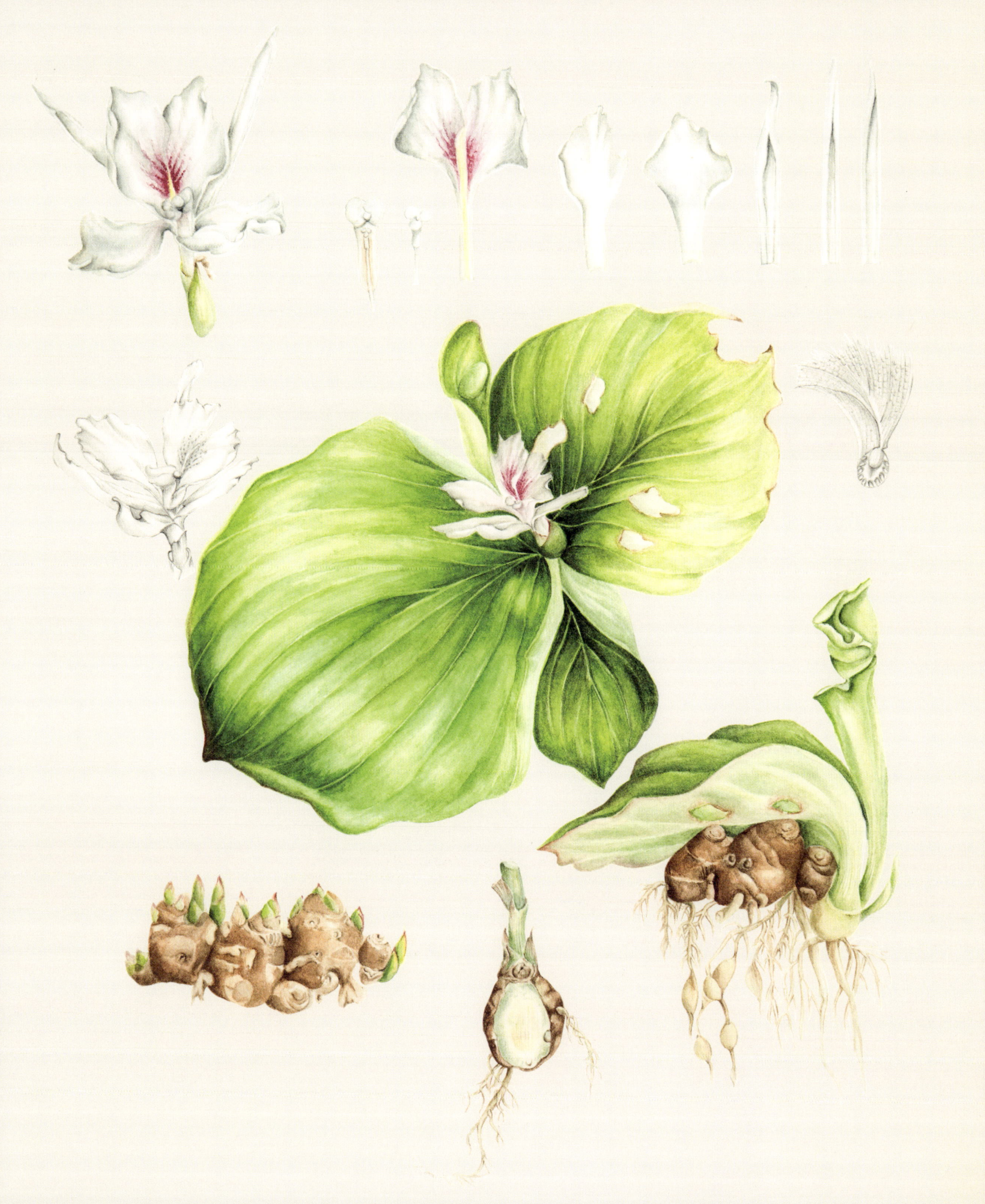

Chapter eight

Non-flowering plants

These include ferns that reproduce by producing spores, and various woody plants – the gymnosperms – that replicate by manufacturing seeds that are not protected within an ovary – for example, conifers, ginkgo and cycads.

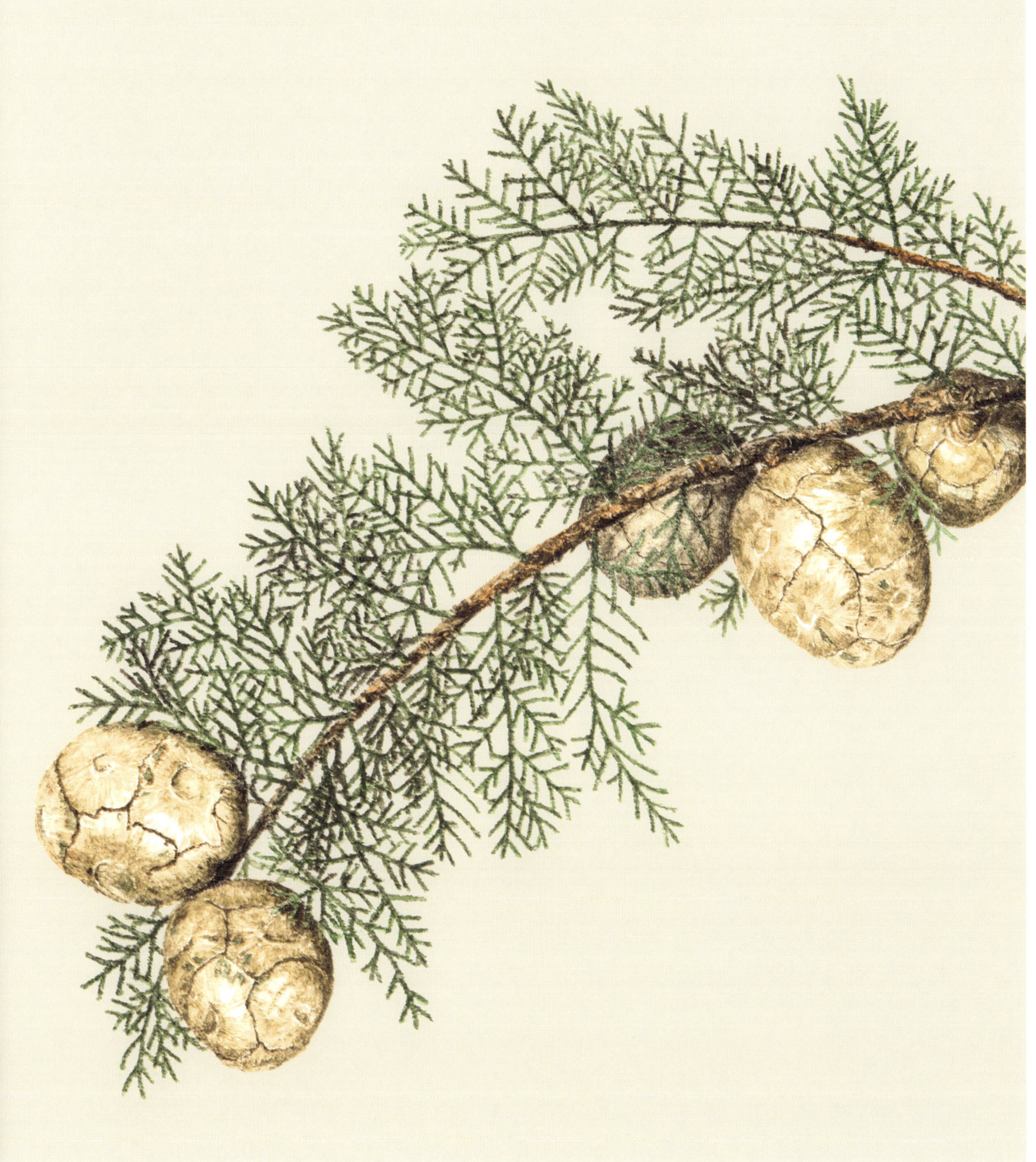

Ferns

A typical fern has frond-like leaves and a branched system of veins. Just like angiosperm leaves, variations depend on their venation, margins, tips, and so on (see chapter 4).

They produce tiny spores in spherical sacs called sporangia. These are found in groups called sori. They make lovely patterns and are often seen on the underside of the leaves. Sori are often covered by a membrane called an indusium. The indusium and sori are diagnostically important in a painting.

Fern leaves are called fronds. They are held on stalks called stipes. A system of scales of various shapes is often attached to them; these also come in various shapes and sizes. Fronds are particularly attractive. The lobed fan-shaped leaflets of a maidenhair fern frond are delicate, others are robust. Some are deciduous, others evergreen. The hart's tongue fern is an example with simple leaves, but leaves of ones such as bracken are divided and subdivided into many pinnae. Most types of ferns have leaves that, when young, emerge as curled-up croziers or 'fiddleheads'. The croziers provide a challenging, yet beautiful, subject for artists. They eventually uncurl into simple or pinnate leaves.

Fern life cycle

The life cycle of a fern is a fascinating subject to draw and paint. It involves two stages. First, the spores ripen on 'fertile' leaves. Some ferns have fertile as well as sterile fronds, which may look different. The spores then drop to the ground for the second stage. There, the spore produces a tiny green, heart-shaped structure called a prothallus, which is no more than 1cm (^{1}/2in) high. Tiny, slender rhizoids anchor it to the ground. On it are both male and female gametes (a reproductive cell). Under damp conditions, the male gamete swims to one of the female ones to fertilize it. Once fertilized, a new fern plant then sprouts on top of the prothallus.

Ferns also have vegetative methods of reproduction, which an artist might include. They can spread by rhizomes (see chapter 7) and might produce new plants on a leaf, or they could develop bulbils. Hickey & King (see Glossary) provide excellent diagrams to help you understand the parts and the different forms they can take.

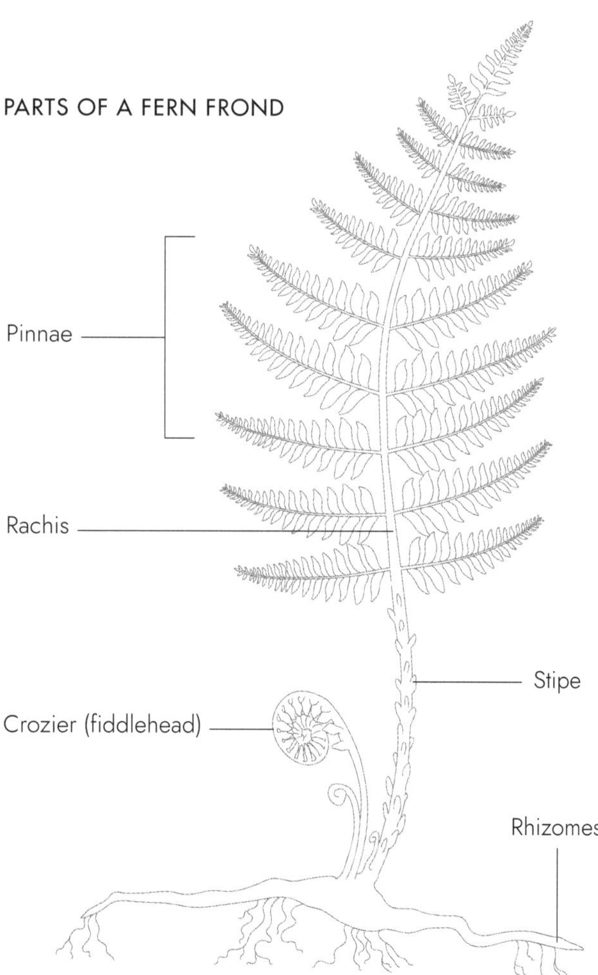

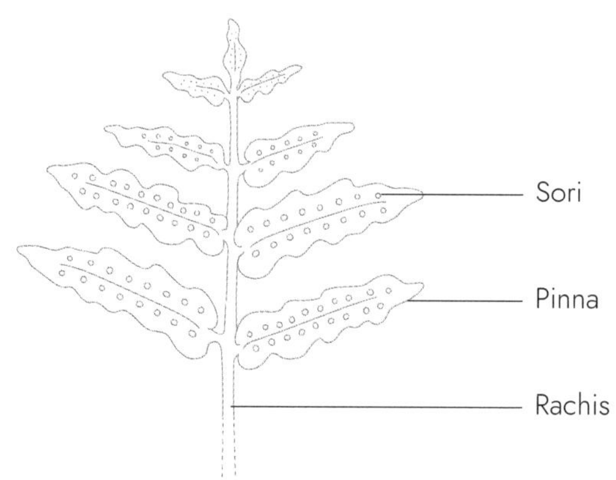

Gymnosperms

Gymnosperms have true seeds, but they are not protected within an ovary, like those of angiosperms. Seeds are typically borne on the upper surface of a hard scale and are only protected by the covering of other overlapping scales to form a cone, for example, in cycads and conifers. However, there are a number of variations, such as the arils of yew (*Taxus*) or the fleshy cones of junipers, both of which resemble small berries. If a tree has scales or needles for leaves as well as cones, it is a conifer.

 Common yew (*Taxus baccata*) by Roger Reynolds. This tree is poisonous except for the red flesh of the female arils.

 Douglas fir (*Pseudotsuga menziesii*) by Mary Frost. This long-lived conifer from the North American Pacific coast holds cones that are easily identified by their three-pronged bracts.

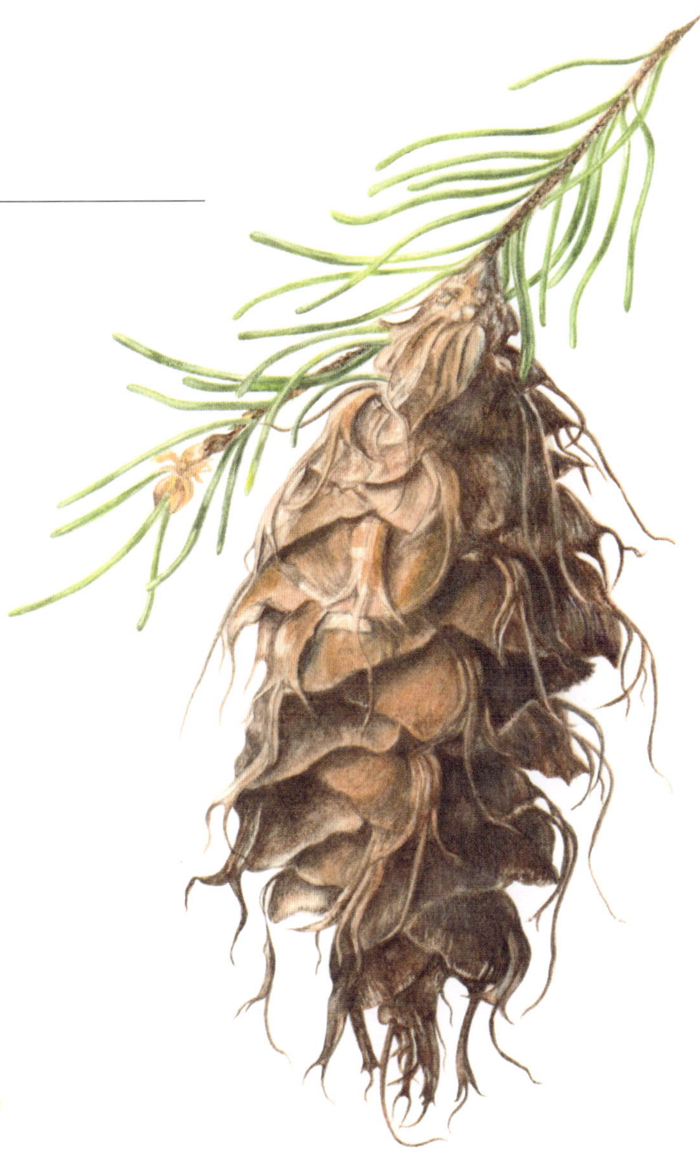

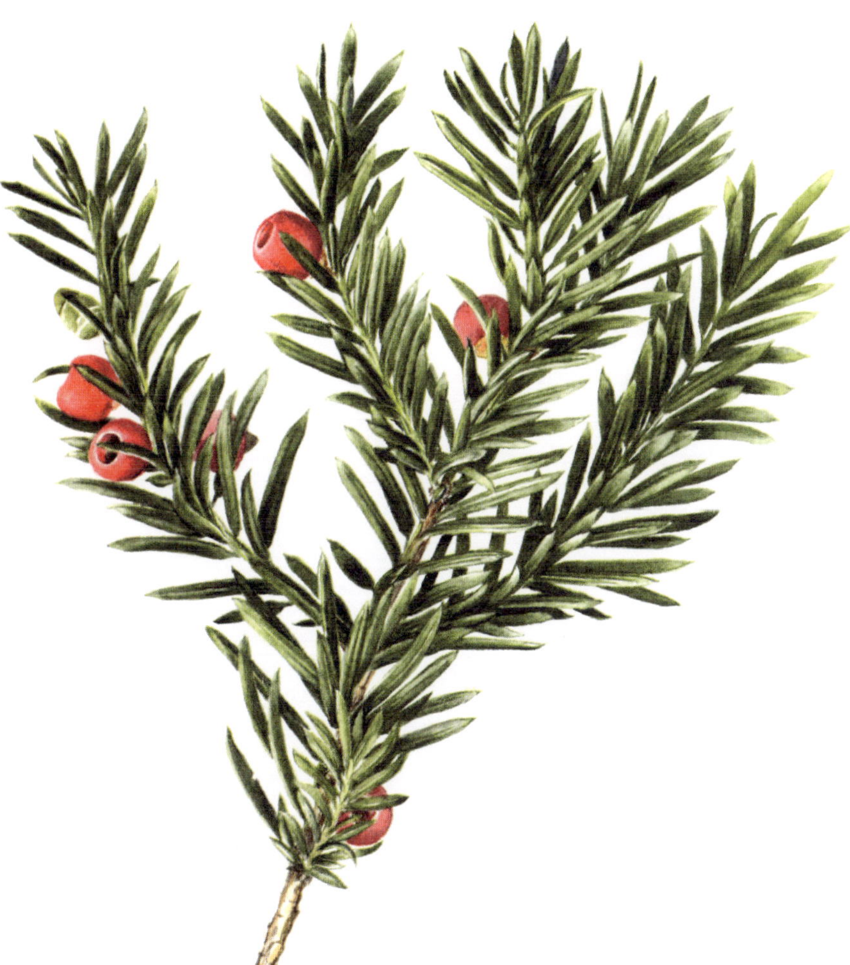

Ginkgo biloba is the only living connection between ferns and conifers. It is often painted for its beautiful bi-lobed leaves with fanned venation; but look out for the short male 'catkins' and the pairs of oval fruit from the female parts. Yews are conifers, with needle-like leaves, but the 'cone' is a soft aril, a red fleshy casing that opens at the top and contains a seed. Separate trees hold the male and female 'flowers'. Study the leaves carefully. Although they are spirally arranged, they are flattened so they look like two rows either side of the twig.

Cycads look like palm trees with fronds for leaves, but they have long cones borne at the tips of the trunks. Both male and female cone scales offer perfect Fibonacci patterns, whilst the fruit inside provides lovely colours.

Conifers

The conifers most likely to be encountered and painted are pine, fir, larch, cypress, cedar and spruce. They all have attractive cones, mostly hard (junipers are the exception) with overlapping scaly arrangements. However, their leaves and bracts also deserve attention.

Leaves are either needles or scales, mostly arranged in a spiral, but some can have whorls of leaves (for example, juniper), as well as alternate or opposite (and decussate) arrangements. How and if leaves are twisted at the base is significant, for it causes them to look flat and pectinate (comb-like) in two rows. It is a feature that enables conifers in dark northern latitudes to maximize exposure to the sun. If they are not twisted, they may have leaves sticking out in all directions, giving the branch a cylindrical look.

Fir leaves are fixed by a cup-like structure at the base. Pines have needle-like leaves in fascicles or bundles of two (for example, Scots pine), three (for example, *Pinus jeffreyi* and Monterey pine) or five (for example, Arolla pine). Cedars and larches have rosettes of short, soft needles along their branches.

The cones alone are fascinating for artists. Their various sizes, shapes and Fibonacci patterning are challenging yet rewarding. Attention should also be given to the scale features and the bracts.

Cones are known botanically as strobili (the singular is strobilus). The male cones are small and ephemeral, whilst the female cones are larger and more diverse. The large cones are made up of woody seed scales, on which the flat seeds sit. They are arranged spirally along the central axis and determine the external pattern. At each point, a bract scale sits beneath a seed scale.

The globular cones of cypress, juniper and yew are very different from the cylindrical shapes of fir cones. Cedar cones are barrel-shaped with short, but horizontally wide, appressed scales. Hemlock (*Tsuga* spp.) cones are pendulous, whereas fir tree cones are upright on the branch.

Unlike in flowers with their ovaries, the bract protects the seed, rather than the flower. Usually, the bract scales are hidden, but occasionally they stick out from behind a scale and can be used to identify the tree. The three-pronged bracts of a Douglas fir (*Pseudotsuga menziesii*) are a giveaway. The long bracts of the North American western larch (*Larix occidentalis*) give their cones a porcupine look, whereas the backward-facing long bracts on a noble fir (*Abies procera*) cone makes them prickly.

BELOW Conifer cones by Leigh Ann Gale. Left to right: cypress (*Cupressus* sp.), pine (*Pinus* sp.) and larch (*Larix* sp.). They make an interesting comparison and a beautiful set.

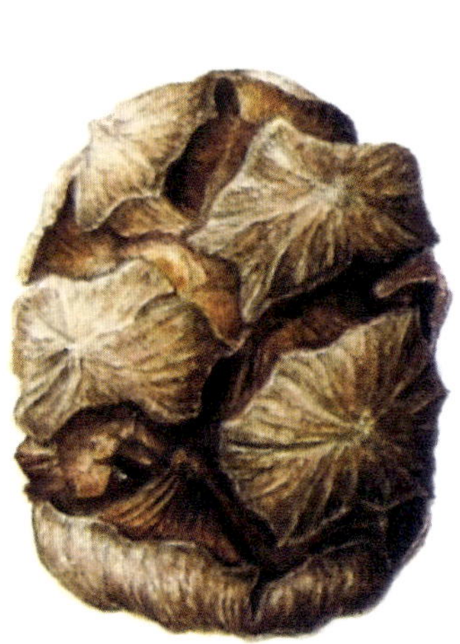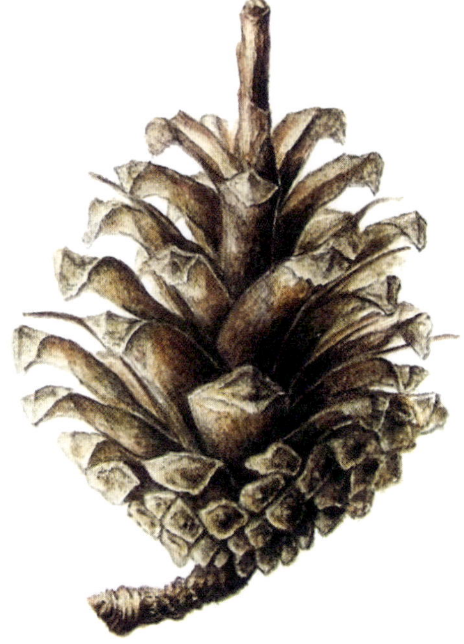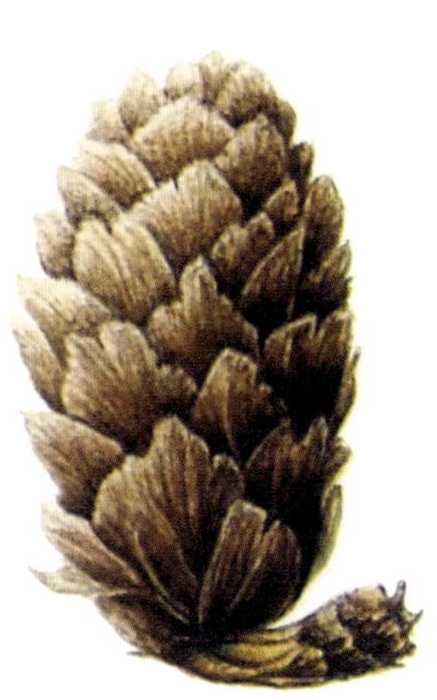

WHAT TO LOOK FOR

The Scots pine (*Pinus sylvestris*) has many interesting features, making it a popular subject for artists.

- Note the pairs of long grey-green leaves, with a well-defined line of pale stomata along the length on the underside.

- Distinguish between male and female cones on the same tree.

- Look for angular, wedge-shaped seed scales on the cones.

- Note a 'boss' on the central ridge of the scale with a prickle, which is small on a Scots pine, but can be long and spiky on other pines.

- The bracts are not obvious because they adhere to the base of the woody seed scale.

- The small seeds inside the cones have a wing, enabling them to blow away.

- The number of Fibonacci spirals should be counted, both clockwise and anticlockwise. You should get two numbers in the Fibonacci sequence. Pines have three steep and five gradual spirals. Where they cross, they form a space for the insertion of a bract. The cone's three-dimensional shape will impact the size of the bract around the shape (perspective) and where the spirals meet (narrow).

FIBONACCI SPIRALS

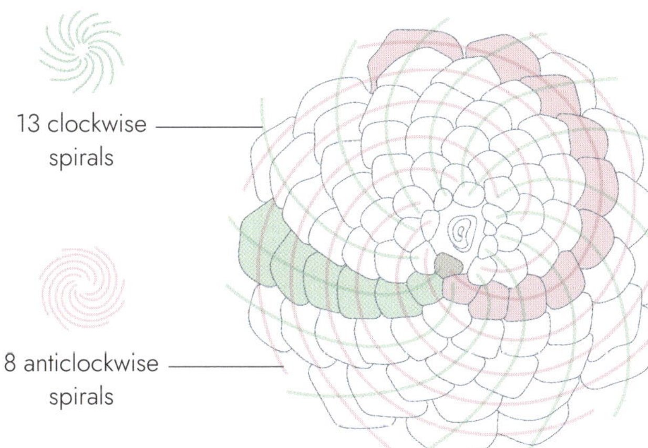

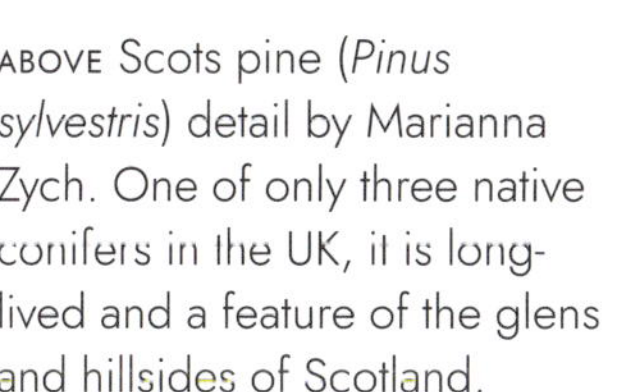

ABOVE Scots pine (*Pinus sylvestris*) detail by Marianna Zych. One of only three native conifers in the UK, it is long-lived and a feature of the glens and hillsides of Scotland.

Painting a pine cone

The pine cone is a study in brown. Although relatively few colours are needed for this painting, the details and application of the shadow colours need careful consideration. A great deal of observation about which area is darker compared to those around it is involved. There is a real difference in brown tones between the inside sections of the scales where the seeds sit, and the ends of the scales. Making sure these appear different helps to show the structure of the cone.

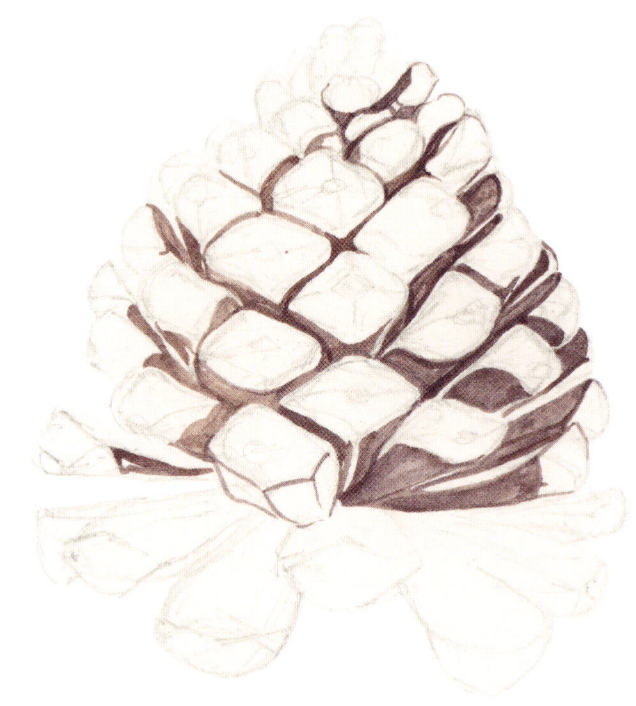

1 | Draw the outline of your pine cone in pencil. Mix a wash of burnt sienna, burnt umber and winsor violet, then carefully apply this to the underside of the long thin sections of the cone scales.

PAINTS USED:

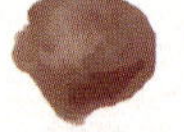 Burnt sienna

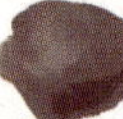 Burnt umber

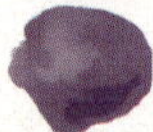 Winsor violet (dioxazine)

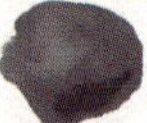 Payne's grey

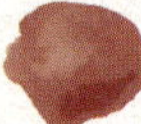 Indian red

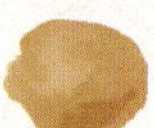 Yellow ochre

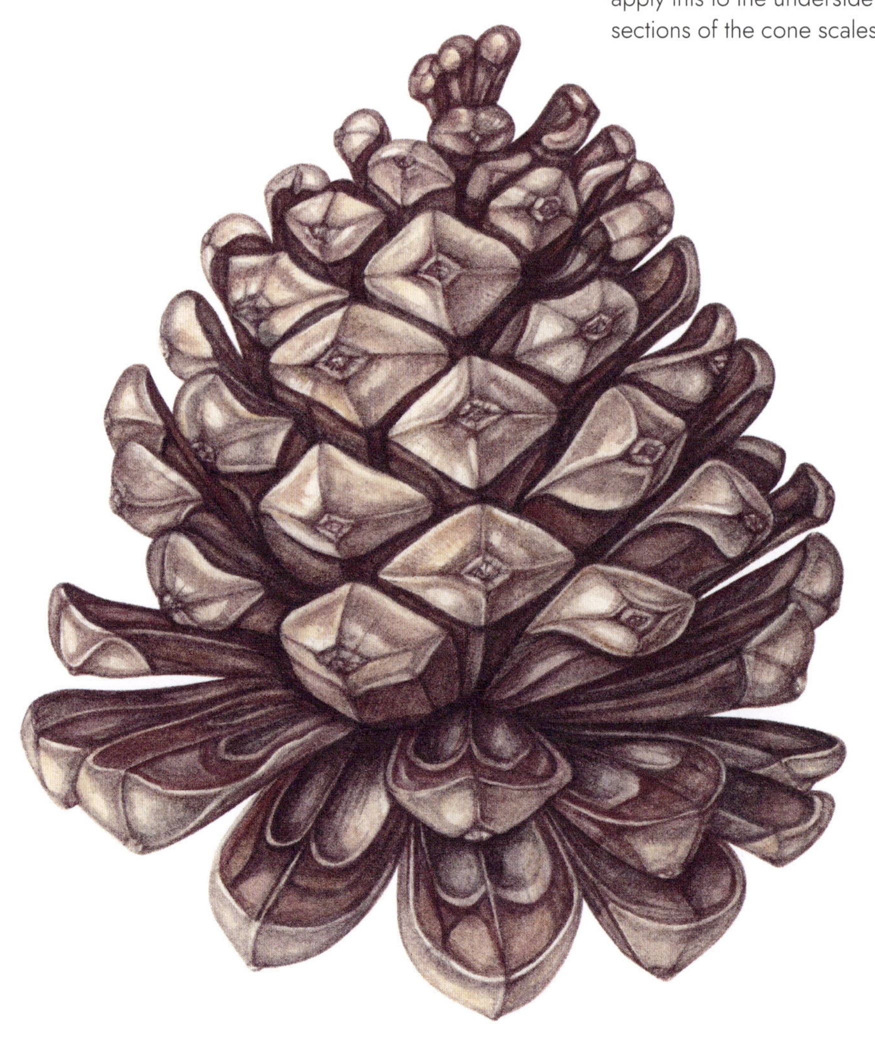

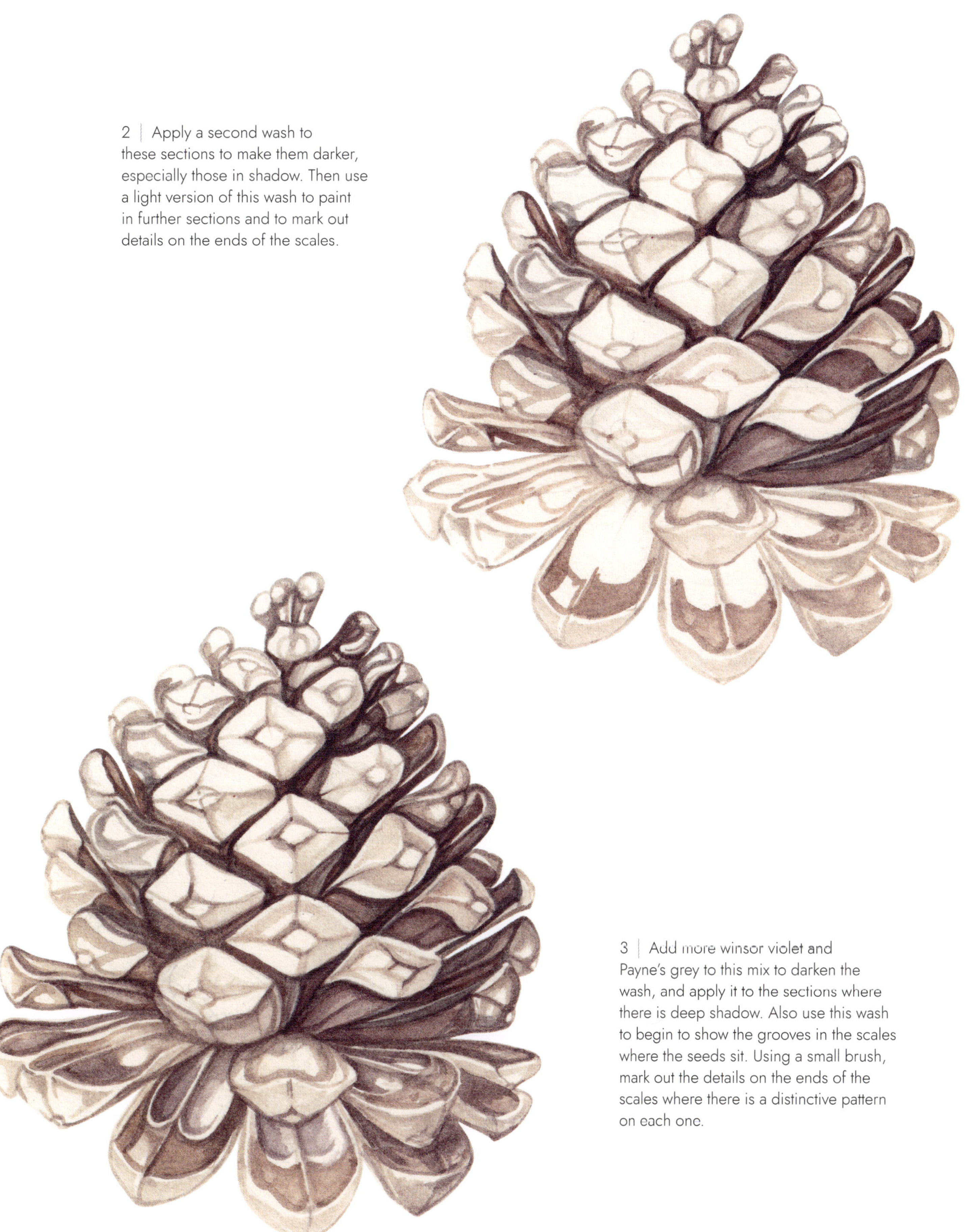

2 | Apply a second wash to these sections to make them darker, especially those in shadow. Then use a light version of this wash to paint in further sections and to mark out details on the ends of the scales.

3 | Add more winsor violet and Payne's grey to this mix to darken the wash, and apply it to the sections where there is deep shadow. Also use this wash to begin to show the grooves in the scales where the seeds sit. Using a small brush, mark out the details on the ends of the scales where there is a distinctive pattern on each one.

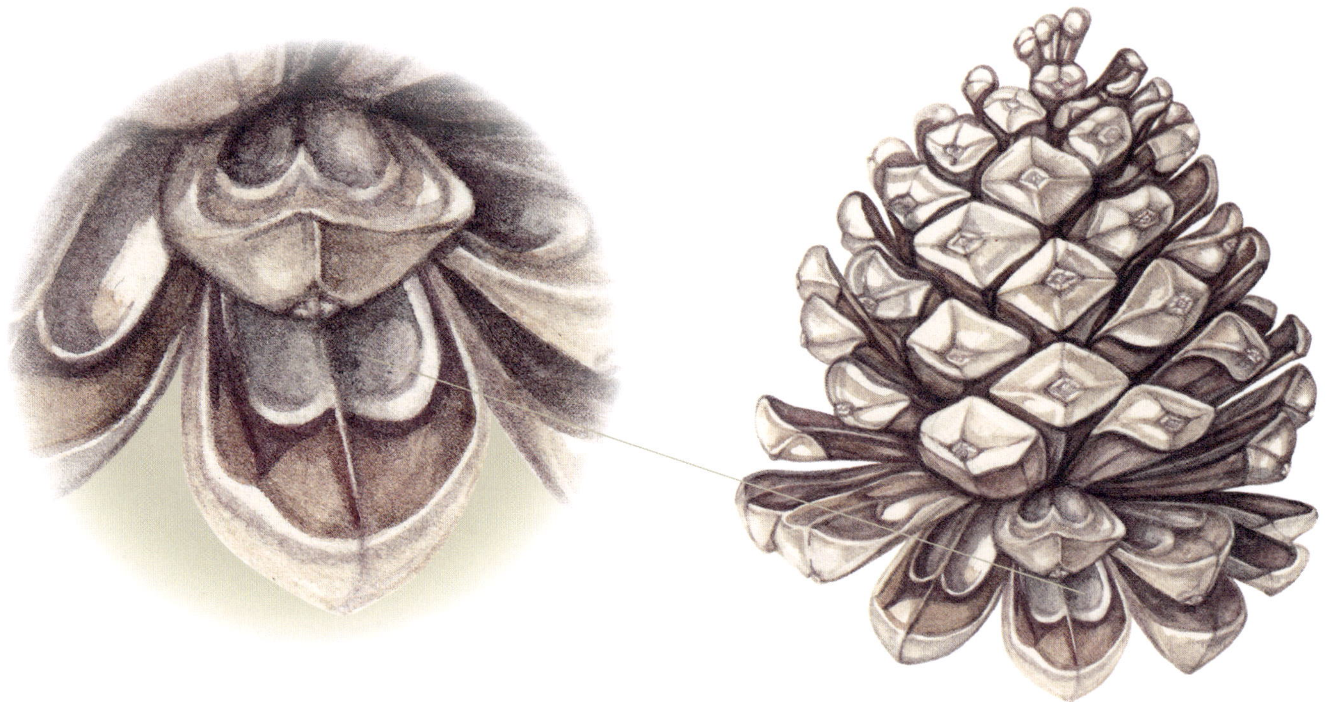

4 | Mix an Indian red and burnt umber wash and apply to the brighter sections of the scales at the bottom of the cone. Use the purple wash from step 3 to further darken the shadow sections and apply details to the ends of the scales, showing the light and dark areas. Make sure to follow the form.

5 | Repeat step 4.

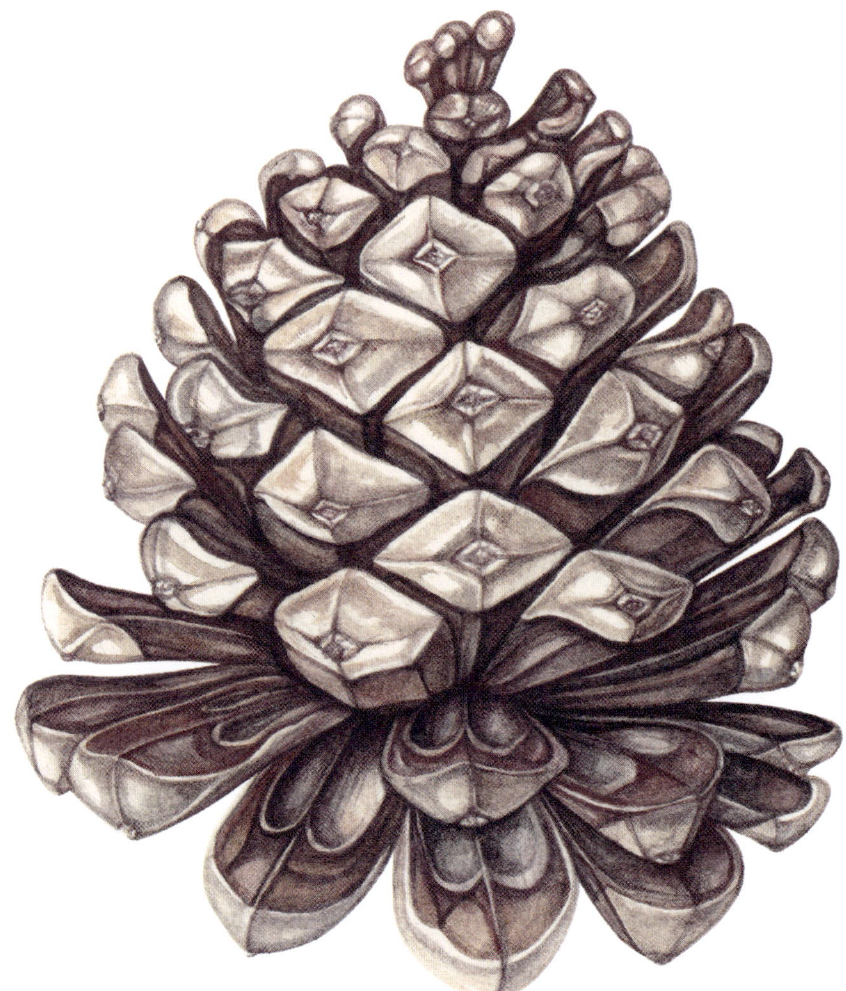

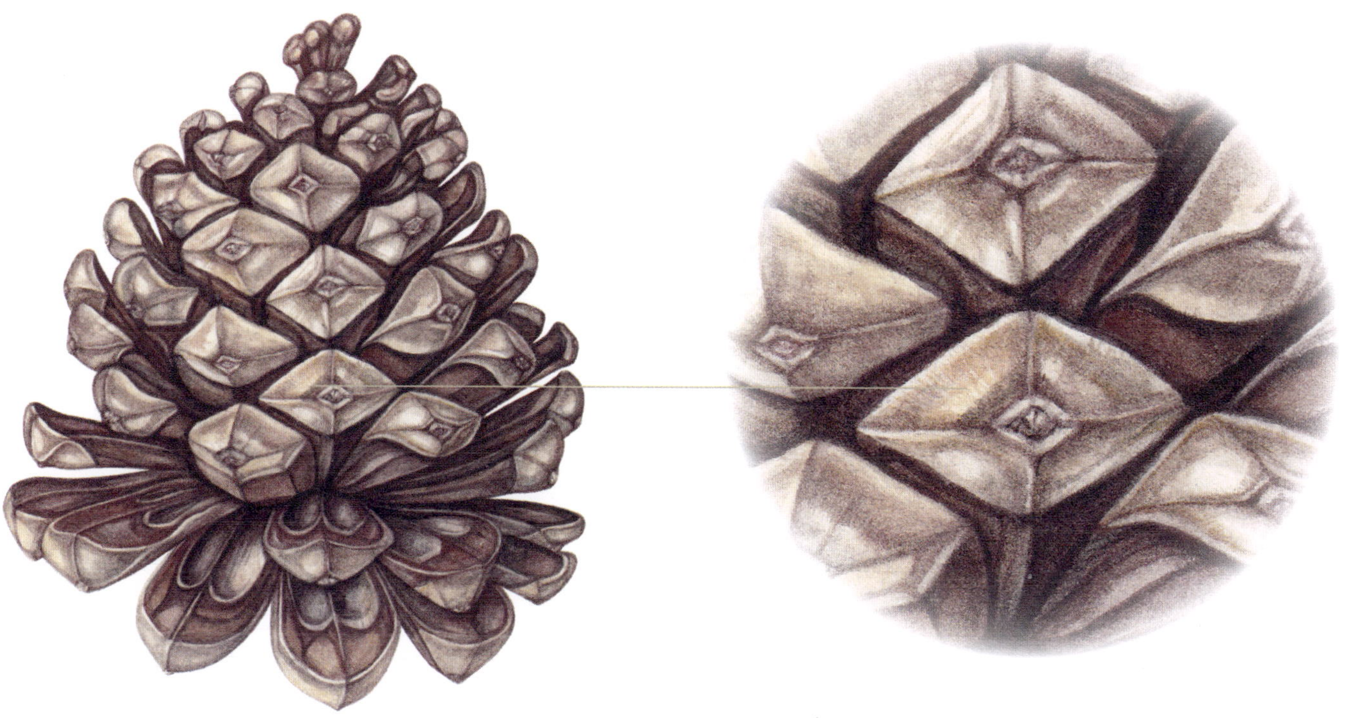

6 | Mix a wash of burnt sienna, burnt umber and a tiny touch of yellow ochre, and apply to the end of the scales. Use a very small brush and small lines to show the form. This should brighten these ends.

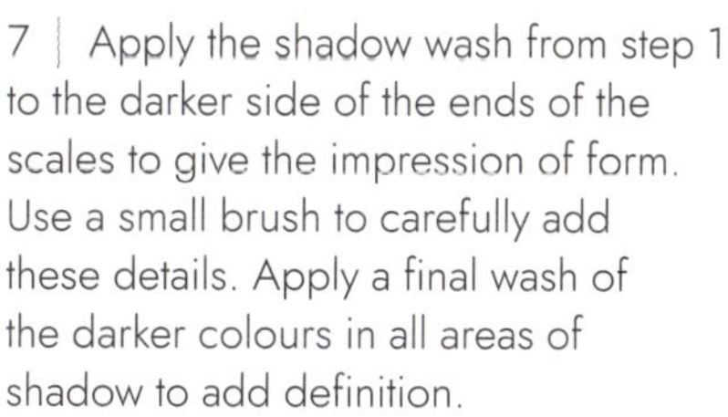

7 | Apply the shadow wash from step 1 to the darker side of the ends of the scales to give the impression of form. Use a small brush to carefully add these details. Apply a final wash of the darker colours in all areas of shadow to add definition.

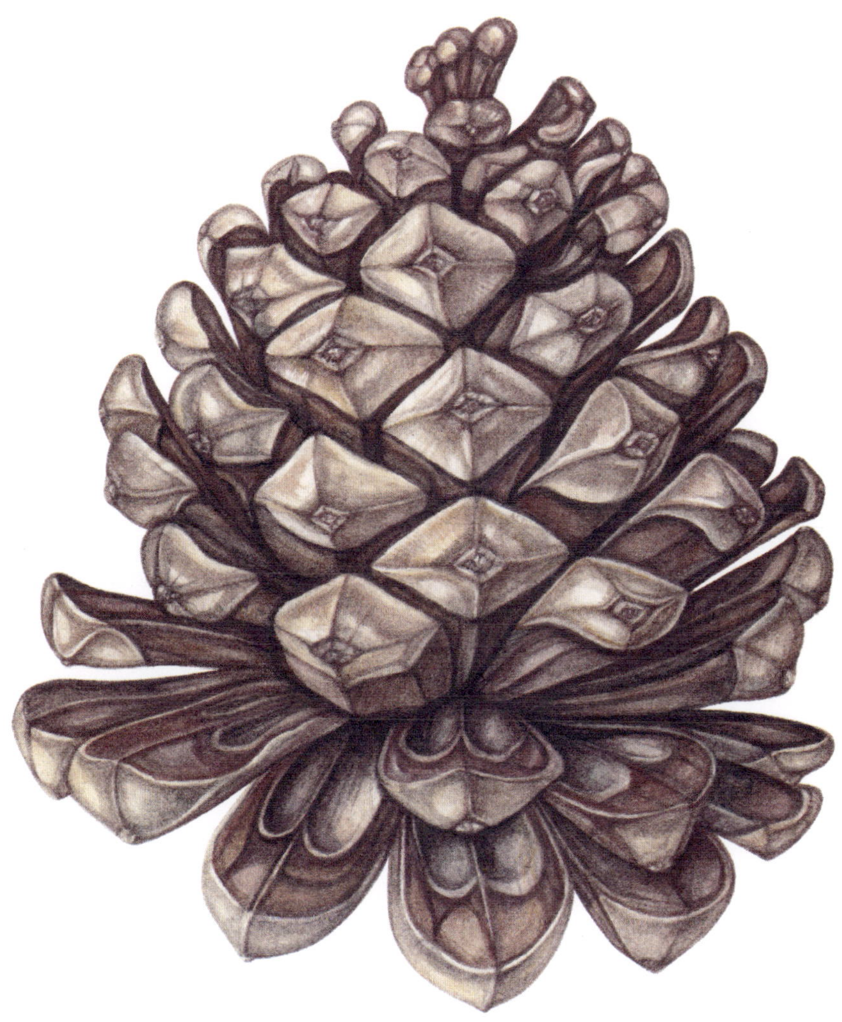

Shaggy shield fern (*Dryopteris cycadina*)

by Laura Silburn

This is a picture that takes one into the heart of a fernery, or a primeval wood, where *Dryopteris* (buckler) ferns are often the dominant ground cover. As a painting, it induces awe for an artist attempting the detail inherent in a fern. And Laura has succeeded, for the picture contains everything required to identify the fern, not as one of the many British buckler ferns, but one that is specifically a temperate East Asian one. It is sold in garden centres as 'shaggy wood fern', 'shaggy shield fern' or 'black wood fern'.

Two aspects of the painting make the viewer feel immersed in the fern. The first is Laura's use of aerial perspective, softening the distant features and enhancing closer ones; the other is placing a cut-off set of fronds in the foreground to make it appear larger than the detailed base of the fern in the background.

Genus features

All the pertinent details are included, from unfurling croziers to slightly falcate, alternately placed pinnae at the top of the fronds. Dark-coloured stipe scales, characteristic of this genus, show up on clearly defined upright stems on the right.

The line of parts at the bottom of the painting, comprising a base margin, makes the picture a botanical masterpiece. From left to right: three prothalli, a cut through of the stipe, two scales, three ripening sori on a leaf, six details of the sporangia ripening in the sorus under the indusium, and two details of a sporangium and front and back of a pinna.

With this picture in hand, anyone tramping through a mixed temperate forest in southern China or Japan between 1,400–3,200m (4,500–10,500ft), will be able to identify a *Dryopteris cycadina*. If you cannot see it in the wild, or you do not have a fernery, it is enough to admire this portrait of one.

RIGHT Shaggy shield ferns have rosettes of fronds borne on scaly stalks with simple, evenly toothed pinnae. Size: 57 x 55cm (22$\frac{1}{2}$ x 22in), watercolour on paper.

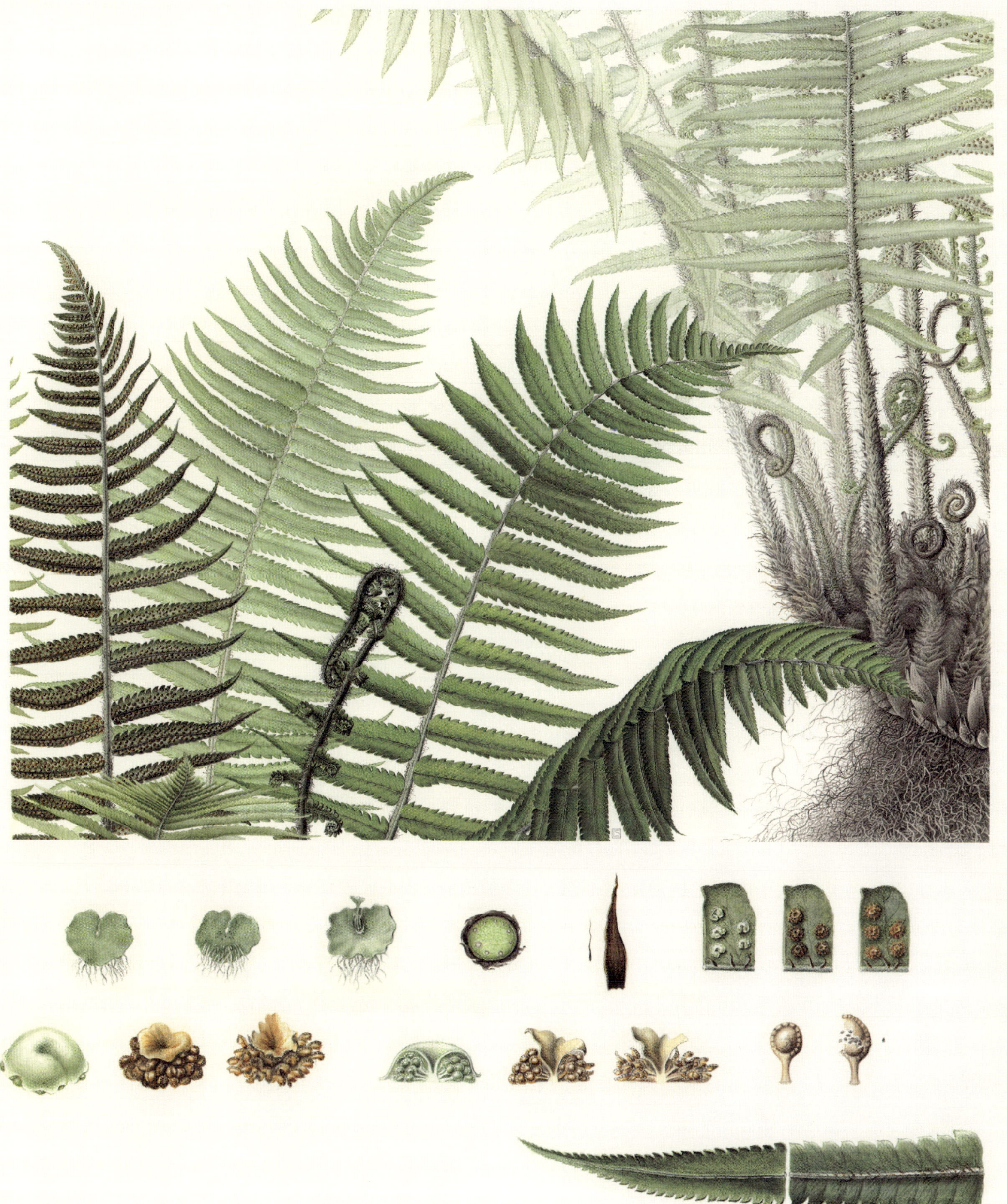

Glossary

Botanical language, like plants, evolves and words in old books can seem arcane. Many everyday words have precise botanical meanings. Sometimes words can have more than one meaning, or botanists disagree about others.

Abaxial: The side that faces away from its axis; the lower leaf

Accessory (fruit): A fruit made up of more than just the ovary; a 'false' fruit

Achene: Dry, non-splitting fruit; small, thin-walled, and containing a single seed

Acicle: A needle-like prickle

Actinomorphic (flowers): Radially symmetric

Acuminate: A tip that narrows gradually to a point

Acute: Sharply pointed

Adaxial: The side that faces towards its axis; the upper leaf

Adventitious: A feature not in its normal place, such as roots or buds

Aerial (roots): Above ground and getting moisture from the air

Aestivation: Arrangement of corolla or calyx in bud

Aggregate (fruit): A fruit made up of many flowers

Androecium: Collective word for the stamens, the male organs

Angiosperm: A flowering plant

Anthesis: When a flower is open and receptive to fertilization by pollen

Apocarpous: Having free carpels

Appressed: Lying flat against a surface

Arachnoid: Like a cobweb, for example, of tangled hairs

Architecture: How a plant looks in terms of branching and habit, dependent on its molecular make-up

Arcuate: Curved, like a bow

Areole: Small spaces in between veins of a leaf; cushion where cactus spines emerge

Aril: An appendage or outer covering of a seed

Arista(e): A stiff bristle, such as in *Compositae*, or the end of a leaf

Asexual: Of reproduction not involving exchanging male and female genes

Auriculate: Like an ear

Awn (grasses): Fine bristle extending from a part

Axil: The angle between stem and leaf

Basal plate: Disc (reduced stem) at base of a bulb

Base/basifixed: At the base of a structure, for example, of leaf rosettes, anthers

Beak (*Asteraceae*): Slender neck of an achene, positioned under the pappus

Berry: Simple, indehiscent fruit with thin skin, in which there may be many seeds within fleshy pulp. No differentiation inside.

Bifid: A tip divided into two

Binomial system: Naming system using two words: genus and species

Blade (leaf): Expanded, main part of a leaf (also called the lamina)

Bract/bracteole: Leafy part associated with flower/inflorescence

Bulb: Underground storage organ comprising scale leaves enclosing buds

Bulbil: Small bulbs, usually in leaf axils or inflorescence

Bulblet: A small bulb or bulb-like structure

Calyx: Combined word for sepals, the outermost whorl of a flower

Campanulate: Bell-shaped

Canaliculate: Grooved, like a canal

Capitate: Head-like

Capitulum: Head of flowers in *Asteraceae* family

Capsule: A dry dehiscent fruit with two or more carpels opening on one side

Carp-: Root word that relates to fruit. Hence, endocarp, mesocarp, epicarp: the three layers of an ovary wall, collectively pericarp. Syncarp, apocarp, monocarp: the nature of fusion and number of carpels

Catkin: A spike of cylindrical racemose unisexual flowers without petals, often pendulous

Cauliflory: Flowers borne directly on a stem or trunk

Cauline/caulescent: Relating to, or growing on, a stem/having an evident stem

Cellulose: The main substance in plant cell walls

Chlorophyll: The green pigment in plant cells

Ciliate: A fringe of hairs

Cladophyll: A branch that looks and functions like a leaf (the phylloclade)

Claw (petal): Narrow base of a petal

Column (orchid): The fused sexual organs that form a solid structure

Composite: A member of *Asteraceae* (daisy family), which have clusters of small flowers composed on a single capitulum

Conduplicate: Folded together so the surfaces face each other

Connate: Fused together, such as leaves, petals or stamens

Contractile root: One that can shorten to pull an underground organ to a particular level

Convergent evolution: Process where similar features evolve independently in unrelated plants, for example, carnivory or succulence

Cordate: Heart shaped

Coriaceous: Leathery

Corm: Short underground storage part of a stem

Cormel: A small new corm

Corolla: Combined term for the petals

Corona: A crown-shaped structure on petals or stamens

Corymb: A flat-topped racemose inflorescence where pedicels arise from different levels

Cotyledon: First leaves arising from a seed

Crenate: Scalloped or having rounded teeth

Crisped: Curled or crumpled, such as the margin of a leaf

Cross pollination: The transfer of pollen between plants

Crozier (ferns): A young, unfurled fern frond; also called fiddlehead

Culm (grass): Stem

Cuneate: Gradual tapering of a leaf base into the petiole

Cupule: Cup-shaped structure at the base of fruits, such as an acorn

Cuspidate: Abruptly tapered, with a sharp point at the tip

Cuticle: Waxy, water repellent surface layer of a leaf

Cyme/-ose: A determinate inflorescence where flower opening follows laterally after the central (top) flower.

Cypsela (*Asteraceae*): Dry one-seeded fruit with appendages for dispersal

Decurrent: Extending downwards, for example, of a leaf base that continues into the stem creating grooves or wings

Decussate: Alternate pairs, such as leaves, which are set at right angles to each other

Dehiscence: Opens when ripe, such as fruit or anthers

Deltate: Triangular, like the Greek letter 'delta'

Dentate: Like teeth, each with a sharp point

Determinate: With a stem whose growth is finite (opposite to indeterminate)

Dichasial: Cymose branching where two new flowers arise on each side of an axis, the older one in between

Didymous (anthers): Paired

Didynamous (anthers): Two pairs, one pair longer than the other

Digitate: Like fingers

Disc floret (*Asteraceae*): The inner, usually fertile, flowers in the centre

Distichous (leaves): Arranged in two vertical rows on either side of an axis

Dorsifixed (anthers): Stamen filaments fixed to the anther halfway between the lobes

Drupe: Fleshy indehiscent fruit containing a stone

Ebracteate: Without bracts

Emarginate: Leaf apex with a distinct notch

Ensiform: Sword shaped; long and narrow, ending in a sharp point

Entire: Smooth and unbroken

Epi-: Prefix. Upon, over, on top of or added to, hence epicalyx, epidermis, epiphytic, epichile (orchids)

Equitant: How the leaf base may clasp the next (opposite) leaf above it, forming a fan-like arrangement

Eudicot: Refers to plants whose seed produces two cotyledons. It is the preferred word in molecular botany to the older word, dicotyledon

Extrorse: Outwards (opposite of introrse)

Falcate: Curved like a sickle

Fall (Iris): Outer tepal that expands into a broad pendulous blade

Family: The classification group below order and above genus

Fascicle: A cluster of leaves or flowers arising from the same point

Fiddlehead: See Crozier

Filament: Stalk which is topped by an anther

Filiform: Thread like

Fimbriate: Margin cut into fringes

Flabellate: Fan-shaped

Floret: Small flower, often applied to individual flowers of inflorescence

Foliole: Leaflet in a compound leaf

Follicle: Monocarpous pod opening on the side where seeds attached

Frond: In ferns and palms the whole leaf

Gamete: Male/female sex cells that unite on fertilization to form a zygote

Genus: Grouping in plant classification which provides the first name in binomial naming system, always written with a capital letter

Glabrous: Smooth and without hairs

Glandular: Covered with glands; hairs with a gland at the tip

Glaucous: Covered in waxy greyish bloom or green-blue colour

Glume (grass): Bracts at the base of a spikelet

Golden section: Section containing the most pleasing part of a composition, which can be related to Fibonacci's sequence of numbers

Growth forms: Grouping of similar habit types, such as Alpine cushion plants

Gynoecium: The female part of the flower, or pistil(s)

Habit: The general appearance of a plant, whether nodding, upright and so on

Hastate: Like the base of a spear, the two basal lobes pointing horizontally

Helicoid: Coiled like a spring

Hesperidium (citrus fruit): Berry with a thick rind and segments inside

Hirsute: With coarse, stiff hairs

Hypanthium: Extension of the receptacle that surrounds the ovary, eventually becoming the fleshy part of a pome

Imbricate (petals): Overlapping

Incised: Cut deeply

Indehiscent (fruit): Not splitting open

Indeterminate: Growth which continues indefinitely as in shoots or inflorescence flowering

Indumentum: Covering of hairs or scales

Indusium (ferns): Thin tissue covering the sorus in some ferns

Inferior (ovary): One that is below its calyx

Inflorescence: A grouping of flowers, including bracts

Infundibular: Funnel shaped

Internode: Length between the nodes on stems

Introrse: Inwards (opposite of extrorse)

Involucre: Group of tightly placed bracts on a capitulum of flowers

Involute: Margins rolled inwards and upwards (opposite is revolute)

Keel: In leguminous flowers, the two lower petals which hide the sexual parts; or a longitudinal ridge along the middle of a leaf

Labellum (orchid): Lowest petal, usually larger and differently shaped to the laterals

Labiate: With lips, often two distinct upper and lower lips (in which case, bilabiate)

Labium: The lip of a labiate flower

Lacerate: Like the torn edges of a piece of paper

Lamina (leaves): a blade

Lanceolate: Egg shape that is elongated towards the tip

Leaflet: Part of a compound leaf; foliole

Legume (fruit): Monocarpic fruit of legume family that opens into two halves

Lemma (grass): Outermost of the two bracts of a floret (palea is inner bract)

Life forms: See growth forms

Ligule (grass/*Asteraceae*): Narrow flap on both sides of leaf sheath at the node/narrow flower petal reduced to one usually retaining five lobes

Linear: Long and narrow with parallel sides

Lip: See Labium

Lobe: Rounded division of the margin of calyx, corolla or leaf

Locule: A cavity in a carpel

Lodicule (grass): Tiny scales between lemma and sexual parts, probably reduced perianth

Lyrate: Like a lyre, such as pinnate leaves at the base with a large, rounded terminal lobe

Margin/marginate: A well-defined margin or edge

Monochasial (cymes): Lateral branching on one side only

Monocot: A flowering plant producing a single cotyledon

Monopodial: Main axis of branching system that continues growing (also racemose); opposite of sympodial

Mucronate: Ending abruptly in a short straight point, a mucro

Multiple (fruit): Fruit formed from more than one carpel (often confused with 'compound' and 'aggregate')

Nectary: Organ where nectar is produced

Node: The point on a stem where a leaf is attached, or an adventitious root

Nodules: Small, rounded structures on roots containing nitrogen-fixing bacteria

Nut: Dry indehiscent fruit with a single seed and very hard shell

Nutant: Nodding

Ob-: Prefix meaning the other way round from the usual e.g. ob-ovate, ob-deltoid

Obtuse: Blunt, rounded apex or base of leaves

Offset: Lateral shoot used in asexual propagation; small bulbs in the outer scales of bulbs

Ovary: The lower part of the pistil or carpels that contains the ovules

Ovate: Egg-shaped

Ovule: Immature seed before fertilization

Palea (grass): The inner bract of a floret

Palmate: In the shape of the palm of a hand

Panicle: Of an inflorescence which has several lateral divisions. Most specifically used for racemes.

Pappus (*Asteraceae*): A series of bristles or hairs around the top of the fruiting part

Parietal (placentation): Where the ovules are set on the periphery of a single-celled syncarp

Pectinate: Like a comb with closely set narrow segments along a margin

Pedate: Further division in the basal lobes of a palmately lobed leaf

Pedicel: Stalk of a single flower or floret

Peduncle: Main stalk of an inflorescence

Peltate: Disc shaped with stalk attached to the centre

Pentamerous: In multiples of five

Perfoliate: A pair of sessile leaves attached to one another around the stem that holds them

Perianth (flowers): Collective name for the calyx and corolla, the non-reproductive parts

Pericarp: Wall of fruit, divided into endocarp, mesocarp and epicarp

Petiole/petiolule: Leaf stalk/leaflet stalk

Phalange (*Pandanus*): Cluster of partly fused drupes forming a unit (also called a key)

Phyllary (*Asteraceae*): One of the bracts that makes up an involucre

Phyllotaxis: Arrangement of leaves on a stem

Pinna(e): The first division(s) of a compound leaf, such as in ferns. Bi-pinna are the second division

Pinnate: Divided into lateral leaflets/veins, like a feather

Pistil: Female organ comprising ovary, style and stigma, or the gynoecium in a syncarpous flower

Placentation: How the ovules are attached inside the ovary

Plicate: With lengthwise folds

Pollen: Small grains containing male reproductive cells

Pollinium (orchids): A mass of pollen grains dispersed as a single unit

Polychasial (cymes): Branching where each axis has more than two branches

Pome: Indehiscent simple fruit in which the hypanthium encloses the ovary

Poricidal (dehiscence): Released through pores

Prickle: Sharp pointed outgrowth from the outer layer of tissue surrounding a stem or leaf

Prop root: Adventitious roots growing out of a stem base or branch, and back into the soil, acting as a support or stilt

Prostrate: Lying flat

Prothallus: A fern's plantlet which carries the male/female sex organs

Pseudobulb: Bulb-like enlargement of the stem in orchids

Puberulent: Covering of minute hairs, hardly visible

Puberulous: Covering of short soft hairs

Pubescent: Covered in soft hairs like down, but often with a wider meaning

Pulverulent: As though dusted with a fine powder

Pulvinus (pulvini): Swelling on the petiole enabling movement of the leaf

Raceme: An indeterminate inflorescence, flowers opening bottom upwards

Rachilla: The secondary axis of a compound leaf

Rachis: The main axis of a compound leaf or of an inflorescence/ultimate axis

Radicle: Young root emergent from the seed

Ray floret (*Asteraceae*): Outermost irregular, often larger and sterile, flowers

Receptacle: Top of the stem that holds the flower

Reflexed: Curving backwards or downwards

Reniform: Kidney-shaped

Revolute: Rolling of a margin towards the underside; opposite of involute

Rhipidium: Fan-shaped cyme with zigzag pattern to flower stalks

Rhizome: An underground stem, often swollen

Root cap: Protective layer of cells on the growing tip of root

Root crown: The point that differentiates root stock from stem

Rostellum (orchid): A beak-like projection of the stigma that separates it from anthers

Runner: Elongated lateral shoot above ground

Sagittate: Shaped like an arrow with two basal lobes pointing downwards

Samara: Dry, indehiscent fruit with wing(s)

Scabrous: Rough to the touch

Scape: Flower stalk without leaves arising from the ground

Secund: Leaves or flowers on one side of the axis only

Self-fertilization: Fertilized by its own pollen

Sepal: Single part of the outermost whorl of the flower, collectively the calyx

Serrate: Like a saw, with angled, pointed teeth

Sessile: Attached direct, without a stalk

Setose: Bristly

Siliqua(e) (*Brassicaceae*): Narrow dehiscent fruit containing a partition on which seeds sit

Sinker: Downward-growing shoot of a bulb/corm that forms a new one at its end

Sinuate: A wavy margin of a leaf

Sinus: The recess between two lobes

Sorus (sori): Structure containing groups of sporangia qv in ferns

Spadix: Inflorescence based on a thick fleshy axis on which sit partially sunken flowers

Spathe: A sheathing bract which surrounds an inflorescence or spadix

Spatulate: Like a spatula: narrow at the base becoming wider and rounded at the tip

Species: Basic group of closely related mutually fertile plants. Always linked with their genus, and written in lower case

Spike: Racemose inflorescence with sessile alternate flowers

Spikelet (grasses): A structure of florets, each with lemma, palea and flower

Spine: Sharp, hard structure derived from vascular part of leaf, stipule, root or branch

Spiral: Coiled along an axis

Sporangi(um)a: Sac or capsule containing spores

Spore: Reproductive unit in ferns

Spur: In flowers, a slender tube usually containing nectar

Stamen: Male organ consisting of filament and anther which holds pollen

Staminode: Sterile stamen, usually short and not having pollen

Standards (*Iris/Papilionaceae*): Narrow, upright inner tepals or upper, often larger, petal in the pea family

Stellate: Star-like

Stigma: The part of the gynoecium receptive to pollen

Stipe: Leaf stalk of a fern frond. Applies to other parts of a flower

Stipule: Leaf appendages, usually in pairs at the base of the petiole

Stoma(ta): Leaf pores that enable transpiration to take place

Strobilus: An inflorescence made up of overlapping scales like cones

Style/style arms: Elongated top of the ovary on which sits the stigma/branching stigma

Sub-: Prefix meaning almost or under, such as in subsessile or subspecies

Succulent: Thick, fleshy and swollen

Sucker: A shoot arising from a root

Superior (ovary): One where the perianth and stamens are set under the ovary

Syconium: Hollow compound fruit such as figs

Sympodial (orchids): Without a single main stem

Synandrium: Anthers joined up, though the filaments are free

Syncarpous: United carpels

Taxon: A unit of species, subspecies, variety or forma

Tendril: Thin, twining structure derived from a leaf or stem

Tepal: Petals and sepals which are undifferentiated

Terminal: At the apex of an axis

Ternate: Arranged in clusters of three

Tessellation: Chequered, like square tiles

Tetradynamous: Four long stamens and two short ones

Tetramerous: In multiples of four

Theca(e): The paired locules of an anther, where pollen is produced

Thorn: Short, pointed woody structure derived from a reduced branch

Thyrse: Inflorescence with racemose main axis and cymose lateral flowers

Tiller (grasses): Sucker or branch emerging from the base of a plant

Tomentose: Short, soft and densely matted hairs

Torose: Cylinder with contractions at intervals

Tridentate: With three teeth

Trifoliate: With three leaves

Trifoliolate: With three leaflets as in a compound leaf

Trimerous: In multiples of three

Tristichous: Arranged one above the other in three rows, for example, of leaf arrangement

Truncate: Looking cut off either at the base or the apex, such as of a leaf

Tuber: Swollen underground stem or root

Tubular floret: Cylindrical flower with lobes representing petals

Tunicate (bulbs/corms): Having a dry outer covering

Umbel: Racemose inflorescence with pedicels set at one point giving a flat-topped or globose flower head

Undulate: A wavy margin

Urceolate: Urn-shaped

Utricle (*Aristolochiaceae*): Swollen basal part of the perianth

Valvate: Petals which meet but do not overlap

Vascular bundle: Specialized cells that carry water and nutrients around the plant

Velamen (orchids): Layer(s) of spongy cells on the outside of epiphytic roots

Verrucose: Bumpy like warts

Verticillaster: Inflorescence where a cluster of cymes are set at intervals on the axis, common in mints

Whorl: A set of similar organs arranged around a stem, also verticil

Wings: Lateral petals of an iris, or extensions either side of a stem or rachis

Zygomorphic: Irregular shaped flowers with bilateral symmetry

Further resources

Books providing an artist's guide to botany

Bebbington, Anne, *Understanding the Flowering Plants: A Practical Guide for Botanical Illustrators*. The Crowood Press, 2014

Leech, Lizbeth, *The Artist's Handbook of Botany*. The Crowood Press, 2023

Leech, Lizbeth, *Botany for Artists*. The Crowood Press, 2011

Simblet, Sarah, *Botany for the Artist: An Inspirational Guide to Drawing Plants*. Penguin Random House, 2010

Systematic and illustrated glossaries

Beentje, Henk, *The Kew Plant Glossary*. Kew Publishing, 2012

Hickey, Michael & King, Clive, *The Cambridge Illustrated Glossary of Botanical Terms*. Cambridge University Press, 2000

Latin

Bayton, Ross, *RHS The Gardener's Botanical*. Mitchell Beazley, 2019

Harrison, Lorraine, *RHS Latin for Gardeners: Over 3,000 Plant Names Explained and Explored*. Mitchell Beazley, 2012

Stearn, William T., *Botanical Latin*. Nelson, 1966

History of botanical art

Blunt, Wilfred & Stearn, William, *The Art of Botanical Illustration*. The Antique Collectors' Club, 1994 edition

Noltie, Henry, *The Dapuri Drawings*. The Antique Collectors' Club, 2002

Noltie, Henry, *Robert Wight and the Botanical Drawings of Rungiah and Govindoo*. Royal Botanic Garden Edinburgh, 2007

Rix, Martyn, *The Golden Age of Botanical Art*. Andre Deutsch/Kew, 2012

'How to' books

A small selection by established contemporary artists:

Gale, Leigh Ann, *Drawing and Painting Leaves*. The Crowood Press, 2025

Güner, Isik, *Botanical Illustration from Life*. Search Press, 2019

Jenkins, Carolyn & Birch, Helen, *New Ideas in Botanical Painting*. Batsford Books, 2022

King, Christabel, *The Kew Book of Botanical Illustration*. Search Press/Kew, 2015

Morrish, Sarah, *Natural History Illustration in Pen and Ink*. The Crowood Press, 2021

Oxley, Valerie, *Botanical Illustration*. The Crowood Press, 2013

Showell, Billy, *Botanical Painting in Watercolour*. Search Press, 2016

Smith, Lucy, *Botanical Sketchbooks: An Artist's Guide to Plant Studies*. The Crowood Press, 2024

Trickey, Julia, *Colour: Observations in Botanical Art* (www.juliatrickey.co.uk)

Trickey, Julia, *Leaves: Observations in Watercolour* (www.juliatrickey.co.uk)

Woodin, Carol & Jess, Robin(eds), *Botanical Art Techniques*. ASBA/Timber Press, 2020

Courses

New York Botanical Garden: Certificate course

Royal Botanic Garden Edinburgh: Diploma course in botanical illustration and Certificate courses, online and in person

Society of Botanical Artists (UK): Distance Learning Diploma course

Many artists around the world also provide in person and on-line courses

Florilegia

A small selection of a growing number of publications from Florilegium societies:

Brown, Andrew, *Botanical Illustration from Chelsea Physic Garden*. The Antique Collectors Club, 1988

Dawson, Sîan, *The Iris Florilegium of Sir Cedric Morris*. Sîan Dawson, UK, 2024

Franklin, Ros, *A coming of Age: Celebrating 18 Years of Botanical Painting by the Eden Project Florilegium Society*, Two Rivers Press, 2018

Morris, Colleen & Murray, Louisa, *The Florilegium: The Royal Botanic Gardens Sydney*. Kew, 2018

Oxley, Valerie, *A Florilegium: Sheffield's Hidden Garden*. The Crowood Press, 2021

Collections

Brooks, Charlotte, *RHS Botanical Illustration: The Gold Medal Winners*. ACC Art Books, 2025

Gardner, Martin, *Plants from the Woods and Forests of Chile*. Royal Botanic Garden Edinburgh, 2022

Sherwood, Shirley, *The Shirley Sherwood Collection: Modern Masterpieces of Botanic Art*. Kew Publishing, 2019

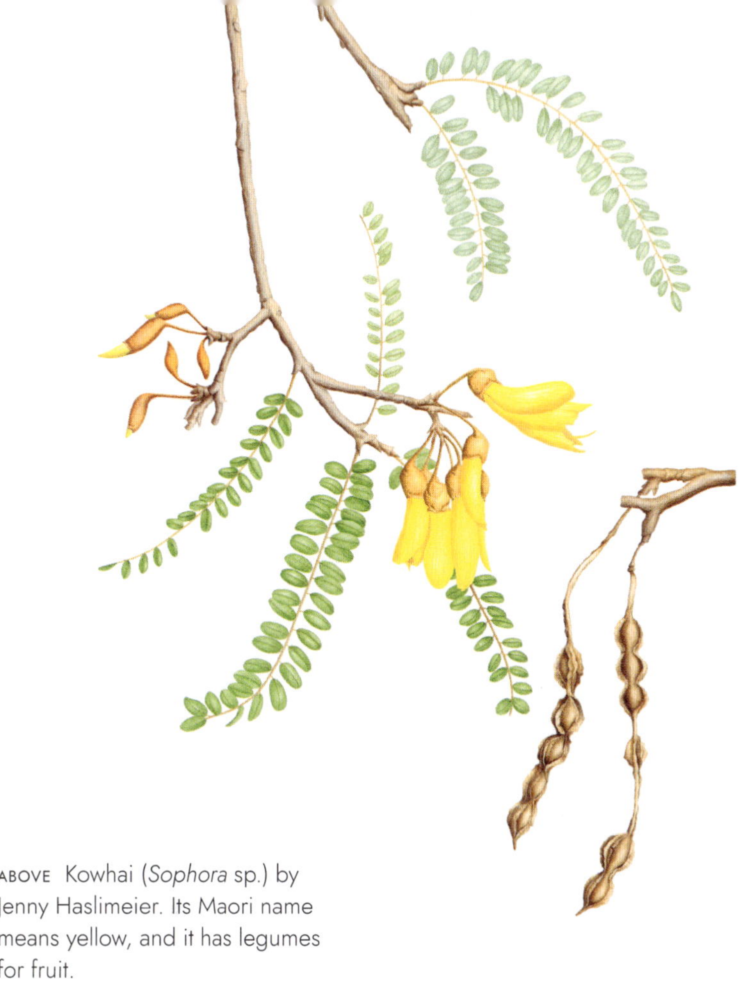

ABOVE Kowhai (*Sophora* sp.) by Jenny Haslimeier. Its Maori name means yellow, and it has legumes for fruit.

Inspiration from artists featured in this book

Trickey, Julia, *Botanical Artistry: Plants, Projects and Processes*. Two Rivers Press, UK, 2019

Li, Fay-Wei & Suissa, Jacob S., *Ferns: Lessons in Survival from Earth's Most Adaptable Plants*, illustrated by Laura Silburn. UniPress, UK, 2025

Hart-Davies, Christina, *The Whole Story, Painting More Than Just the Flowers*. Two Rivers Press, UK, 2020

Online help

Katherine Tyrrell: A mine of information. https://www.botanicalartandartists.com

Lizzie Harper: Lots of botany and advice. https://lizzieharper.co.uk

Julia Trickey: Regular online talks by established artists. www.juliatrickey.com

Index

Picture credits

The author and publisher would like to thank the following artists for generously allowing their artworks to be used on the following pages.

© **Julie Ah-Fa** julieahfa.com / @julieahfa
20: *Dombeya elegans*

© **Ingrid Arthur** @ingridarthurart
144 (top right): *Fagus sylvatica* (Common Beech), 56 × 38 cm, watercolour on Saunders Waterford 140lb (300gsm) HP paper

© **Gillian Barlow** gillianbarlow.com
16: *Eryngium bourgatii*, watercolour on paper, by kind permission of Chelsea Physic Garden Florilegium Society

© **Isobel Bartholomew**
87 (bottom right): Common stork's bill (*Erodium cicutarium*), 30 × 38 cm, watercolour on Arches HP, from *Breckland Wild Flowers – Heaths and grasslands*, by Iceni Botanical Artists, 2016
88 (top): *Iris* 'Edward of Windsor', 72 × 56 cm, watercolour on Arches HP, by kind permission of *The Iris Florilegium of Sir Cedric Morris*

© **Auriol Batten / RBGE**
19: *Zaluzianskya maritima*, by kind permission of the Trustees of the Royal Botanic Garden Edinburgh

© **Tina Bone**, Wildlife & Botanical Illustrator
www.tinasfineart.uk / @tinasfineartuk
124: Hollyhock (*Alcea cv.*)
133 (top): Three-banded passion flower (*Passiflora trifasciata*)
205: Water chickweed (*Callitriche palustris*)

© **Victoria Braithwaite**
167 (top): Solomon's Seal (*Polygonatum multiflorum*)

© **Lyn Campbell**
53: Trillium (*Trillium sulcatum*)
57 (top right): Ramson (*Allium ursinum*)
120 (left): Wake Robin (*Trillium chloropetalum*)
125 (top): Fawn lily (*Erythronium revolutum*)

© **Lucilla Carcano** www.lucillacarcano.com / @lucillacarcano
73 & 87 (top right): *November* (*Fraxinus ornus* and *Ginkgo biloba*), 45 × 35 cm, watercolour on paper

© **Giovanni Cera / RBGE**
217: *Cupressus sempervirens*, by kind permission of the Trustees of the Royal Botanic Garden Edinburgh

© **Ruth Cox**
86 (bottom right): Ornamental pear (*Pyrus* sp.) and blueberry (*Vaccinium* sp.)
87 (top left): Red Japanese maple (*Acer palmatum*)

© **Mary Crabtree** marycrabtreebotanicalartist.com
157: Switchgrass (*Panicum virgatum*)

© **Toni Dade** toni-maria-design.com / @tonimariald
108 (bottom left): Spanish mallow (*Malva hispanica*)
161 (top left): Fading allium (*Allium* sp.)
181 (top): Avocado (*Persea americana*)

© **Gaynor Dickeson** gaynorsflora.com / @gaynorsflora

168 (right): *Far from Common Time* (*Thymus vulgaris*), 30 × 21 cm, watercolour and graphite on HP paper

© **Sandra Doyle**
105: Cyprian plane (*Platanus orientalis* var. *insularis*), 63 × 44cm, watercolour and coloured pencil on paper

© **Janet Dyer**
46 & p204 (left): Few-flowered garlic (*Allium paradoxum*)
90 (bottom): White stemmed or Chinese bramble (*Rubus cockburnianus*)
107 & 134 (centre): *Rhododendron ponticum*
119 (box) & p182: Crab apple (*Malus*)
121 (top): 'Paul's Himalayan musk' rose
144 (bottom): Butterfly bush (*Buddleja davidii*)
161 (bottom right): Himalayan balsam (*Impatiens glandulifera*)
190 (bottom): Horse chestnut (*Aesculus hippocastanum*), from *Autumn Fragments*
207 (top): Montbretia (*Crocosmia × crocosmiiflora*)

© **Gülnur Ekşi Bona / RBGE**
117: Ulmo (*Eucryphia cordifolia*), 49.5 × 35cm, watercolour on paper, by kind permission of the Trustees of the Royal Botanic Garden Edinburgh

© **Maggy Fitzpatrick**
89 (top): *Horse Chestnut Full Leaf and Flower*, No 7 in series of 12, 50 × 50 cm, watercolour on HP paper

© **Jenny Ford**
110 (top): Japanese wineberry (*Rubus phoenicolasius*)

© **Mary Frost**
219 (top right): Douglas fir (*Pseudotsuga menziesii*), 14 × 10cm, watercolour on Arches Aquarelle Grain Satine HP 140lb.

© **Betty Gadsby** www.bettygadsby.co.uk
156: *Brassia* 'Eternal Wind'
204 (right): Wild strawberry (*Fragaria vesca*)

© **Leigh Ann Gale** www.la-botanicalart.co.uk / @leighann3518
58 & 98 (left): Lilium bulbs, 33 × 23 cm, watercolour on paper
61 (centre): Winter Twigs 2, 26 × 19 cm, watercolour on paper
62 (bottom): Common stinging nettle (*Urtica dioica*), 30 × 20 cm, watercolour on paper, by kind permission of the artist and Michael Kopinski
99 (top): Pigface (*Carpobrotus* sp.), 23 × 23 cm, watercolour on paper
220: *Coniferae* cones, 8 × 17 cm, watercolour on paper

© **AJ Harley**
45: *Antirrhinum*, full painting size 38.5 × 56 cm, aperture 31 × 47.5 cm, watercolour on HP 300g/m² Fabriano Artistico
118: Pansies, full painting size 37.5 × 49.5cm, aperture 25.5 × 37.5cm, watercolour on HP 300g/m²

© **Jean Harlow SSBA** (Scottish Society Botanical Artists) https://jeanharlowartist.yolasite.com / www.facebook.com/jeanharlowartist
49 (top): *Ananas comosus*, pineapple (origin Costa Rica), 35 × 24cm, watercolour on paper

© **Lizzie Harper** lizzieharper.co.uk
24 (bottom left and top right): Sketchbook page, Giant Hogweed
24 (top right): Leaf study
32: Sketchbook page, lesser periwinkle (*Vinca minor*)
33 (top left & top right): Colour gradients
33 (bottom): Building up colour
86 (box): Blackberry (*Rubus fruticosus*) leaf
88 (box): Snowdrop (*Galanthus nivalis*)

© **Mayumi Hashi** @mayumihashiart
120 (right): *Iris* 'Benton Evora', 56 × 41 cm, watercolour on Fabriano Artistico Extra White 600g, by kind permission of *The Iris Florilegium of Sir Cedric Morris*

© **Jenny Haslimeier** www.jennyhaslimeier.com
9: *Rhododendron arboreum* ssp. *Cinnamomeum*
86 (bottom left): Hazel (*Corylus avellana*)
90 (top): *Clianthus puniceus*
154 (left): Thistle
168 (left): Jasmine (*Jasminum officinalis*)
233: Kowhai (*Sophora* sp.)

© **Marianne Hazlewood**
153: Himalayan cobra lily (*Arisaema consanguineum*), 45 × 32 cm, watercolour on Fabriano 5 paper

© **Sarah Howard** www.sarah-howard.co.uk
25: *Aloe elegans*
26: Sketchbook page, Sargent's Cherry (*Prunus sargentii*)
34: *Iris* 'Strathmore' tracing
35 (top left and right): *Iris* 'Strathmore', by kind permission of *The Iris Florilegium of Sir Cedric Morris*
47 (top): Red hot poker (*Kniphofia foliosa*)
51 (left) and 109 (right bottom): *Magnolia* 'Galaxy'
59 (top left): Onions (*Allium cepa*)
60 (top): Moth orchid (*Phalaenopsis* sp.)
63: Crown of thorns (*Euphorbia milii* var. *splendens*)
74 (bottom) & p192 (top & centre): Spindle (*Euonymus europaeus*)
76 (left) & p163 (right): *Kniphofia schimperi*
76 (right) & p163 (left): Curled dock (*Rumex crispus*)
79 & 192 (bottom left): Opium poppy (*Papaver somniferum*)
80 (top): Daisy (*Bellis perennis*)
83 (top & box): *Aloe schelpei*
84 (top right): African violet (*Streptocarpus* sect. *Saintpaulia*)
109 (left): Drooping star of Bethlehem (*Ornithogalum nutans*)
142 (bottom): *Delphinium wellbyi*
155 (left): Smooth sow thistle (*Sonchus oleraceus*)
160: *Aloe* aff. *percrassa*
164: Wild teasel (*Dipsacus fullonum*)
166: Kashmir rowan (*Sorbus cashmiriana*)
179 (top): *Crotalaria rosenii*
191: Coffee (*Coffea arabica*)
193: Bere barley (*Hordeum vulgare*)

© **Waiwai Hove** www.waiwaihove.com / @waiwaihove
175: *Papilionanda* Tan Chay Yan, 31 × 41 cm, watercolour on Lanaquarelle HP 300gsm paper, by kind permission of Singapore Botanic Gardens, National Parks Board

© **Mariko Ikeda / RBGE**
201: Common screwpine (*Pandanus utilis*), 57.3 × 39.4 cm, watercolour on vellum, by kind permission of the Trustees of the Royal Botanic Garden Edinburgh

© **Mieko Ishikawa / RBGE**
108 (top right): Foxglove tree (*Paulownia tomentosa*), by kind permission of the Trustees of the Royal Botanic Garden Edinburgh

© **Carolyn Jenkins** www.carolynjenkins.co.uk / @jenkinscarolyn
4: *Quercus coccifera*
142 (top): *Helleborus* × *hybridus*, by kind permission of RHS Lindley Collections
165 (top): Holm oak (*Quercus ilex*)
165 (bottom): Globe artichoke (*Cynara cardunculus*)
178 (left): Hybrid Lenten rose (*Helleborus* × *hybridus*) by kind permission of RHS Lindley Collections

© **Agita Keiri** www.keiriart.uk / @ akeiri_art
82 (top): *Nuphar lutea* (L.) Sm. Yellow water-lily, 43 × 38 cm, watercolour on Fabriano Artistico
180 (top): Hornbeam, 57 × 55 cm, watercolour on Fabriano Artistico

© **CF King**
48 (top): *Senecio jacobaea*, watercolour
48 (bottom): *Dendrosenecio adnivalis*, watercolour, by kind permission of Dr & Mrs R Polhill. From *The Kew Book of Botanical Illustration* (Search Press, 2016)

© **Hulya Korkmaz / RBGE**
119 (right) & 134 (top): *Lobelia bridgesii*, by kind permission of the Trustees of the Royal Botanic Garden Edinburgh

© **Nicola Macartney**
nicolamacartney.com / @nicola.macartney.art
50 (right): *Pinus jeffreyi*, 40 × 40 cm, watercolour on paper
52: *Malus domestica* 'James Grieve', 35 × 45 cm, watercolour on paper
89 (bottom): *Sorbus aucuparia*, 60 × 40 cm, watercolour on paper

© **Vicki Malone** vickimaloneartist.com
131: New Mexico Grasses, 61 × 45.5 cm, watercolour on Arches Aquarelle HP 300gsm

© **Alister Mathews**
57 (top left): Dandelion (*Taraxacum officinale*), by kind permission of Chelsea Physic Garden Florilegium Society

© **Mhorag McDowall**
81 (top): Apple 'Bloody Ploughman' (*Malus domestica*)

© **Annie Morris**
51 (right): Common vetch (*Vicia sativa*), 25 × 18 cm, watercolour on paper
85 (bottom): Common whitebeam (*Sorbus aria*), 23 × 19 cm, watercolour on paper

© **Gloria Newlan** gloria-newlan.pixels.com / glorianewlan
36: Christmas cactus (*Schlumbergera* cv.), 28 × 38 cm, watercolour on Arches Aquarelle watercolour paper HP 300 gsm, 2021
207 (right): Desiree potatoes (*Solanum tuberosum*), 56 × 37 cm, watercolour on Arches Aquarelle watercolour paper HP 300 gsm, 2024

© **Simonetta Occhipinti** www.simonettaocchipinti.it
181 (bottom): *Citrus medica* 'Mano di Buddha', 53 × 44 cm, watercolour on paper (private owner: Stefano Casciu, Regional Director of the Museums of Tuscany)

© **Rachel Pedder-Smith**
38—41, 64—69, 92—95, 100—103, 112—115, 126—129, 136—139, 146—151, 170—173, 184—189, 194—199, 208—213, 222—225

© **Jacqueline Pestell**
145: Foxglove (*Digitalis purpurea*)

© **William D Phillips** @wdphillipsart
49 (bottom): Orchid (*Vanda* 'Blue Magic')

© **Linda Pitkin** www.lindapitkin.net
62 (top): Horse chestnut (*Aesculus hippocastanum*)
77 (left): Arrowhead (*Sagittaria sagittifolia*)
78 (top): Yellow archangel (*Lamium galeobdolon*)
84 (bottom left): Borage (*Borago officinalis*)
109 (right top) & 135 (top): Bogbean (*Menyanthes trifoliata*)
180 (bottom): Stinking iris (*Iris foetidissima*)

© **David Reynolds** www.davidreynoldsart.com.au / @botanicalarttv / @davidreynoldsfineart
141: *Banksia serrata*, 63 × 59 cm, watercolour on paper, by kind permission of Geoff and Elizabeth Clear

© **Roger Reynolds**
111: Japanese quince (*Chaenomeles japonica*)
134 (second from top): *Aconitum carmichaelii*
162: Carmichael's monk's hood (*Aconitum carmichaelii*)
219 (bottom left): Common yew (*Taxus baccata*)

© **Maria Alice de Rezende** @mariaalice.rezende
97: Arrowhead plant (*Syngonium podophyllum*), 69 × 50cm, watercolour on Fabriano Artistico watercolour paper
125 (bottom): *Archytaea triflora*

© **Gillian Rice**
37: Indian Blanket, *Gaillardia pulchella*, 26 × 36 cm, watercolour on paper
155 (right): Mexican Hat, *Ratibida columnifera*, watercolor on vellum, 18 × 18 cm

© **Pamela Richardson**
www.pamelarichardsonartist.com
123 (left): *Clematis* 'Princess Diana'

© **Sarah Roberts**
75: Sea buckthorn (*Hippophae rhamnoides*)

© **Daleen Roodt** www.daleenroodt.com / @artist.daleenroodt
77 (right): Fever tree (*Vachellia xanthophloea*)
159: *Erythrina affra* and Cape white-eye, 35 × 24.8 cm, watercolour on paper, 2016

© **Laura Silburn** www.laurasilburn.co.uk / @laura.silburn
122: Dutchman's pipe (*Aristolochia cathcartii*), left: 21 × 30 cm, watercolour on paper; right: 40 × 50 cm, watercolour on paper
227: Shaggy shield fern (*Dryopteris cycadina*), 57 × 55 cm, watercolour on paper

© **Lucy T Smith**
91 (top): Cabbage-tree palm (*Livistona australis*)

© **Lilian Snelling**
21: Jersey lily (*Amaryllis belladonna*), RHS Lindley Collections / Lilian Snelling

© **Liudmyla Spodin** www.liudmylaspodin.art / @liudmylaartist
123 (right): *Physalis: Autumn Lantern*, 46 × 33 cm, watercolour on paper

© **Mary Ellen Taylor FLS DipEGS CPGFS**
www.maryellentaylor.co.uk / @metaylor10

50 (left): *Asplenium trichomanes*, 27 × 37 cm, watercolour and graphite on HP Arches, by kind permission of Chelsea Physic Garden Florilegium Society

© **Vivienne Taylor**
99 (bottom): Mitchell's pitcher plant (*Sarracenia* x *mitchelliana*), by kind permission of Chelsea Physic Garden Florilegium Society

© **Julia Trickey**
www.juliatrickey.co.uk / @ juliatrickeyART
71: *Vintage Amaryllis*, 47 × 50cm, watercolour on Fabriano Artistico paper
169 (right): Russian comfrey (*Symphytum* × *uplandicum*)

© **Pauleen A Trim HS SBAF GM**
@pauleentrim8
7: Common hornbeam (*Carpinus betulus*), 27 × 34 cm, watercolour on Fabriano Artistico extra white HP 640 gsm

© **Sarah-Jane Tweed**
sjtweed.com / @sjtweed.art / www.soc-botanical-artists.org/artist/sarah-jane-tweed
206: Red onions (*Allium cepa*), 42 × 29.7cm, watercolour on HP watercolour paper, original painted life-size

© **Nataya Udompat / RBGE**
215: East Indian galanga (*Kaempferia galanga*), 42 × 29.7 cm, watercolour on paper, by kind permission of the Trustees of the Royal Botanic Garden Edinburgh

© **Emma van Klaveren DipCSBA, SBAF**
www.emmasbotanicals.com / @emmasbotanicals
31: Emma painting in her studio
110 (bottom): *Narcissus* 'Pheasant's Eye', A4, watercolour on Arches HP

© **Fiona Ward Dip BI RBGE**
47 (bottom): Ivy (*Hedera helix*), original watercolour on paper
121 (bottom): Herb Robert (*Geranium robertianum*), original watercolour on paper

© **Sunanda Widel / RBGE**
132 (right): Dancing lady ginger (*Globba winitii*), by kind permission of the Trustees of the Royal Botanic Garden Edinburgh

© **Michie Yamada / RBGE**
13: *Rosa sambucina*, by kind permission of the Trustees of the Royal Botanic Garden Edinburgh
183: *Rosa luciae*, by kind permission of the Trustees of the Royal Botanic Garden Edinburgh

© **Marianna Zych**
221 (right): Scots pine (*Pinus sylvestris*)

Pictures from other sources
Alamy Stock Photo / Old Images 177; iStock / Irina Trumpe 33 (top centre); James Lawrence 28; New York Public Library Digital Collection 59 (bottom); Rawpixel 17 (left), 43, 55, 143, 203; Shutterstock /Sergey Skleznev 27 (second from top), /M. Thanaphum 27 (third from top), /Elena Lar 29, /Dasha Lapshina 30; Wikimedia Commons 12, 13 (bottom), 14, 15, 17 (right), 18, 27 (second from bottom), 27 (bottom) CC BY-SA 3.0 FR / Rama.

All line drawings by **Sarah Skeate**

Acknowledgements

A complex book of this sort is simply not possible without a wonderful team of people to make it happen. I hope you agree the result is stunning and useful.

A big thank you to the many gifted and generous artists who allowed us to use their work, and often reduce them to details. A special thanks goes to those who went the extra mile to re-scan or put us right about nomenclature and the artistry involved. In the end we could not use everyone's contributions, but I hope you will visit the many exhibitions of botanical art to discover even more beautiful works. Rachel Pedder-Smith wowed the botanical art world with her PhD paintings of herbarium specimens at Kew, and is now a tutor; her projects added greatly to the usefulness of the book.

The keepers and curators of collections have been most helpful and encouraging. Mary Ellen Taylor and Gillian Barlow of the Chelsea Physic Garden Florilegium and Lorna Mitchell, Librarian at the Royal Botanic Garden Edinburgh, allowed me to peruse their original paintings. The RHS Lindley Library helped me navigate their online catalogue. Singapore Botanic Gardens kindly put us right about Waiwai Hove's work. Thank you.

Botanical art societies are a rich resource. Some of their members are or have been my colleagues, others I hadn't met. They include the Iceni Botanical Artists, the Scottish Society of Botanical Artists (where Valerie Gordon helped me locate images), the Edinburgh Society of Botanical Artists, the Society of Botanical Artists, the Institute for Analytical Plant Illustration, the Association of Botanical Artists, and the artists of the Iris Florilegium of Sir Cedric Morris – all in the UK. The American Society of Botanical Artists also proved a good source of pictures from its international membership. It helped enormously that many had online galleries, or personal websites, where I could easily peruse work and contact artists. I am proud to know you and am very grateful.

Errors are strictly my responsibility, even though various experts agreed to cast their eye over the text, unquestioningly and in their free time. Special thanks to Dr Sabina Knees, who checked various aspects, Dr Axel Paulsen for identifying a ginger, and Dr Greg Kenicer, who helped with the vexed question of inflorescences – all of the RBGE; Dr Ross Bayton; and RHS experts James Lawrence, Simon Maughan and Mike Grant. My husband, Charles, provided explanations for Latin and Greek words. Thank you for your knowledge.

My final thanks are for my production colleagues at Bright Press, who were unfailingly patient: Kathy Steeden, picture researcher, was gentle with my clumsy use of software; May Corfield, editor, reduced my text to a readable format; and Lindsey Johns, designer, arranged my picture suggestions into lovely spreads. Joanna Bentley, senior editor, cast a professional eye over the project, and James Lawrence, art director, ensured high standards for illustrations. There were undoubtedly others in the background, known only to the publishing team. I hope you enjoyed working on this project.

This book was a joint effort by all the above, who, together with encouraging friends, inspired progress and gave weight to the result. Thank you all.

Sarah Howard

Author biographies

Sarah Howard is a botanical artist based in Scotland, where she works on commissions and projects, including exhibitions and book illustrations, and teaches botanical illustration. Sarah's work has been selected for exhibitions and institutions including the Florilegium of the Royal Botanic Garden, Edinburgh; American Society of Botanical Artists; Flora Scotia; and the Hunt Institute, Carnegie Mellon University, Pittsburgh. She has been awarded medals by the Royal Horticultural Society, including a Gold Medal for her Horn of African Aloes. She received a Diploma with Distinction in Botanical Illustration from the Royal Botanic Garden, Edinburgh, and has studied and painted the flora of Ethiopia and Romania, as well as contributing to collections in the UK.

Rachel Pedder-Smith is a botanical artist with a PhD in Natural History Illustration from the Royal College of Art. She has spent many years illustrating pressed and dried specimens within the Herbarium at the Royal Botanic Gardens, Kew, and she has been awarded four gold medals by the Royal Horticultural Society (RHS). Rachel's work has been exhibited in three solo shows in London and featured in numerous exhibitions worldwide, including Tate Britain's 'Watercolour' exhibition in 2011. One of Rachel's botanical paintings (Small Gourd) is held in the RHS's Lindley Library in London. She is the author of *Botanical Art* and *Flowers* in the RHS *Watercolour Art Pad* series.

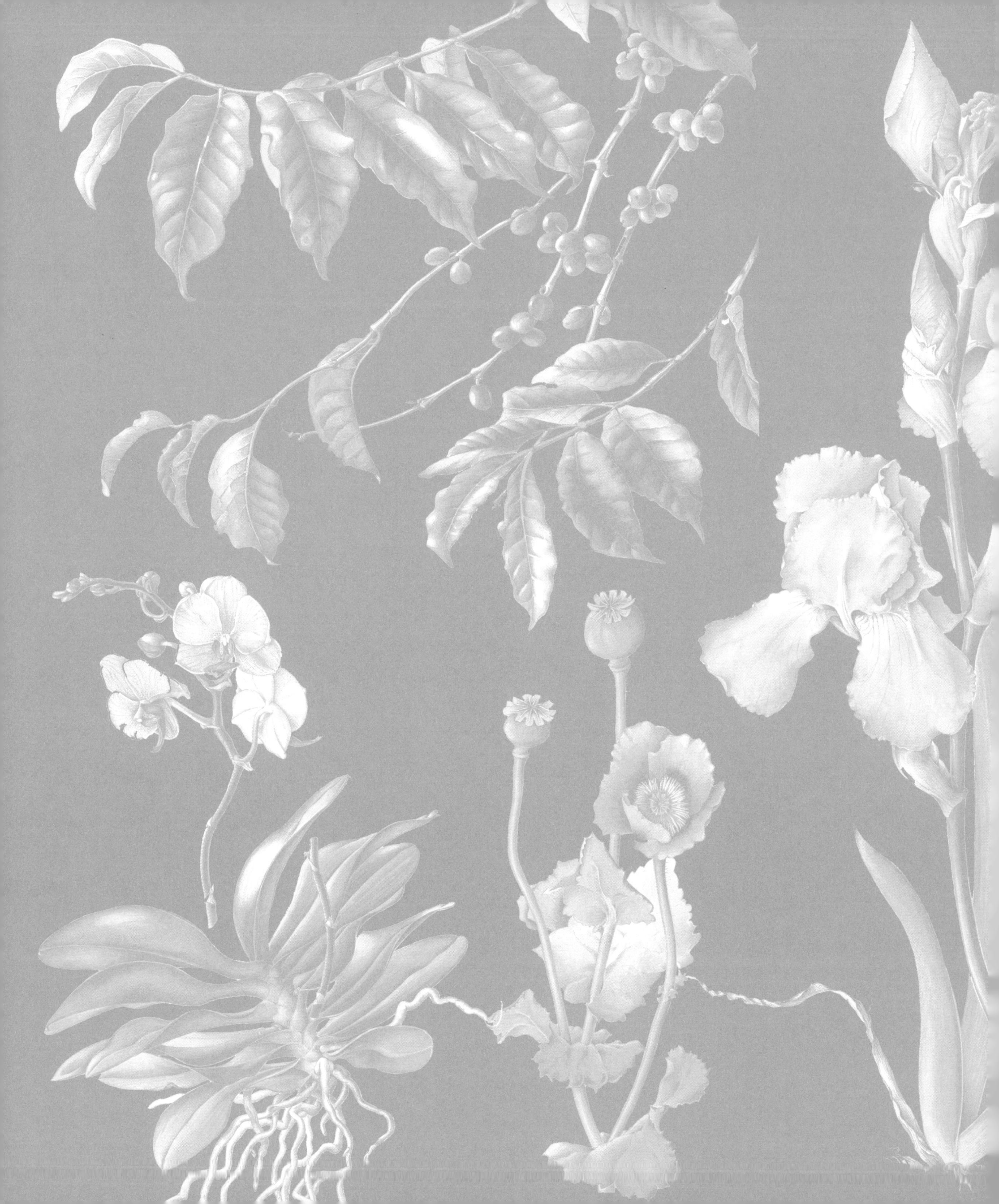